III International Conference on Small Angle Neutron Scattering Dedicated to the 80th Anniversary of Yu.M. Ostanevich (YuMO 2016)

Journal of Physics: Conference Series Volume 848

Dubna, Russia
6 – 9 June 2016

ISBN: 978-1-5108-4242-7
ISSN: 1742-6588

Printed from e-media with permission by:

Curran Associates, Inc.
57 Morehouse Lane
Red Hook, NY 12571

Some format issues inherent in the e-media version may also appear in this print version.

Printed by Curran Associates, Inc. (2017)

For permission requests, please contact the Institute of Physics
at the address below.

Institute of Physics
Dirac House, Temple Back
Bristol BS1 6BE UK

Phone: 44 1 17 929 7481
Fax: 44 1 17 920 0979

techtracking@iop.org

Additional copies of this publication are available from:

Curran Associates, Inc.
57 Morehouse Lane
Red Hook, NY 12571 USA
Phone: 845-758-0400
Fax: 845-758-2633
Email: curran@proceedings.com
Web: www.proceedings.com

Table of contents

Volume 848

III International Conference on Small Angle Neutron Scattering dedicated to the 80 anniversary of Yu.M. Ostanevich

6–9 June 2016, Dubna, Russian Federation

Accepted papers received: 12 May 2017
Published online: 8 June 2017

Preface

Papers

Preface

"Among the many well-known fundamental and applied consequences of neutron discovery, one of the first was the emergence of structural neutron diffraction - one of the methods for studying the spatial structure of condensed matter at the atomic level. Investigating the dynamics of biological macromolecules by neutron diffraction methods is an area that is almost untouched. We believe that also in this direction one should expect rapid progress."

Yu.M.Ostanevich and I.N.Serdyuk
Uspekhi Fizicheskih Nauk (1982)

More than 30 years after Ostanevich and Serdyuk reviewed the then 50 years from the neutron discovery, we are happy to support their statement and prediction by reporting on the further progress in the field of neutron scattering research. Frank Laboratory of Neutron Physics has organized the III International Conference on Small Angle Neutron Scattering (YuMO2016) that was – not coincidently – dedicated to the 80th anniversary of Ostanevich.

Yuriy Mechislavovich Ostanevich (1936–1992) has had a determinative and crucial contribution to the construction of spectrometers at the pulsed reactor IBR in Dubna, Russia. He contributed in particular to the development of time-of-flight small-angle neutron scattering (SANS) technique, and the selection of advanced scientific areas for its application. The SANS instrument at the IBR-2 reactor is called YuMO in his honour. The Ostanevich's leadership and outstanding scientific achievements in SANS studies of polyelectrolytes, small molecules, fractals, metallic glasses, macromolecules, polymers, etc., were recognized also by a number of awards including the State Prize of the Russian Federation in 2000. To this end of course, we should not forget mentioning the contribution of close collaborators of Ostanevich: Laszlo Cser, Josef Plestil as well as Alexander Kunchenko, Vadim Bezzabotnov and Nikolay Gorski.

The YuMO2016 conference focused on providing opportunities to discuss various possibilities of exploiting the SANS technique in the many aspects of condensed matter research as once appreciated and predicted by Ostanevich, and perhaps even surpassing his predictions. The FLNP have had an opportunity to welcome more than 110 participants from 14 different countries and 3 continents comprising Europe, North America and Australia. The scientific program was filled with 43 oral presentations extending over 930 minutes, while more than 60 posters were presented during the poster sessions and breaks.

The content of the conference and its proceedings are fully commensurate with world-wide trends in neutron scattering applications. The contribution dedicated to the SANS methods and instrumentation describes for example a role of polarized neutrons in the localization of free radicals in the structural biology. It is hard to imagine another probe than neutron, and another technique than small-angle scattering that have such many utilization possibilities in the studies of drug delivery systems, neurodegenerative diseases, and other aspects of biomembranes and soft matter in general, to name but a few.

Besides the neutron's sensitivity to the light elements, and H/D differentiation making the hydrogen-rich biomaterials an innate target, magnetism is accessible to neutrons too. Our proceedings document studies where neutron scattering techniques come to the rescue when other methods have failed, and/or provide examples of synergy between the various methods. Atomic force microscopy, NMR, Raman spectroscopy, molecular dynamics simulations, and X-ray scattering are clearly the complementary methods of choice for condensed matter research. It is a matter of time, and already happening future in some cases, for the studies concerning nanoparticles and nanocomposites to spin-off into advanced and industrial applications.

It is more than intriguing to realize the role of neutrons and/or small angle scattering method in such studies across various scientific fields. The neutron history is written by those who touched the neutron. Let us pay a tribute to one of them. We are fortunate to open our collection of YuMO2016 conference dedicated to the 80th anniversary of Yu.M. Ostanevich by the very personal commemorations of his close colleague and friend Dr. Laszlo Cser.

NORBERT KUČERKA, MARIA BALASOIU,
AND ALEXANDER I. KUKLIN
Frank Laboratory of Neutron Physics,
Joint Institute for Nuclear Research,
Dubna, Russia
April 2017

IOP Conf. Series: Journal of Physics: Conf. Series **848** (2017) 011002 doi:10.1088/1742-6596/848/1/011002

Conference Photo

III International Conference on Small Angle Neutron Scattering dedicated to the 80-th anniversary of Yu. M. Ostanevich, June 6-9, 2016, Dubna

IOP Conf. Series: Journal of Physics: Conf. Series **848** (2017) 011003 doi:10.1088/1742-6596/848/1/011003

Conference Poster

Acknowledgements

The Organizers of the III International Conference on Small Angle Neutron Scattering dedicated to the 80-th anniversary of Yu. M. Ostanevich, acknowledge the financial support from the Grants of the Governmental Plenipotentiary Representatives of Romania, Slovakia, Czech Republic, scientific projects of the Cooperation Programmes JINR-Romania. Special gratitude to the JINR and FLNP administration and staff is expressed.

The conference is also dedicated to the 60th Anniversary of the Joint Institute for Nuclear Research, Dubna.

IOP Conf. Series: Journal of Physics: Conf. Series **848** (2017) 011005 doi:10.1088/1742-6596/848/1/011005

Peer review statement

All papers published in this volume of *Journal of Physics: Conference Series* have been peer reviewed through processes administered by the proceedings Editors. Reviews were conducted by expert referees to the professional and scientific standards expected of a proceedings journal published by IOP Publishing.

IOP Conf. Series: Journal of Physics: Conf. Series **848** (2017) 012001 doi:10.1088/1742-6596/848/1/012001

In Memoriam Yu.M. Ostanevich

László CSER

Wigner Research Centre for Physics Hungarian Academy of Sciences

cser.laszlo@wigner.mta.hu

Abstract. Ten years ago, Joint Institute Nuclear Research released a book containing commemoration written by co-workers, colleagues and several guest scientists who had had close and fruitful companionship with Yuri M. Ostanevich. In their papers, the authors presented many of his scientific ideas and achievements proposed and even accomplished by his hands. When I was asked to give this short memorial speech, I got confused. It is without any reason to repeat what can be already found in the book. Thus, I had to make the decision to tell you some rather personal and somehow minor impression of mine, and due to this approach, it is tackled from a subjective point of view.

1. Introduction

In 1963, I was delegated for a three years period to JINR – more precisely to Neutron Physics Laboratory – aiming to learn the application of Mössbauer effect to solid-state physics.

Entering the lab where I was directed by the administrators, I saw a tall blond young man being busy with soldering some rather complicated electronic scheme. He was whistling a beautiful tune giving a proof of perfect ear for music. This whistling accompanied his deep meditation on a given problem. He kept this habit for a long time afterwards. Several years later he had started to smoke a pipe.

After mutual introduction, he immediately got me acquainted with the aim of his actual activity: working on building a new system for actuation the Co^{57} source. The idea of the so-called resonant stabilization proposed by him. The explanation I got was compact and clear without any unnecessary details.

I felt immediately that I had arrived at the right place and a right person with balanced temper. Indeed, Yura had a very good character. He was calm, never shouted, and my Russian dictionary was compiled by a collection of dirty words from many other sources but him.

2. Einstein and Fermi

To work with him was easy and even informative. Yura immediately understood what I was going to tell and quickly found both the mistakes and accuracy in my words.

We soon became friends. Funny, we both deeply respected Einstein's theory of relativity, yet after reading the book about Fermi's biography, the style of Fermi's group in Rome stood closer to our hearts. We started to call each other by the title "dottore" both at the lab and later on in letters.

Our colleagues at the neutron physics lab were surprised by our habit not being on first name terms with each other. We expressed our respect for each other by this uncommon way.

3. The theory of relativity

Despite of our attraction to Fermi's way of conducting physical research, the interpretation and experimental proving of the theory of relativity was one of our permanent topics almost every evening when I was asked to accompany him to his place.

For example, the theory of relativity of space and time was a basic statement. Space was considered empty and no material was endowed with any material properties like ether. On the other hand, the structure of this emptiness is influenced by gravitation. In fact, the geometry of this emptiness can be modified by presence of gravitation. From this statement it follows that space should possess material properties.

You can see that trying to solve the above-described contradiction was a good theme for long discussions. Amongst other issues, diminishing our ignorance was also a way of our thinking.

4. Another intriguing problem

Another intriguing, unsolved for us problem, was the following. A charged body emits electromagnetic radiation if it is accelerated. On the other hand, a weak gravitation field is equal to accelerated motion at the first approximation. (See the "gedanken experiment" with the free falling elevator). If it is so, then an electron placed on the table has to emit radiation. Yet another problem to be discussed.

Once, Academic Ginzburg visited the Neutron Physics Laboratory and obviously presented a talk at the seminar of the Laboratory. Yura grabbed the opportunity and brought up our problem to Ginzburg. He stared at the ceiling for a while and promised to consider the problem. Half a year later appeared in journal Uspekhi of Physical Science a paper submitted by Ginzburg dealing with some problems of the theory of general relativity, and a solution to our problem was described in it. However, at this time, the Mössbauer epoch has ended and our interest in science has changed.

5. Interest to neutron scattering

At the end of 1960's I left JINR and the Yura's interest turned to neutron scattering.

When I went to JIRN again for a longer period, Yura has had finished a series of studies of the water behavior at critical point using neutron scattering. About that time, Ilya Mihaylovich Frank - the director of Neutron Physics Laboratory got motivated with the implementation of investigations of biological samples using neutron scattering.

The neutron source available at that time was the so-called IBR-30. It produced regular pulses of neutrons but the average flux was rather low for many types of neutron scattering experiments. Trying to satisfy Frank's desire Yura came to employ the small angle scattering method, because the time of flight method allowed us to use a very large wavelength range, which made the pulsed source competitive with the method using single wavelength techniques. I was involved in the preparation of the construction of small angle scattering instrument. Our first task was to build the shielding around the neutron guide, which was an evacuated tube of 80 cm diameter, and to do this at the detector as well. It was a boring and hard physical work because we had to build a wall out of several tons of concrete and lead blocks around a track in order to move the detector.

During the breaks, Yura was busy with calculating the resolution power of the small angle device, planning and elaborating on a method to evaluate the optimal relative position between the sample and detector. By doing that, his mathematical erudition was revealed. Back then, computer techniques have been in rudimentary state.

6. Computer calculations and program languages

At the Laboratory of Neutron Physics, a computer "M-20" was installed. The necessary software for such kind of a computer was uneasy to find, program languages were far below the present level.

Yura was the first physicist learning the rules of communication with the computer at so high level that he was able to teach colleagues. He conducted a series of seminars on the topics of the way of writing programs for data evaluation by using the computer "M-20". His explanations were logical, clear, and the way, which made his explanations even very exciting. He picked up the then-most recently released program languages like Algol, Fortran, and Assambler quickly on a user level.

7. Teaching of students

His mathematical talent was employed for teaching students at the filial of university placed in Dubna. His lectures dealt with Statistical physics, which is a branch of physics that uses methods of probability theory and statistics, and particularly the mathematical tools for dealing with the behavior of large populations and finding new approximations for solving physical problems. It can describe a wide variety of fields with an inherently stochastic nature. Its applications include many problems in the field of physics, biology, chemistry, neurology and others. Its main purpose is to clarify the properties of matter generally, in terms of physical laws that govern atomic motion.

After we lost him, a thick dossier was found on his table. In that dossier, there was a manuscript. At first sight, they seemed to be the lecture notes. However, in fact, it was full of numerous complex formulae; therefore, it is very probable that he tried to write a book on this topic.

8. Style of his job

A few words about the Yura's style of doing his job.

Firstly, he arrived at the lab exactly at nine o'clock morning using a bicycle as the mean of transport during both the summer and winter.

He worked twelve hours a day. Typically, at seven or eight o'clock evening he sat down and accurately wrote a short résumé about achievements and problems that need to be solved next day. At nine o'clock he took his velo and went home. At home, before going to bed, he worked on improving his knowledge of English language while reading often books dealing with children topics such as Winnie the Pooh and so on. He did that in order to get his own children involved in learning English.

Nota bene, Yura was proud of the young Ostaneviches. All three of them became nice, clever and promising people.

9. Conclusion

He usually spent his holidays either hiking in the mountains, or more often rowing on rivers, where the riverside was almost uninhabited. He liked the NATURE very much and it was a bitter response from nature that he was grasped at the place where he expressed his feeling of love.

We lost a talented scientist, a teacher, and a good friend as well, a strictly moral person who was a torch in our laboratory.

Time-resolved proton polarisation (TPP) images tyrosyl radical sites in bovine liver catalase.

Oliver Zimmer[1], Hélène M Jouve[2], and Heinrich B Stuhrmann[2,3]

[1] Institut Laue-Langevin, F-38042 Grenoble, CEDEX 9, France.
[2] Institut de Biologie Structurale Jean-Pierre Ebel, CEA/CNRS/UJF, F-38027 Grenoble CEDEX 1, France.
[3] Helmholtz-Zentrum Geesthacht, D-21494 Geesthacht, Germany.

heinrich.stuhrmann@orange.fr

Abstract. A differentiation between dynamic polarised protons close to tyrosyl radical sites in catalase and those of the bulk is achieved by time-resolved polarised neutron scattering. Three radical sites, all of them being close to the molecular centre and the heme, appear to be equally possible. Among these is tyr-369 the radial site of which had previously been proven by EPR.

1. Introduction

At first sight, it appears that an enzyme like bovine liver catalase does not have much in common with polarized proton spin targets. The latter are used in experiments of high energy physics, whereas a redox enzyme like catalase is treated in life sciences. We will show that it is polarized neutron scattering which connects both worlds. Before entering into this complex undertaking, it seems to be appropriate to describe some aspects of both worlds, which may bear a tremendous potential of synergies in neutron scattering [1].

Let us start with polarized protons. It is widely established that abundant spin-½ nuclei, and protons in particular, in a paramagnetically doped insulator, like most hydrogenous organic materials are, reach thermal equilibrium with the lattice *via* interaction with paramagnetic impurities, e.g. unpaired electrons [2]. The theoretical foundation of this nuclear relaxation process has been formulated by Bloembergen [3] who postulated that nuclear spins *close* to paramagnetic centers are relaxed by fluctuations in the magnetic field of the paramagnetic center (direct relaxation) and that *bulk* spins are relaxed by diffusion of Zeeman energy to *close* nuclei (spin diffusion). The number of close protons in direct contact with the non-Zeeman electron-electron spin interaction reservoir (or bath) is limited by the fact that the local spin lattice relaxation rate is inversely proportional to the sixth power of the distance between the unpaired electron and the proton [2] [4]. The result is a gradient of proton polarization reflecting a proton spin diffusion barrier which separates the close protons from those of the bulk. As the local magnetic field near an unpaired electron is very large, the Larmor frequencies of the close protons differ appreciably from those of the bulk. Consequently close protons hardly contribute to the NMR signal of the bulk (Fig.1). It is this picture which provides the environment of dynamic nuclear spin polarization (DNP).

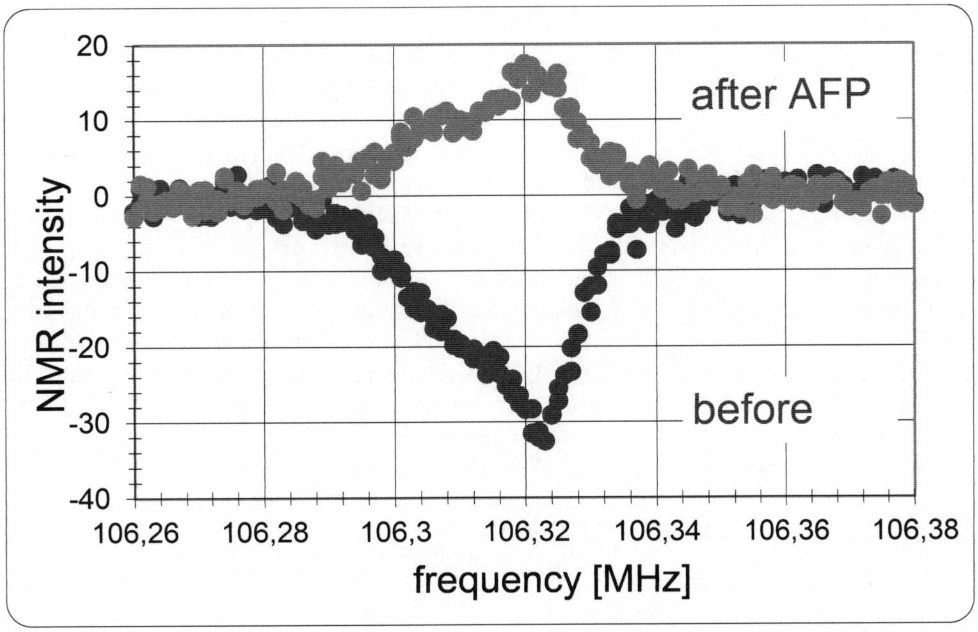

Fig. 1. The NMR intensity from tyrosyl doped catalase is due to the dynamically polarised protons of the bulk.. • negative proton polarization with microwave irradiation at a frequency slightly above the EPR line of the tyrosyl radical, • after adiabatic fast passage (AFP) using an rf sweep across the NMR profile. B= 2.5 T, T = 0.1K.

Dynamic nuclear spin polarization is a widely used method to polarize nuclear spins. The macroscopic aspects of DNP are well understood in the frame of the spin temperature theory [5]. The thermodynamic model of DNP assigns two heat reservoirs to various degrees of freedom of the electronic and nuclear spin system via mutual and external interactions[5] [6]. The mechanism of DNP then is described as a two-step process: the cooling of the non-Zeeman reservoir by a non-saturating microwave field at frequencies close to the electronic resonance and the subsequent transfer of entropy from the nuclear Zeeman system consisting of the close protons. Depending on the type and density of the unpaired electrons in the sample and the external conditions (temperature, applied magnetic field) DNP proceeds either by the so-called solid effect or by thermal mixing or a combination of them [5] [7]. Note that the exact type of the process is not relevant for the evaluation of the experimental data. In any case, it is assumed the magnitude of the mixing is strongly influenced by the microscopic structure of the material and in particular the nuclear spins close to the unpaired electron [8] [9]. The transfer of the polarization of the close protons to those of the bulk is controlled by the magnetic spin diffusion barrier as it has been mentioned above. Hence the polarization of the close protons is expected to be different from those of the bulk, particularly at the onset of the microwave irradiation.

Now let us turn to the radical proteins. Radicals play an important role in living cells. Radical intermediates allow enzymes to perform a wide variety of chemically challenging reactions. Catalase is a redox enzyme which converts hydrogen peroxide at an incredibly fast diffusion limited rate into water and molecular oxygen by offering peroxyacetic acid, a derivative of hydrogen peroxide with more steric hindrance, to bovine liver catalase the reaction differs in at least two points: first the response is slow, and second, after some intermediate steps one of the amino acids, a tyrosine, is converted into a tyrosyl radical. It is assumed that one and only one of the 20 tyrosines of each of the four subunits of catalase becomes a tyrosyl radical [10].

Finding the exact location of a (tyrosyl) radical is an important step in the understanding of the mechanism of an enzyme. Electron paramagnetic resonance is the method of choice for detecting a radical, and in favorable cases the site of the radical inside a protein molecule can be obtained from a detailed analysis of the EPR line [11]. A more general approach to determine the location of an unpaired electron is provided by magnetic neutron scattering. However, a radical doped protein can be considered as an extremely dilute paramagnet, with a correspondingly small magnetic amplitude.

If it comes to the size of the polarization dependent scattering length, then polarized nuclei, and certainly protons, are more rewarding targets. This has been realized forty years ago, when the dynamically polarized proton spins in a Nd^{3+} doped crystals of lanthanum magnesium nitrate were studied by polarized neutron scattering [12]. It took another ten years until dynamically polarized protons in non-crystalline targets have considerably widened the range of applications of neutron scattering [13] [14] [15]. These and later experiments of nuclear spin contrast variation (e.g. [16] [17] [18]) which profited from a high proton polarization dependent scattering amplitude, prepared the ground for a more detailed study of the build-up of proton polarization by polarized neutron scattering. The starting signal came from a problem of structural biology concerning the location of a tyrosyl radical in bovine liver catalase [10]. In an initial step, it led to a collaboration of groups of the IBS, Grenoble, of the CEN Grenoble, and of the PSI, Villigen.

2. First experiments with tyrosyl radicals of catalase.

Once it had been shown that free tyrosyl radicals in frozen solution do support DNP [1], similar studies were carried out with samples of catalase containing tyrosyl radicals. Microwave irradiation at frequencies slightly below the EPR line of tyrosyl leads to a slow increase of positive proton polarisation, far below the polarisation rate known from polarised targets used in high energy physics experiments. At $T = 0.1$ K a maximum proton polarisation of $P = 3\%$ was reached after several hours of microwave irradiation. Microwave irradiation at frequencies slightly above the EPR line leads to negative values of proton polarisation. At $T = 0.1$ K the proton polarisation reaches $P = -0.98\%$ (Fig. 2). With higher temperatures the difference ΔP between the extreme values of proton polarisation decrease. At T=0.5 K and 1.0 K we obtain $\Delta P = 0.6\%$ and 0.2%, respectively. In the latter case, the proton polarisation remains positive and close to the proton polarisation at thermal equilibrium, $P_e = 0.25\%$, encountered in a magnetic field strength $B = 2.5$ T, whatever direction of dynamic nuclear polarisation is used. There is hardly any change of the initial speed of proton polarisation with the temperature. The implications of this observation will be discussed below.

The method of adiabatic fast passage (AFP) is a powerful method to invert the sign of nuclear polarisation. Using a radio frequency sweep at an appropriate speed and power [19], a negative polarisation of protons in tyrosyl doped catalase is obtained within a second (Fig.1). Impressive this method is [20], it suffers from two drawbacks: first, a considerable part of the polarisation is lost, and second, the preparation preceding the jump of proton polarisation is long compared with the time of nuclear relaxation after AFP, which results in a poor use of neutron beam time. A solution to this problem is offered by frequently alternating the direction of dynamic polarisation. The build-up of a polarisation gradient then is observed after each change of the direction of DNP allowing for a high duty cycle. Another merit of this method is that it will focus on the polarisation of the close protons which are expected to image the electron spin density.

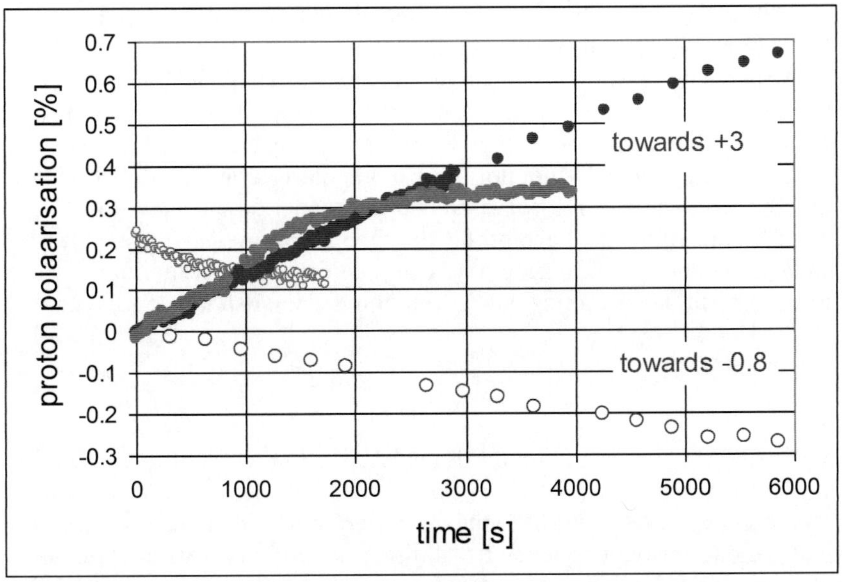

Fig. 2. The evolution of proton polarization in frozen solution of tyrosyl doped catalase with the time of microwave irradiation. • Positive DNP at T = 0.1 K, ○ negative DNP at T = 0.1 K. • Positive DNP at T = 1.0 K, ○ negative DNP at T = 1.0 K.

3. Experimental procedures

A detailed description of the instruments and their use for time resolved neutron scattering from dynamic polarised protons has been given earlier [21] [22]. Here we present experimental details that are necessary in order to achieve a self-contained presentation

3.1. Sample preparation

The preparation of the sample started from tetramer bovine liver catalase (BLC) in crystalline suspension (Boehringer Mannheim, Germany, lot 106828). 12 mL of the suspension were dissolved in 2 L Tris maleate buffer (50 mM, pH 7.5), purified, concentrated to 5 ml and dialyzed against 150 mL Tris maleate in D_2O. The buffer had been exchanged three times. The content of the dialysis-tube had partly precipitated, but a clear solution (0.515 mM heme) was obtained by addition of an equal volume of deuterated glycerol. Note that the BLC molecule is a tetramer enzyme with one heme as part of each of its four identical subunits. At a temperature only slightly above 0° C, to 1 mL of the clear solution of BLC 28 μL peroxyacetic acid (32% solution, Merck, diluted to 3% solution in 0.1 M Tris, and adjusted to pH 4.3) were added. The immediate formation of a tyrosyl radical can be followed by a change of the colour to deep red. After a few minutes the tyrosyl concentration reaches its maximum and the sample is frozen to liquid nitrogen temperature. From EPR measurements a tyrosyl concentration of 0.58 per heme has been determined. Samples with a lower catalase concentration were prepared in a similar way.

3.2. Neutron small-angle scattering (SANS)

The SANS measurements were performed at the D22 instrument of the Institut Laue-Langevin (ILL) Grenoble, France. We used polarized incident neutrons of wavelengths λ=4.6 Å with a wavelength spread of $\delta\lambda/\lambda = 0.1$ (FWHM). The solid sample was inserted into the NMR coil, placed in microwave cavity and mounted into a ^{4}He refrigerator, which was operated at a temperature of about 1K. A

longitudinal magnetic field of 3.5 T was provided by a superconducting split-coil solenoid. Samples could be changed and cooled down to 1K in less than half an hour.
Two IMPATT-diodes, tuned to frequencies required for positive and negative DNP, 97.2 and 97.5 GHz, respectively, could be connected to the sample cavity by a waveguide switch (total switching time 100 ms comprised of shutting within 5 ms, no microwaves for 90 ms, opening in 5 ms). Continuous wave (cw) NMR was used in parallel to monitor the proton polarisation. The direction of DNP was reversed each 5 s, and the acquisition of the polarised neutron scattering intensity was triggered synchronously in time intervals of 50 ms. Each of the 200 time frames was averaged over several thousand cycles to obtain sufficient statistical accuracy. The time dependent scattering intensity then was obtained as usual by radial averaging, and correction for polarisation dependent transmission.

4. Mathematical formalism

The evolution of the proton polarisation in space and time can be described in terms of rate equations that govern the flow of proton polarisation between the reservoirs coupled in series [22]. The reservoirs can be identified as follows: As the source of DNP, we have the electronic spin-spin interaction reservoir R0. Depending on the choice of the microwave frequency it will drive the proton polarisation towards $P = +0.98$ or $P = -0.98$, i.e. the reservoirs R0 will assume the values $P_0 = \pm 0.98$ (B=3.5 T, T = 1 K). R1 is the reservoir of the close protons that are presumably inside a spherical magnetic spin diffusion barrier with a radius of 5 Å [23]. In order to simulate a diffuse barrier, an additional reservoir, R2, of not so close protons inside a hollow sphere with an outer radius of R2 = 10 Å was assumed. R3 contains all the protons of the catalase molecule except those in R1 and R2. The protons of the deuterated solvent define R4. The protons in R3 and R4 are bulk protons, while those of R2 may be a mixture of close and bulk protons.

The build-up of proton polarisation $P_1(t)$ in R1, $P_2(t)$ in R2, $P_3(t)$ in R3 and $P_4(t)$ in R4 is governed by a set of four coupled first order differential equations:

$$\begin{aligned}
\frac{dP_1}{dt} &= \frac{W_{01}}{N_1}(P_0 - P_1) + \frac{W_{12}}{N_1}(P_2 - P_1)\\
\frac{dP_2}{dt} &= \frac{W_{12}}{N_2}(P_1 - P_2) + \frac{W_{23}}{N_2}(P_3 - P_2)\\
\frac{dP_3}{dt} &= \frac{W_{23}}{N_3}(P_2 - P_3) + \frac{W_{34}}{N_3}(P_4 - P_3)\\
\frac{dP_4}{dt} &= \frac{W_{34}}{N_4}(P_3 - P_4)
\end{aligned} \qquad (1)$$

As for the catalase molecule itself, N_i are the number of protons in Ri. N_4 is the number of solvent protons per catalase molecule. The rate constant W_{ij} are defined as probabilities of a mutual spin flip per time, e.g. the contact term W_{01} describes the flow of polarisation from R0 to R1.

The thermodynamic description of the evolution of nuclear polarisation given above has to be extended by the structural aspects which are relevant in neutron scattering. Using the expansion of the scattering amplitude as a series of spherical harmonics, we obtain the intensity of coherent small-angle scattering [1] [23]:

$$I_{coh.}(Q,t) = 2\pi^2 \sum_{l=0}^{L} \sum_{m=-l}^{m=l} \left| A_{l,m}^{(0)}(Q) + \sum_{i=1}^{3} P_i(t) B_{l,m}^{(i)}(Q) - P_4(t)\, B_{l,m}^{(4)}(Q) \right|^2 \qquad (2)$$

where $A_{l,m}(Q) = \sqrt{\frac{2}{\pi}}\, i^l \sum_n b_n\, j_l(Qr_n)\, Y_{l,m}^{*}(\theta_n, \varphi_n)$

Y and j are the spherical harmonics and spherical Bessel functions, respectively. The atomic positions are given by the polar co-ordinates r,θ,φ. The upper indices (i) > 0 refer to the reservoirs defined above. Apart from intensity of coherent scattering the experimental data of neutron scattering will contain incoherent scattering and background scattering.

$$I(Q,t) = C\,[I_{coh.}(Q,t) + I_{inc.}(t)] + I_{background}(Q,t) \qquad (3)$$

The factor C takes into account how the instrument is operated (e.g. incident neutron flux, measuring time, detector efficiency) and some properties of the sample, like concentration of the solute and the transmission of the sample.

The intensity of incoherent scattering is mainly due to the protons [24]

$$I_{inc.}(t) = \sum_{i=1}^{4} N_i\, \sigma_{incoh.}^{(i)}(t) = 8.48 \sum_{i=1}^{4} N_i \left(\frac{3}{4} - \frac{1}{2} P_i(t) - \frac{1}{4} P_i^2(t)\right) 10^{-24}\, cm^2 \qquad (4)$$

A first estimate of the background scattering intensity can be made by the measurement of the neutron scattering intensity of the empty refrigerator and the time dependent transmission of the polarised sample. There may be an additional Q- and t-dependent scattering intensity which could be due to spatially non-uniform proton polarisation, or to an intrinsic property of the instrument. These are taken into account in two ways.

1) The time-dependent background is described by a model, like the following
$$I_{background} = (c_1\, P_{i<4}(t) + c_2\, P_4(t))/Q^{c_3} \qquad (5)$$
2) The time dependent background scattering is substituted by that from a sample with a very low concentration of catalase.

After subtraction of $\boldsymbol{I}(Q) = 1/T_m \int_{t=0}^{T_m} I(Q,t)dt$ we obtain $Z(Q,t) = I(Q,t) - \boldsymbol{I}(Q)$. The difference between the calculated $Z_{calc.}(Q,t)$ and the measured $Z_{exp.}(Q,t)$ is minimized in an iterative procedure using as variables the scale factor C, and the rate coefficients W_{ij} and – only in case 1 – the coefficients c_1, c_2 and c_3.

$$\int_{Q=0.03}^{Q=0.09} \int_{t=0}^{t=10s} \left(Z_{exp.}(Q,t) - Z_{calc.}(Q,t)\right)^2 = Min \qquad (6)$$

The result is stabilized by including two more conditions, concerning the total scattering intensity and the variation of the NMR intensity during one cycle of DNP, respectively [18].

The equations (1) to (5) define the analysis of data from time –resolved polarised neutron scattering from protons in dynamic polarised targets. We replace this somewhat lengthy term by time-resolved proton polarisation and abbreviate it as TPP.

As the data are very noisy we use the leading members of a development as a Fourier series. $Z_{exp}(Q,t)$ then is defined as a two-dimensional array of intensities at $[m_1\, \Delta Q,\ m_2\, \Delta t]$ with $1 \leq m_1 \leq M_1$ and $1 \leq m_2 \leq M_2$. Omitting ΔQ and Δt we have

$$f(m_1,m_2) = \sum_{n_1=-N_1}^{N_1} \sum_{n_2=-N_2}^{N_2} F_C(n_1,n_2)\cos(z) + F_S(n_1,n_2)\sin(z) \qquad (7)$$
$$z = 2\pi\,(n_1 m_1/M_1 + n_2 m_2/M_2)$$

Reducing the Fourier analysis to the time axis we have

$$f(m,Q) = \sum_{n=-N}^{N} F_C(n,Q)\cos(z) + F_S(n,Q)\sin(z) \ \text{ with } \ z = 2\pi\, nm/M \qquad (8)$$

The resolution of the time-dependence will be restricted to |N|=1 throughout.

5. Results from time-resolved neutron scattering.

SANS from frozen solutions of catalase in mixtures of glycerol and D_2O as recorded by the area detector of D22 is shown in Fig. 3. The most striking feature is an unexpected important leaking out of the primary beam intensity, most likely due to an enlargement of the primary beam in the horizontal direction due the neutron spin polariser half a meter upstream the sample position. No amendment to this anomaly was made as it had been believed that these intensity peaks could well serve as measure of the time-dependent transmission of the sample.

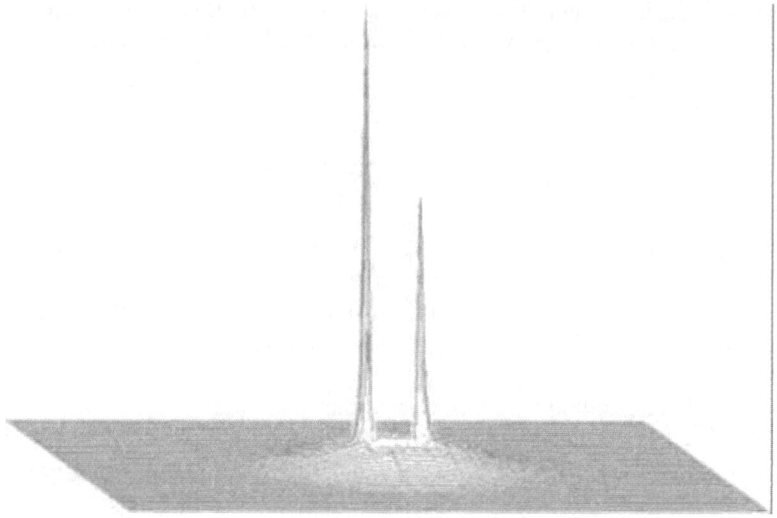

Fig. 3. Intensity of neutron scattering from catalase as recorded by the detector of D22. An important leaking out of the primary neutron beam is observed on the left and right side of the beam stop [1].

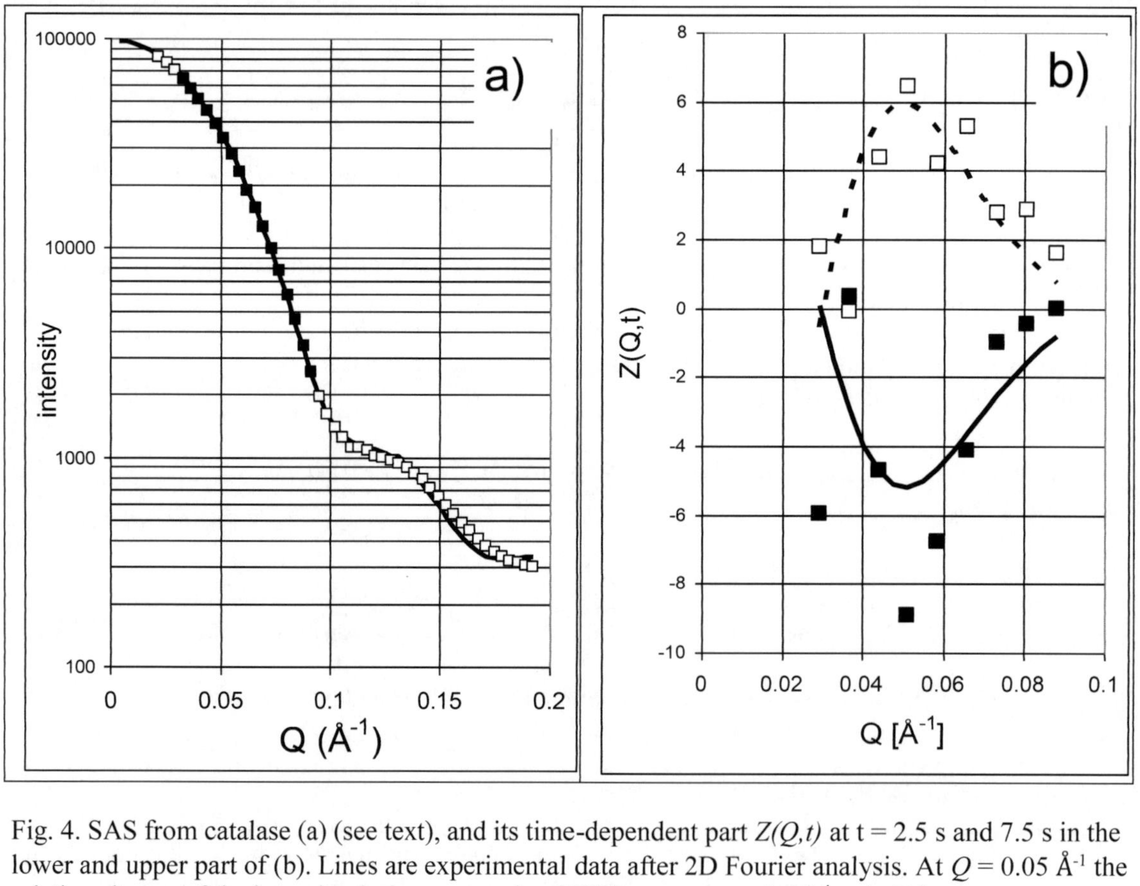

Fig. 4. SAS from catalase (a) (see text), and its time-dependent part *Z(Q,t)* at t = 2.5 s and 7.5 s in the lower and upper part of (b). Lines are experimental data after 2D Fourier analysis. At $Q = 0.05$ Å^{-1} the relative change of the intensity during one cycle of DNP amounts to 4×10^{-4} only [1].

In fact, the initial increase of the peak intensity at the onset of positive polarisation due to a decreasing transmission of the sample is compatible with the simultaneous decrease of SANS from catalase [1]. The relative change of the intensity is slightly below 1/1000 in both cases. From time resolved NMR measurement to be presented in section 6 it appears that the large change of the peak intensity cannot be reconciled with change of the transmission of the sample.

A Fourier analysis of the time dependence for each pixel contributing to the large peaks shows a regular shift of the phase defined as $\varphi = \arctan(F_S(1,Q)/F_C(1,Q)$ of 50° on passing from one to the next to the row of the area detector, i.e. the three pixels of the column on each side of the beam stop show an artefact which must be a property of the instrument rather than a physical property of the sample [1]. The question arises how to deal with this artefact and in particular how to separate it from SAS of catalase. Two approaches have been tried:

1) A 2D Fourier analysis is used with N_1=3 for the Q-axis. The background is described by (5).
2) A 1D Fourier analysis is used. SAS from catalase then is given as the difference between the intensity from a sample with a high concentration of catalase diminished by that from a sample with low concentration catalase. $F_C^{(high)}(1,Q) - F_C^{(low)}(1,Q)$ and $F_S^{(high)}(1,Q) - F_S^{(low)}(1,Q)$.

The measured intensity profile of SAS from catalase agrees with that of the obtained from the crystallographic model [25] (line in Fig. 4a). The experimental data at Q values between 0.03 and 0.09 Å^{-1} (■ in Fig. 4a) are used for a further analysis. The decrease of the time-dependent scattering $Z_{exp}(Q,t)$ at $Q < 0.05$ Å^{-1} is the signature of an instrument dependent response function mentioned above.

Using the background scattering defined by (5) and taking the radical site at tyr-369 for granted [10], $Z_{exp}(Q,t)$ in terms of the 2D Fourier analysis is reasonably well approximated by $Z_{calc.}(Q,t)$ derived from the model (Fig. 5a). There is an increasing fraction of background scattering at smaller Q (Fig. 5b).

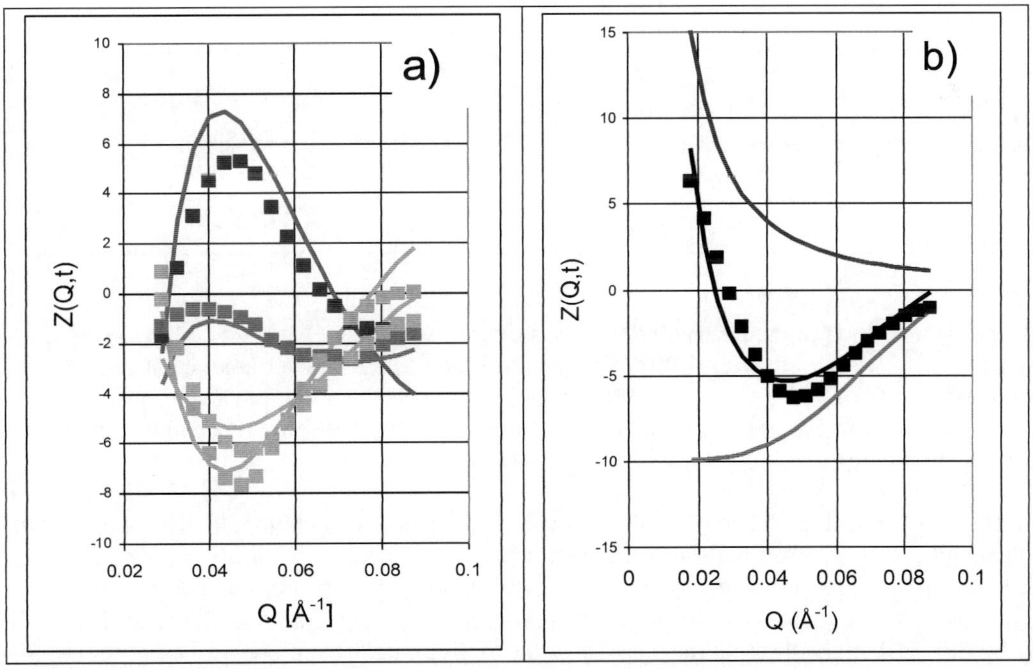

Fig. 5. Comparison of $Z_{exp}(Q,t)$ with $Z_{calc.}(Q,t)$ for various times of the cycle of DNP (a), and its decomposition into SAS from catalase and background scattering (b). a) $Z_{exp}(Q,t)$ at ■ t=0, ■ t=1.25 s, ■ t=2.5 s, ■ t=3.75 s, after 2D Fourier analysis. The corresponding $Z_{calc.}(Q,t)$ are presented as lines. b) Decomposition of $Z_{calc}(Q,t)$ (———) into SAS from catalase (———) and $I_{background}(Q,t)$ (———) at t = 2.5 s [1].

Starting from a 1D Fourier analysis as described under 2) a similarly good agreement between experimental and calculated $F_C(1,Q)$ and $F_S(1,Q)$ is obtained [1]. The scattering intensity for a sample with a low concentration of catalase appears to be an acceptable substitute for a blank sample containing no catalase at all.

An important spin-off of the fit of $Z_{calc.}(Q,t)$ to $Z_{exp}(Q,t)$ is the evolution of the proton polarisation $P_i(t)$ in each of the reservoirs Ri (Fig. 6).

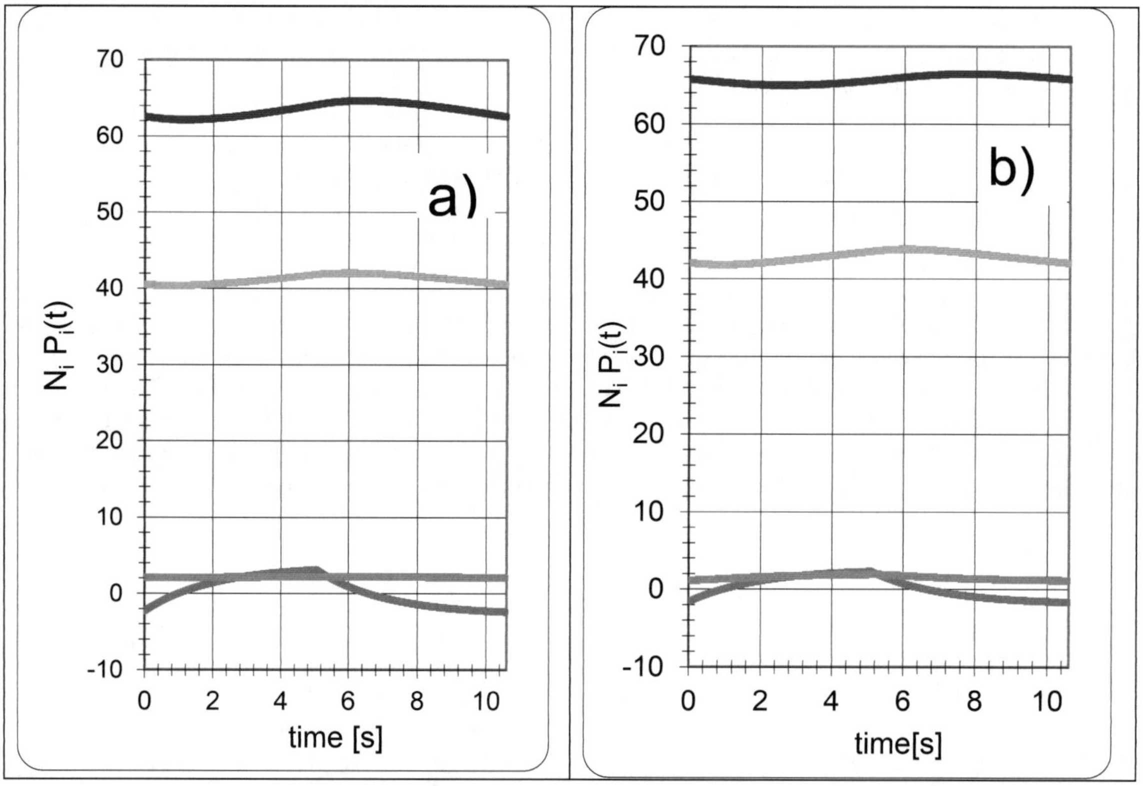

Fig. 6. The product Ni Pi(t) of the reservoirs Ri versus time. ——— i=1, ——— i=2, ——— i=3, ——— i=4. a) from a 2D Fourier series. b) from a 1D Fourier series. For the tetramer catalase molecule we have N_1 = 92(62), N_2=612(410), N_3 = 11700, N_4 = 18000. Taking into account the occupancy by tyrosyl in catalase of the 'active' catalase molecules N_1 and N_2 are reduced by a factor 0.67 (relevant to this figure).

The number of tyrosyl radicals per catalase molecules varies according to the binomial distribution $p_4(k)$ where k is the number of tyrosyls. Catalase with a single tyrosyl molecule will hardly contribute to dynamic polarisation because of the large inter-radical distance [1]. Catalase molecules with not more than a single radical will be disregarded. Moreover the occupancy of the radical sites determined by EPR is 0.58. Restricting the analysis to the 'active catalase molecules, i.e. those with more than one tyrosyl, the number of protons N_1 =92 in R1, and N_2 =612 in R2, is reduced to two third, i.e. 62 and 410 respectively (Fig.6). The number of polarized protons in excess, i.e. $N_P = N(up) - N(down)$, equals to the value of the ordinate. We are interested in the variation of polarized close protons N_P in R1 and R2, during one cycle of DNP, because these will image the electron spin density at the site of the radical. From Fig. 6a we obtains $N_P^{(1)}$ = 5.36 and $N_P^{(2)}$ = 0.1 polarised protons in R1 and R2, respectively. Taking into account the occupancy of 0.67 and sharing his value among four radical sites we obtain 5.37/0.67/4 = 2.0 polarised close protons per tyrosyl. An increase of $N_P^{(2)}$ in R2 at the expense of $N_P^{(1)}$ in R1 is obtained from data subjected to a 1D Fourier analysis.(Fig.6b). In detail we have $N_P^{(1)}$ = 3.85 and $N_P^{(2)}$=0.79 which results in 1.73 close protons per tyrosyl. An explanation for this lower value can be found under case 2 in section 4.

A final point concerns the detection of the tyrosine converted into a radical. So far our analysis of time-resolved neutron scattering started from tyr-369 as the most likely radical site. What is the answer from neutron scattering? The answer comes from a test subjecting all the 19 remaining tyrosines as possible radical sites to the same analysis as for tyr-369. The minimum rms values as defined by (6) increase with the distance of the hypothetical radical site from the centre of the catalase molecule [18]. This is true for both sets of input after 1D and 2D Fourier analysis, respectively. Tyr-369 together with tyr-357 and tyr-342 are equally good candidates for a radical site. Tyr-136 with a slightly higher rms deviation might be included as well. TPP selects four tyrosines out of 20 per subunit as possible radical sites (Fig. 7).

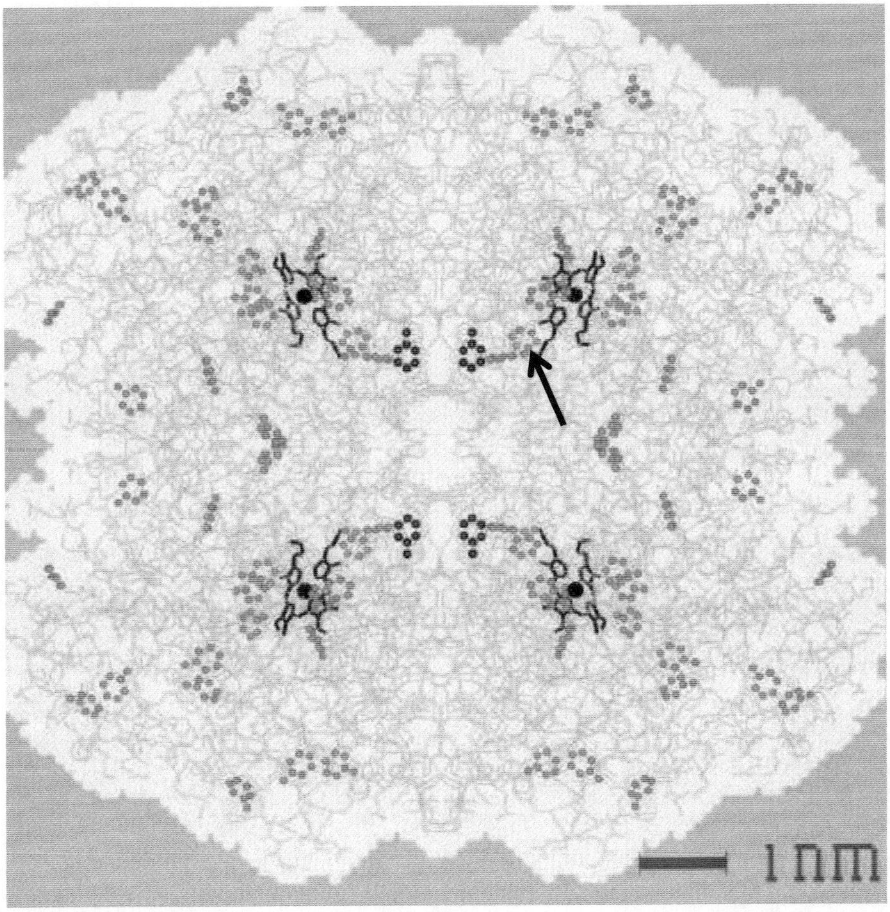

Fig 7. The tetramer bovine liver catalase molecule with its tyrosines presented by their phenyl rings, most of them in orange colour. From neutron scattering it appears that one of the following tyrosines may have become a radical: tyr-324, tyr-357, and tyr-369. They are marked in blue colour. Tyr-136 (marked by an arrow) is another slightly less possible candidate. From EPR [10] tyr-369 is suggested (in dark blue). Tyr-357 is very close to the iron atom (•) of the porphyrin ring [1].

6. Results from time resolved NMR.

The variation of the intensity of proton polarisation during each cycle of DNP has been measured simultaneously with the intensity of neutron scattering. As the sampling of the proton NMR signal takes nearly one second, four points were taken during each half cycle of DNP (Fig. 8). The change of the proton polarisation during one cycle of DNP is largely obscured by electronic noise. Selecting those cycles of DNP supposed to be less contaminated by the noise as described in the Appendix A, we observe a relative change of the proton polarisation close to $\Delta P = 10^{-4}$ which agrees remarkably well with ΔP from time-resolved neutron scattering. Moreover, the Fourier analysed measured proton polarisation follows fairly closely the proton polarisation calculated from the model. We note in passing that the change of the transmission of the catalase sample obtained from $\Delta P = 10^{-4}$ is of the order of 10^{-5}, far below the change of the peak intensities discussed in the preceding section. .

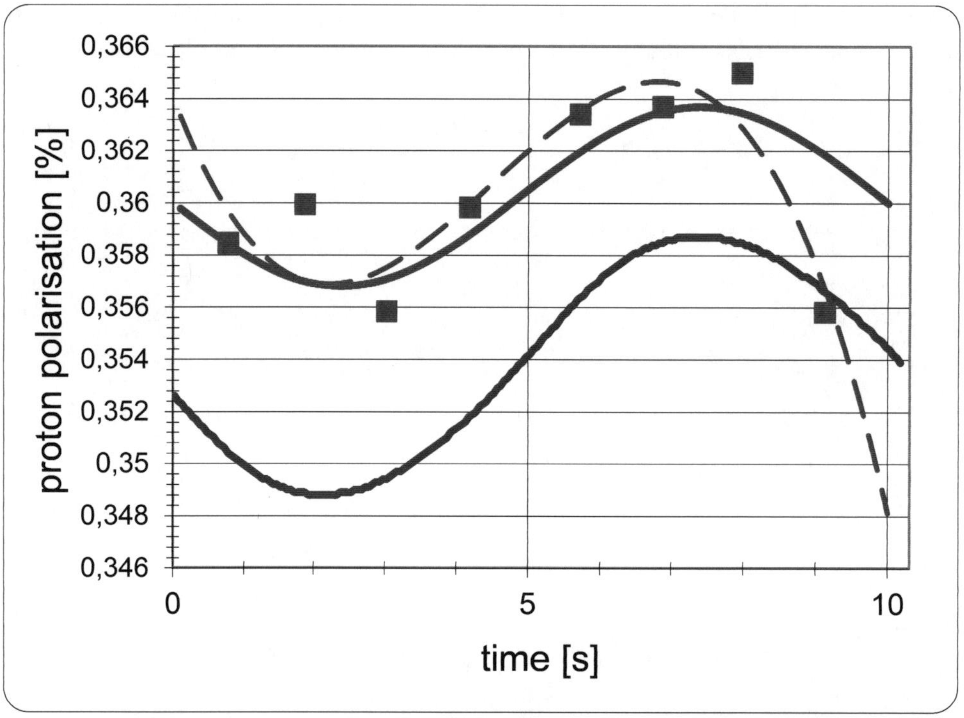

Fig. 8. Comparison of the measured proton polarisation with that obtained from the analysis of time-resolved neutron scattering. ■ measured proton polarisation, ——— its approximation by a Fourier series, and — — — by a polynomial of 3rd degree. ——— proton polarisation obtained from time-resolved neutron scattering. .

7. Outlook.

Two points might be considered in future applications of TPP: A technical one which concerns the optimization of the conditions for DNP. The other, strategic one, which concerns the future use of TPP.

The technical aspect is inspired by Fig. 2, which shows the increase of the proton NMR intensity with DNP at T = 0.1 K and T=1K. In spite of the change of the temperature by an order of magnitude there is nearly no change in the speed of proton polarisation. However the increase of the polarisation comes to an end after half an hour of DNP at T=1K, whereas a continuous increase of polarisation for several hours is observed at T=0.1K. Prolonged irradiation at microwave frequencies below and above the EPR will lead to final values of proton polarisation which become more different at lower temperatures (Fig. 9). An extrapolation to higher temperatures leads to $\Delta P = 0.01\%$ at T =15 K. What would be the outcome of time-resolved neutron scattering at T = 10 K? The reservoir R0 then would operate at P_0 = -0.23 and P_0 = +0.23 for negative and positive dynamic polarisation, respectively. The expected change of proton polarisation of 0.015 would be reached within 10 s, approximately. Assuming that nuclear spin diffusion will not depend on the temperature, the method of TPP might still be feasible at higher bath temperatures.

The method of TPP might well find greener pastures in protein crystallography. The determination of the radical density map of proteins then can be achieved in a fairly straight way. The application of the difference Fourier method to time-resolved neutron diffraction allows to follow the evolution of the radical density map in space and time [18]. The change of the intensity of some low–order reflections under the conditions discussed above by 10% looks promising [1].

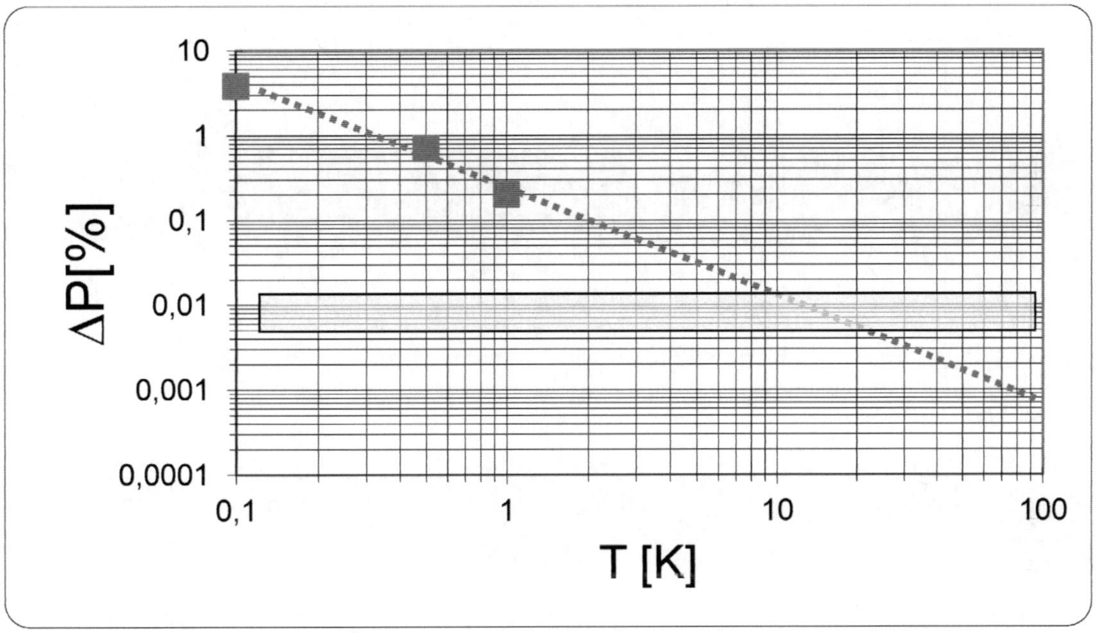

Fig. 9 The difference between the extreme values proton polarisation obtained by positive and negative DNP, respectively, versus the bath temperature. The yellow bar indicates the change of polarisation observed with TPP from catalase at T=0.1 K (this experiment).

8. Acknowledgements

This work was performed by the French-Swiss-German collaboration on the proton polarisation build-up by DNP, the team members of which are B. van den Brandt, H. Glättli, I. Grillo, P. Hautle, J. Kohlbrecher, E. Leymarie, S. Mango, R.P. May, A. Michels, and the authors. We thank the Paul-Scherrer-Institut (PSI) for the use of the facility of dynamic nuclear polarisation, the Institut Laue-Langevin for allocating beam time at the small-angle diffractometer D22, and last not least J. Gaillard (CEN Grenoble) for the EPR measurements. We are grateful to Salvatore Mango with whom we conceived the project at the PSI and to Ben van den Brandt for the successful continuation of the collaboration.

9. Appendix A

The detection of a periodic change of the proton polarisation by $\Delta P \leq 10^{-4}$ during one cycle of DNP is a technical and a numerical challenge. Here we show how the result shown in Fig. 8 has been obtained. The NMR intensity has been measured for 5 hours. At a first glance, there appears to be no periodic change of the NMR intensities with the alternating directions of DNP (Fig. 10). The time frame generator worked correctly. In a next step we calculate the rms of the NMR intensities for each of 1891 cycles of DNP (Fig. 11). A major part of the rms values is a broad band, and a small fraction must be considered

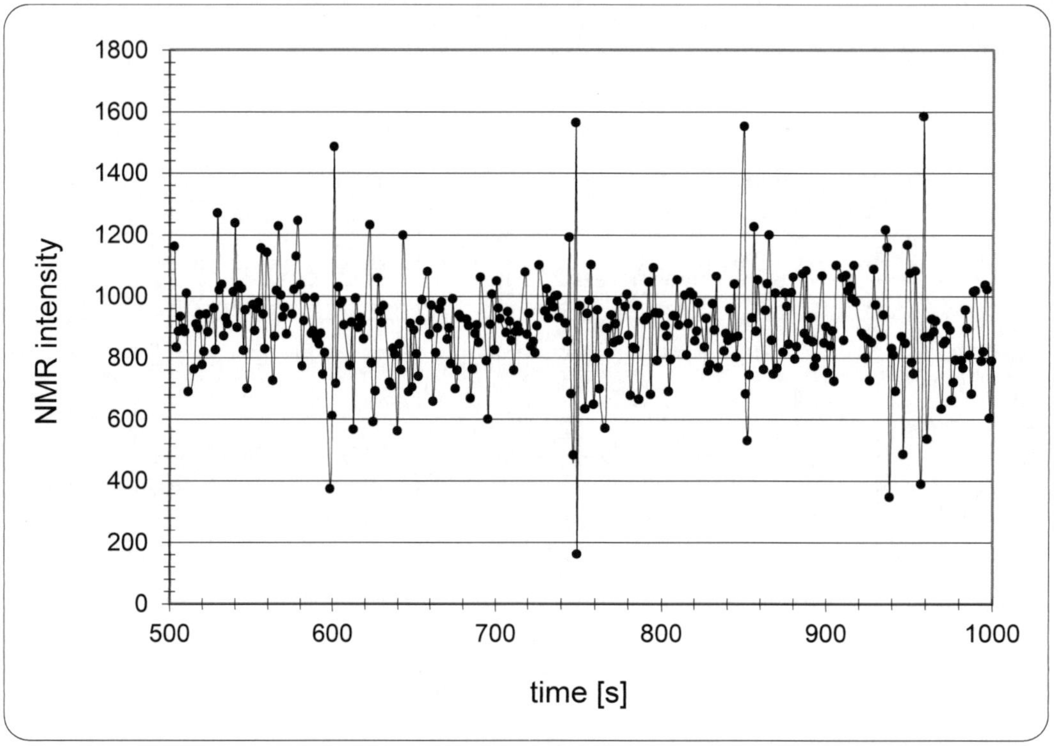

Fig. 10. A part of the NMR data. During one cycle of DNP 8 points were taken. Total time: 5 hours.

as outliers. At this point we cannot exclude 'hidden' outliers inside the majority of cycles represented by their rms values. To test this supposition we subdivide the broad band of rms values into stripes as indicated by the orange bars in Fig. 11. The NMR intensities of each of the 10 subsets are averaged, which results in 8 points for each of the 10 subsets (Fig. 12). The time dependence of NMR has been developed as a Fourier series for each subset. There are large differences among the 10 Fourier representations. We now look for the subsets which present coherent results. They appear to lie in the middle of the 10 subsets: the 3rd, 4th , 5th and 6th are quite similar. We assume that these 4 subsets which comprise the majority of cycles are closest to an unperturbed variation of the NMR intensity during one cycle of DNP.

The NMR intensities reflect proton polarisations close to the thermal equilibrium, i.e. P = 0.35% at B=3.5 T, and T=0.1K. As the NMR intensities are quite small, they are also very sensitive to an off-set due to instrument effects. After inspection of all our NMR data from different samples of catalase it appears that the NMR intensities should be crowded at a value lower than that shown in Fig.12. Using an off-set of 400 we obtain the proton polarisation of the 5th subset shown in Fig. 8 (in brown, both in Fig. 8 and 12).

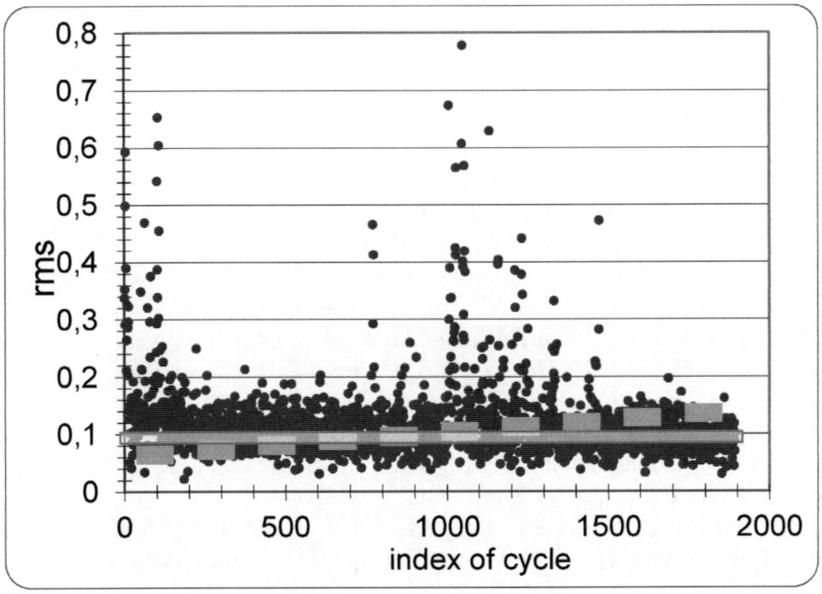

Fig. 11 The rms of the 8 NMR intensities for each cycle of DNP. The orange bars denote limits of intervals of rms. 10 intervals have been chosen. The rms values comprised by the 5th interval (from the left) are within the yellow semi-transparent area, for instance.

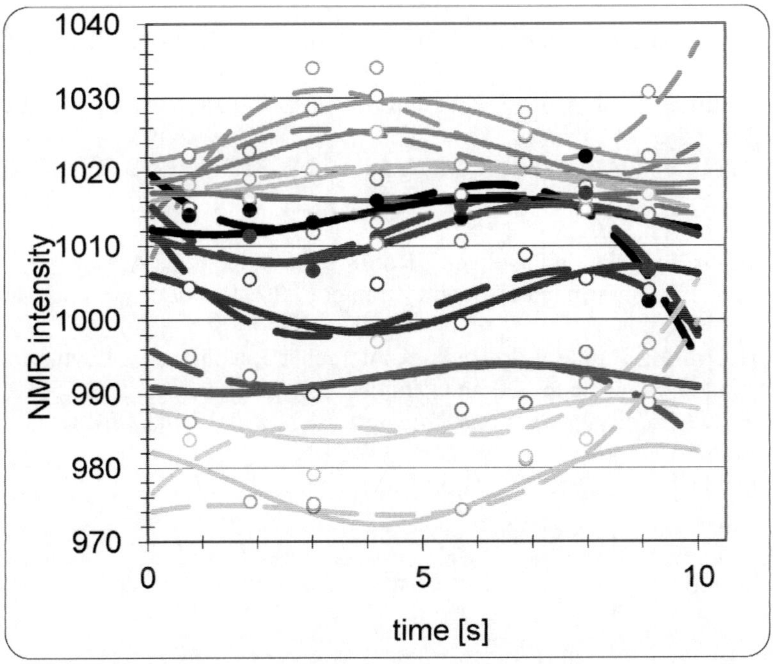

Fig. 12. Time dependent NMR from ten intervals of measured NMR intensities as described in Fig. 11. Interval numbers from lowest to highest intensity. NMR after 1D Fourier analysis: ——— 1, ——— 2, ——— 3, ——— 4, ——— 5, ——— 6, ——— 7, ——— 8, ——— 9, ——— 10. Intervals 5 and 6 are shown in Fig. 13. Broken lines represent a fit by a polynomial of 3rd degree. Symbols are NMR data averaged within a given interval.

References

[1] Zimmer O, Jouve H M, and Stuhrmann H B 2016 IUCrJ, 3, 326-340
[2] Wolfe J P 1973 *Phys. Rev. Lett.* 31 907-910
[3] Bloembergen N (1949 *Physica (Utrecht)* 15 386-426
[4] Furman G B, Goren S D 2002 *Z. Naturforsch.* 57a 307-314
[5] Abragam A, and Goldman M 1978 *Rep. Prog. Phys.* 41 395-468
[6] Abragam A. and Goldman M 1982 *Nuclear Magnetism, Order and Disorder* Oxford University Press
[7] Wenckebach W Th 2016 *Essentials of Dynamic Nuclear Polarisation* Amazon
[8] Cox S F J, Read S F J, and Wenckebach Th 1977 *J. Phys. C: Solid State Phys.* 10 2917-2936
[9] Jannin S, Comment A, and Klink J J 2012 *Applied Magnetic Resonance* 43 59-68
[10] Ivancich A, Jouve H M, Sartor B, and Gaillard J 1997 *Biochemistry* 36 9356-9364
[11] Svistunenko D A, and Cooper C E 2004 *Biophysical J.* 87 582-595
[12] Hayter J B, Jenkin G T, and White J W 1974 *Phys. Rev. Lett.* 33 696-698
[13] Knop W,Nierhaus K H, Novotny V, Niinikoski T O, Krumpolc C, Rieubland J M, Rijllart A, Schärpf O, Schink H-J, Stuhrmann H B and Wagner R 1986 *Helvetica Physica Acta* 50 741-746
[14] Koghi M, Ishida M, Ishikawa Y, Ishimoto S, Kanno Y, Masaike A, Masuda Y, and Morimoto K 1987 *J. Phys. Soc. Japan* 56 2681-2688
[15] Glättli H, Fermon C, Eisenkremer M and Pinot M 1989 *J. Physique* 50 2375-2388
[16] Willumeit R, Burkhardt N, Diedrich G, Zhao J, Nirhaus K H, and Stuhrmann H B (1996) J. Mol. Struct. 383 201-211
[17] Kumada T, Noda Y, Koizumi S, and Hashimoto T 2010 The Journal of Chemical Physics 133 054504
[18] Noda Y, Yamaguchi D, Hashimoto T, Shamoto S, Koizumi S, Yuasa T, Tominaga T and Sone T 2013 *Physica Procedia* 42 52-57
[19] Hautle P, Grübler W, van den Brandt B, Konter J A, Mango S, and Wessler K 1992 *Phys. Rev.* B 46 6596-6599
[20] Buckingham A D 2003 *Chem. Phys. Lett.* 371 517-521
[21] Van den Brandt B, Glättli H, Grillo I, Hautle P, Jouve H, Kohlbrecher J, Konter J A, Leymarie E, Mango S, May R P, Michels A, Stuhrmann H B, and Zimmer O 2004 *Nucl. Instr. and Meth.* A 526 81-90
[22] Van den Brandt b, Glättli H, Grillo I, Hautle P, Jouve H, Kohlbrecher J, Konter J A, Leymarie E, Mango S, May R P, Michel A, Stuhrmann H B and Zimmer O 2006 *Eur. Phys. J.* B 157-165
[23] Stuhrmann H B 2015 *Journal of Optoelectronics and Advanced Materials* 17 1417-1424
[24] Glättli H, and Goldman M 1987 *Methods Exp. Phys.* 23 C 307-314
[25] Fita I, and Rossmann M G 1985 J. Mol. Biol. 185 21-37

The contribution of small angle and quasi-elastic scattering to the physics of liquid water

José Teixeira

Laboratoire Léon Brillouin (CEA/CNRS) CEA Saclay F-91191 Gif-sur-Yvette Cedex France

jose.teixeira@cea.fr

Abstract. Many properties of liquid water at low temperature show anomalous behaviour. For example, density, isothermal compressibility, heat capacity pass by maxima or minima and transport properties show a super-Arrhenius behaviour. Extrapolations performed beyond the homogeneous nucleation temperature are at the origin of models that predict critical points, liquid-liquid transitions or dynamic cross-overs in the large domain of temperature and pressure not accessible to experiments because of ice nucleation.
A careful analysis of existing data can be used to test some of these models. Small angle X-ray or neutron scattering data are incompatible with models where two liquids or heterogeneities are present. Quasi-elastic neutron scattering, taking advantage and combining both coherent and incoherent scattering show that two relaxation times are present in liquid water and that one of them, related to hydrogen bond dynamics, has an Arrhenian behaviour, suggesting that the associated dynamics of the bonds, similar to the β relaxation of polymers, determines the glass transition temperature of water.

1. Introduction

The importance of water doesn't need to be emphasized. Liquid water is almost the only liquid present on the Earth surface, although it represents an extremely small fraction of the mass of the planet. It is generally accepted that life is only possible in its presence, reason why the research for life in the Universe coincides with research of water molecules in liquid state. Temperature and pressure conditions at the Earth surface allow the natural existence of the three states, liquid, ice and vapour, and associated phase transitions which feed a permanent water cycle between states through evaporation, condensation, melting and freezing processes. As well, there are circulation of big masses of liquid water in oceans with periodicities of the order of several thousand years.

Behind such behaviours there are physical processes which, contrary to the case of other liquids are complex, specific, eventually not yet well understood despite the huge amount of data available. Indeed, most of the thermodynamic properties of water show an "anomalous behaviour", an expression that means only that such properties depend on temperature and pressure much differently than for other liquids.

Knowing that a water molecule is chemically described by the simple expression H_2O, corresponding to the arrangement of one atom of oxygen and two atoms of hydrogen, the obvious question is: why is liquid water so complex? The simplest answer is that water cannot be reduced to an assembly of molecules similar to that of hard spheres or of many other simpler liquids. Between two

IOP Conf. Series: Journal of Physics: Conf. Series **848** (2017) 012003 doi:10.1088/1742-6596/848/1/012003

neighbouring water molecules both the interaction of dipolar moments and sharing of part of electronic clouds generate an attractive interaction: the intermolecular hydrogen bond, which is very directional.

Concretely, the attractive energy between two water molecules is maximum when the hydrogen atom of one molecule is almost aligned with the oxygen of a neighbouring molecule, a situation described by the notation O-H ... O. The bond length is the largest distance but the hydrogen atom can jump from one position to the symmetrical one along the bond direction.

At room temperature, at a given moment, most of hydrogen atoms are engaged in so-called "intact" bonds, meaning that, at very short times of observation, liquid water is like a gel far above its percolation threshold [1].

Due to hybridisation of molecular orbitals, the water molecule has the shape of a V with an angle HOH very close to the tetrahedral angle, that formed by the diagonals of a cube ($\cos^{-1}(-1/3)$). Together with the directionality of the bonds, this feature generates a local structure where each molecule is surrounded by 4 neighbours, sitting at opposite vertexes of a cube. This coordination number, that characterises most of the crystalline forms of water, is very small as compared to the much more compacted structures of atomic liquids. In liquid water, because of thermal energy, the coordination number is larger than 4 but always affording a large empty volume and room for large amplitude fluctuations of intermolecular bonds which damp rapidly the extension of the oscillations of the pair correlation function. Despite the existence of a large connectivity length, the extension of the radial correlation function is small.

The energy of a hydrogen bond formed between two water molecules is relatively high: of the order of 8 $kJ.mol^{-1}$ a value more than 3 times above the kinetic energy at room temperature. However, the lifetime of an hydrogen bond, whatever is the definition taken to evaluate it, is short - of the order of 1 ps. This is due to the large amplitude librations of the hydrogen atoms in the direction perpendicular to that of the bond which bring the light hydrogen atom out of the narrow cone where the attractive energy is high. In liquid water at room temperature, even if the number of "intact" bonds is larger than 60% of the possible bonds between neighbouring molecules, their lifetime is very short, what explains the low viscosity of water.

The temperature and pressure dependences of most of water properties are very different from those of other liquids, particularly at low temperature when the number of intact bonds is higher. Qualitatively, some properties can be understood. For example, larger the average number of bonds per molecule and smaller the coordination number. Indeed, this number decreases from 4.6 at room temperature to 4.3 at 0° C [2]. In crystalline ice I_h it is equal to 4. This temperature dependence of the coordination number explains the existence of a temperature of maximum density of the liquid and the lower density of ice, the effects due to the formation of bonds compensating the normal thermal expansion.

Shortly, the anomalous properties of liquid water are due to three features:
- large number of intermolecular directional hydrogen bonds
- local tetrahedral arrangement
- short lifetime of hydrogen bonds

Among the unusual thermal and transport properties of water, isothermal compressibility and thermal capacity pass through minima at 46 and 30 °C, respectively. They represent enhanced density and entropy fluctuations.

Numerical extrapolations of such temperature dependences diverge around -45° C, which is close to the temperature of homogeneous nucleation of ice as first noticed by G. S. Kell [3]. As well, R. Speedy and C.A. Angell [4] have shown that several thermodynamic and transport properties of supercooled water (metastable state) could be well represented by power laws similar to those describing critical phenomena, through a reduced temperature parameter $(T-T_s)/T_s$, with $T_s = -42°$ C. Because nucleation prevents experiments at temperatures below -30° C, in some cases -35° C, data are scarce in the interesting temperature domain where the anomalies are larger and show apparent divergences.

Instead a rapid quenching of liquid water yields formation of a glassy state with a structure similar to that of amorphous ice obtained by vapour deposition on a cold surface, what confirms the continuity between room temperature and glassy water. Consequently, the properties of water between the lowest temperatures accessible to experiment (around -35 °C) and the temperature of the glass transition (135 K) are unknown and have been object of many speculations since almost 50 years!

The discovery of a high density form of amorphous ice [5] contributed substantially to the discussion of the properties of liquid water in the large temperature range extending over 100°, from 240 K down to 135 K, in some cases the resurgence mixture models left out since 40 years.

Very schematically, the approaches based either in "critical" phenomena or in mixture models are the following:

1) Existence of a spinodal line on the low temperature side of the phase diagram of water, along which a real divergence of water properties does exist. This line is an extension of the normal spinodal that starts from the critical point on the high temperature side [4].

2) Assuming that there is a first order transition between the two forms of amorphous ice (high and low density), the coexistence line can be extrapolated till a "low temperature critical point" (LTCP). The position of this point is extremely diverse depending on extrapolations and on effective potentials used in molecular dynamic simulations but it would be placed at pressures above atmospheric pressure. From this point, another extrapolation of the coexistence line beyond the LTCP is the loci of extrema of the values of water properties. For example, compressibility or heat capacity, don't diverge but show "bumps" when crossing this extrapolated line (sometimes called "Widom line") [6] .

3) Similar ideas about existence of a line, loci of extrema but due to a dynamic crossover: the temperature dependence of the transport properties of supercooled water are non-Arrhenian or Arrhenian, at temperatures above or below this line, respectively. Crossing the line by decreasing temperature corresponds to what is called a "fragile to strong liquid transition" where the words fragile and strong correspond to non-Arrhenian and Arrhenian behaviour, respectively [7].

4) A pure mixture model that supposes that, with decreasing temperature, water goes from a high density homogeneous liquid at high temperature to a low density form at very low temperature. In between, i.e. room temperature and supercooled water are mixtures of the two hypothetic liquids, with clusters of the low density phase immersed in the high density phase at higher temperatures and the opposite at lower temperatures [8].

There is a large and controversial literature about these models, and others, which are mostly based on several computer simulations of molecular dynamics of water, a few of them being compatible with the existence of a LTCP. In other cases, extrapolations are done from data obtained either in aqueous solutions or under confinement. Indeed, in such conditions, water doesn't freeze or freezes at very low temperatures. However, because one or more of the three features at the origin of the specific behaviour of water, which are mentioned above, most of these results are irrelevant for the case of bulk water. Actually, under confinement or in aqueous solutions, the anomalies vanish.

The purpose of this paper is to show the contribution of a few neutron scattering experiments within the context of the proposed models. A precise analysis of data, even at room temperature, show that some models are hardly incompatible. In section 2, Small angle Scattering data are discussed. In section 3, arguments will be presented about different interpretations of quasi-elastic neutron scattering. Section 4, mostly based on the previous discussion, presents a possible framework for the explanation of water anomalies, free of critical behaviour.

2. **Small Angle Scattering**

Small angle X-ray and neutron scattering probes spatial correlations at scales of the order of 1 nm. One of the more frequent applications of these techniques is the evaluation of sizes of nano-particles, aggregates or inclusions, in situations where two phases can be identified and contribute to the scattered intensity through a "contrast", which is a difference between the density scattering length of each phase. In the case such entities have small interactions, the scattered intensity is described by a "form factor", square of the amplitude of the Fourier transform of the individual particles. The other contribution to the scattered intensity is due to the number fluctuations which is determined by the isothermal compressibility of the sample.

At small values of the momentum exchange Q, the general expression of the scattered intensity S(Q) is:

$$S(Q) = k_1(\Delta\rho)^2 + k_2(\Delta c)^2 + k_3(\Delta\rho\Delta c) \qquad (1)$$

where $\Delta\rho$ and Δc represent number and concentration fluctuations, respectively, k_i are constants and the last term takes into account an eventually cross term.

In the case of a homogeneous liquid, only the first term exists and the previous equation reduces to:

$$S(Q) = k_1\,(\Delta\rho)^2 = Nk_BT\chi_T f(Q) \qquad \text{with } f(Q=0) = 1 \qquad (2)$$

In this expression, N is the number of molecules, k_B is the Boltzmann constant, T the temperature and χ_T the isothermal compressibility.

Far from the liquid-vapour critical point, the magnitude of $S(Q)$ is small what implies careful measurements of the absolute intensity. For example, in the case of deuterated water (D_2O) at room temperature, measured by neutron small angle scattering, $S(Q=0) = 0.0071$ cm^{-1}, to be compared with $S(Q=\infty) = 0.0407$ cm^{-1}. This value is very small even compared with other liquids because the compressibility of water is also very small.

Showing that the experimental value of $S(0)$ coincides with the contribution of number fluctuations, in 1974, R.W. Hendricks et al [9] demonstrated the homogeneity of liquid water, excluding contributions due to another phase of different density, which would imply a contrast and a non-zero value of k_2 in equation (1).

The temperature domain of observation has been extended in the supercooled region, down to -20° C [10] confirming the absence of heterogeneities in liquid water, thus of a mixture of two liquids with different densities (figure 1).

Figure 1
X-ray scattering of liquid water [10]. Green circles: 65° C; blue squares: 25° C; red triangles: -20° C. Points at Q=0 are obtained from eq. (2).

Independently, the evaluation of the density, compressibility or thermal capacity of the eventual coexisting phases necessary to fit the observed extrema results suppose implicitly aberrant temperature dependences of the properties of each constituent [11]. Instead, it is worth noticing that, at low temperature, the function *f(Q)* in equation (2) has a small *Q* dependence denoting a small correlation length of the order of 5 to 8 Å, depending on the assumptions of the performed analysis [10]. It is important to emphasize that, even for large values of the correlation length as observed in liquids with high compressibility at the vicinity of critical points, a correlation length of number fluctuations corresponds to a decay of density fluctuations seen around the origin of the pair correlation function. In any case can it be associated to heterogeneities visible in real space.

In the case of water, an increase of the correlation length of number fluctuations associated to a parallel increase of isothermal compressibility and to some form of critical behaviour with decreasing temperature cannot be excluded. However, nucleation of the crystalline phase prevents large density fluctuations.

3. Quasi-elastic neutron scattering

Quasi-elastic neutron scattering (QENS) is one of the rare techniques that probes individual dynamics of atoms and molecules. Due to the extremely large incoherent cross section of protons, the scattered intensity of liquids containing hydrogen atoms probes the ensemble average of the time-dependent correlation function $<r(0).r(t)>$ where *r(t)* is the position of each hydrogen atom at time *t*. Analysis of QENS of bulk water can be done assuming a decoupling between motions of the whole molecules and motions of hydrogen atoms relative to the molecular centres of mass [12], but other strategies are possible, namely in conjunction with molecular dynamics simulations [13].

In the first case, coupling between global and internal motions of hydrogen atoms does exist but its contribution is often neglected. Relaxations are described in energy space by Lorentzian functions which are the Fourier transforms of single exponentials $exp(-t/\tau)$.

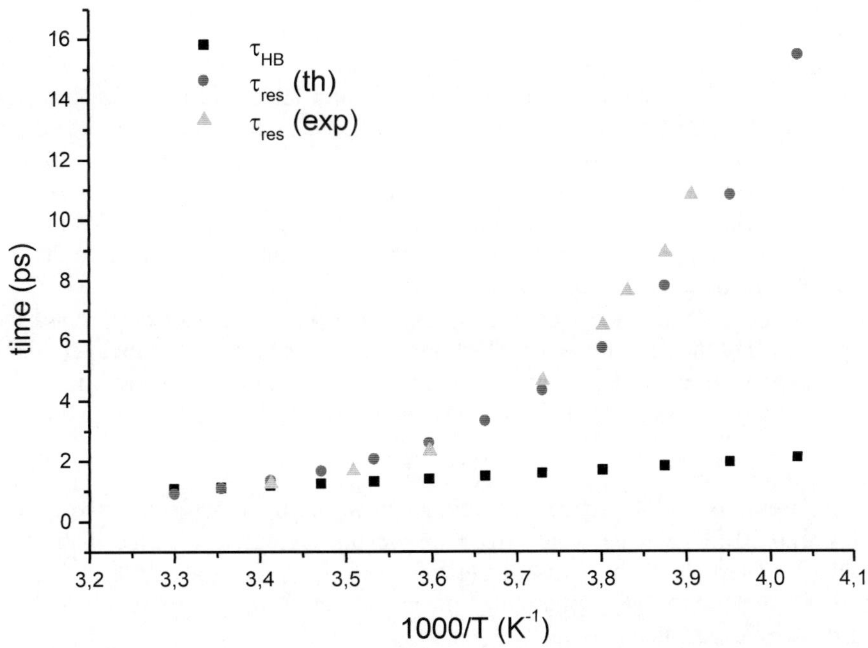

Figure 2
Black squares and green triangles: respectively, hydrogen bond lifetime and residence time of liquid water evaluated by QENS [12], plotted in function of the reciprocal temperature. Red dots: The same residence time calculated from the hydrogen bond time and the number of intact hydrogen bonds [15].

Analysis of the QENS spectra of supercooled water separates a narrow component describing the molecular self-diffusion at different length scales and a wider one that corresponds to motions of hydrogen atoms around the oxygen atom, similar to large amplitude librations. Because of the high directionality of intermolecular motions, this component is associated to the dynamics of the bonds, what means that, in some way, it measures hydrogen bond lifetimes. The remarkable result of this analysis is that, while the self-diffusion component follows the strongly non Arrhenius behaviour of transport properties of water, the temperature dependence of the hydrogen bond lifetime is Arrhenius at all measured temperatures, i.e. -20° C, with an activation energy that corresponds to the energy of the bonds. This result is also found in recent computer simulations [14].

Moreover, it was shown that the anomalous temperature dependence of the molecular residence time can be evaluated from the hydrogen bond lifetime taking into account the temperature dependence of the number of intact bonds. [15]

Figure 2 depicts the temperature dependence of the residence time and of the hydrogen bond lifetime as well as the evaluation of the residence time based on the temperature dependence of the number of intact bonds.

When water is confined either in tiny pores or sitting on hydrophilic surfaces, a variety of relaxation times coexist depending on local environment and structure. Discrimination between several populations is not or not always possible, what may justify a heuristic method that describes all the relaxations by a global stretched exponential, $exp\ (-t/\tau)^{\beta}$, with $0<\beta<1$. From this analysis, an average value $<\tau>$ of the ensemble of relaxation times can be extracted and interpreted.

Obviously, the same analysis can be applied to bulk water and the quality of the fits does not allow to discriminate between the consideration of two components and the heuristic approach.

The analysis of the dynamics of coherent scattering of water gives the solution of the conundrum. For this purpose, a sample of deuterated water has been measured by a triple-axis instrument and by a spin-echo spectrometer. In the first case, one measures the function $S(Q,\omega)$ while the second technique measures the intermediate function $I(Q,t)$, Fourier transform in time domain of $S(Q,\omega)$.

Both functions are linear combinations of the three partial pair correlation functions $S_{ij}(Q,\omega)$ where the indexes i and j refer to oxygen and deuterium atoms. Two Q values where selected. To one of them (Q_1=1.95 Å^{-1}) corresponds approximately the maxima of the three partials, thus the well-known maximum of the total $S(Q,\omega)$ around 2 Å^{-1}. To the larger value Q_2=3.54 Å^{-1} corresponds a local maximum of the partial DD, and negligible contributions of the other partials. With this choice, the measurement of $S(Q,\omega)$ at Q_1 gives information about the molecular dynamics while the measurement at Q_2 is dominated by the dynamics of deuterium atoms.

In both cases, i.e. measurements of $S(Q,\omega)$ and of $I(Q,t)$, it was clearly established that the temperature dependence of the linewidths (or relaxation times) was very important at Q_1 and hardly seen at Q_2, justifying experimentally the decomposition of the incoherent scattered intensity into two components discriminating non-Arrhenius residence times from Arrhenius hydrogen bond lifetimes.

4. **Discussion and conclusion**

Experimental data presented in the preceding sections allow a critical analysis of the different models that have been proposed to describe water properties in the temperature range that, at atmospheric pressure, extends from -30° C to the glass transition temperature. Necessarily based either on a few effective potentials extended to low temperatures or on simpler numerical extrapolations, they are not always compatible with experimental data.

This is the case of all mixture models which suppose that liquid water is a mixture of two liquids of different densities in relative concentrations that change with temperature.

The two liquids, would correspond to two of the different forms of amorphous ice: low and high density amorphous ice (LDA and HDA, respectively) with densities that differ by about 20%.

Such large difference of density generates a X-ray or neutron contrast that would be detected as a "concentration fluctuation" contribution to the scattered intensity similarly to what happens in any liquid mixtures, for example water-ethanol [10, 16].

Moreover, as pointed out above, assuming that the density of two liquids have a monotonous temperature dependence, the maximum of water density at 4° C cannot be explained without accepting very unusual temperature dependences of the densities of the two liquids [11].

Similarly, the results obtained by QENS exclude any complexity of the water dynamics that could generate distributions of residence times and relaxations described by stretched exponentials. Indeed, the separation into two components is the right way to take into account the dynamics of the hydrogen bonds and that of the whole molecule.

This result doesn't necessary contradict QENS of water under confinement [7]. In such cases, indeed, several relaxation times are present. For example, the formation of bonds and their lifetime depend on the interface [17] and molecular diffusion is not the same for molecules more or less close to the interfaces. The main merit of analysis of such situations using stretched exponentials to account for the variety of relaxation times is to define an average time $<\tau>$ that can be followed in function of external parameters, namely the temperature, although its numerical value is not necessarily very significant.

It was established [7] that the temperature dependence $<\tau>$ is super-Arrhenius above 220 K and Arrhenius at lower temperatures, what was interpreted as dynamic cross-over of water under confinement, in this case hydration water of lysozyme. Trying to relate this observation to the dynamics of supercooled water, it is possible that, at temperatures below 220 K, all translational motions are frozen and that the only contribution to the incoherent scattering is due to the hydrogen bond dynamics, with an Arrhenian dependence on temperature. In other words, dynamics of bulk and confined water are similar below 220 K.

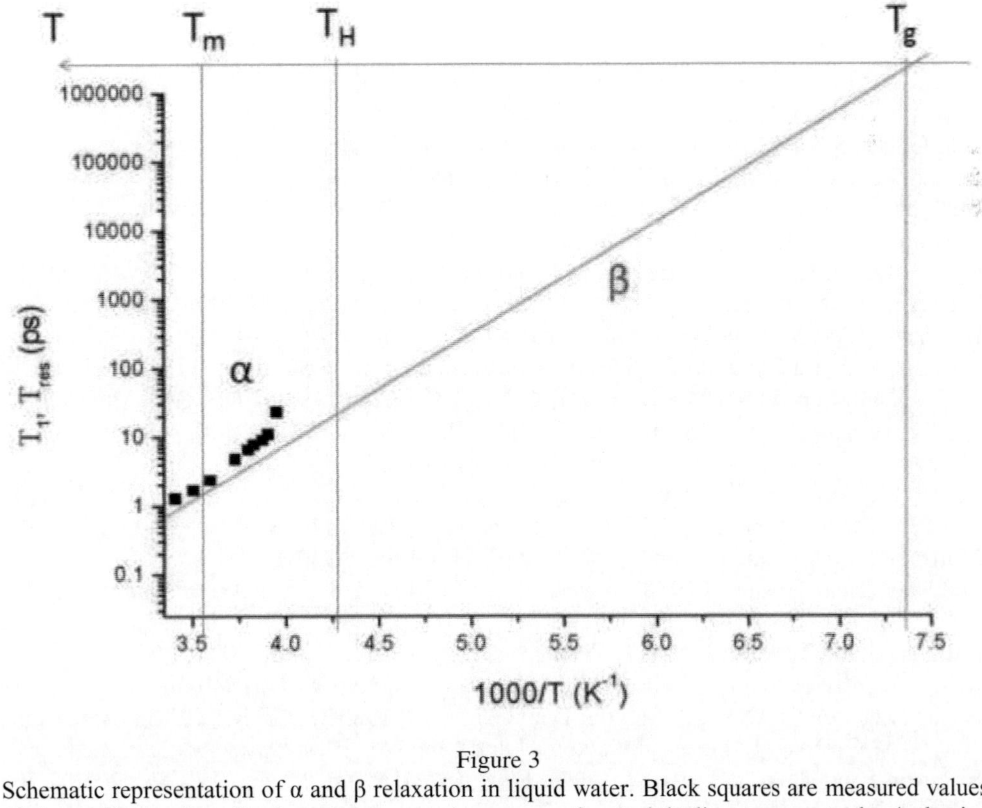

Figure 3

Schematic representation of α and β relaxation in liquid water. Black squares are measured values of the residence time. In the same temperature range, the straight line represents the Arrhenius temperature dependence of hydrogen bond life-time which extrapolates to the glass transition at T_g. T_M is the melting temperature and T_H is the temperature of homogeneous nucleation of crystalline ice.

Taking into account these results, mixture models or postulating the existence of heterogeneities are not compatible with the available experimental data. Likely, more attention should be paid to the fact that both the extrapolated values of criticality and dynamic crossover coincide with the experimental value of homogeneous nucleation temperature, T_H, within a few degrees. Actually, it has been established since a long time [18] that the local structure of liquid water converges to that of LDA at T_H, far above the glass transition temperature, T_g , what means that at T_H, the remaining disorder is due only to hydrogen bonds and rotations of hydrogen atoms around the centres of mass of water molecules. Similarly to polymer dynamics, it is possible to call α relaxation, that related to translational motions of molecules and β relaxation that of hydrogen bonds, the first being strongly non-Arrhenius while the second is Arrhenius [19] (figure 3).

However, contrary to the case of polymers, at the macroscopic level it is the β relaxation that imposes the glass transition. Between T_H and T_g, small amplitude rotations of hydrogen atoms allow crystallisation of the stable crystalline form of ice, what is consistent with the relatively small enthalpy change observed at the glass transition.

This relatively simple picture, free of heterogeneities or critical behaviour is compatible with available experimental data and also with the majority of effective potentials used in computer simulations.

Because of the large extension of the domain of stability of liquid water it is unlikely that a general consensus about its properties can be reached. At the best, experiments performed in the accessible regions of temperature and pressure can reduce the number of postulated models. The present work based in two neutron scattering techniques is presented in this context. Other experiments and comparisons will be necessary to solve, whenever possible, what is called the "water conundrum". [20].

References

[1] Geiger A, Stillinger FH and Rahman A 1979 *J. Chem. Phys.* **70** 4185
[2] Némethy GN and Scheraga HA 1962 *J. Chem. Phys.* **36** 3382
[3] Kell GS 1975 *J. Chem. and Engineering Data* **20** 97
[4] Speedy RJ and Angell CA 1976 *J. Chem. Phys.* **65** 851
[5] Mishima O, Calvert LD and Whalley E 1984 *Nature* **310** 393 and 1985 *Nature* **314** 76
[6] Mishima O and Stanley HE 1998 *Nature* **396** 329
[7] Chen S-H, Liu L, Fratini E, Baglioni P, Faraone A and Mamontov E 2006 *PNAS* **103** 9012.
[8] Pettersson LGM and Nilsson A 2015 *J. Non-Crystalline Solids* **407** 399
[9] Hendricks RW, Mardon PG and Shaffer LB 1974 *J. Chem. Phys.* **61** 319
[10] Bosio L, Teixeira J and Stanley HE 1981 *Phys. Rev. Lett.* **46** 597; Bosio L, Teixeira J and Bellissent-Funel M-C 1989 *Phys. Rev.* **A39** 6612 (1989); Clark GNI, Hura GL, Teixeira J,Soper AK and Head-Gordon T 2010 *PNAS* **107** 14003
[11] Johari G and Teixeira J 2015 *J. Phys. Chem. B* **119** 14210
[12] Teixeira J, Bellissent-Funel M-C, Chen S-H and Dianoux AJ 1985 *Phys. Rev.* **A31** 1913
[13] Qvist J, Schober H and Halle B 2011 *J. Chem. Phys.* **134** 144508
[14] Martiniano HFMC and Galamba N 2013 *J. Phys. Chem. B* **117** 16188
[15] Stanley HE and Teixeira J 1980 *J. Chem. Phys.* **73** 3404; Teixeira J 2008 *Pramana -J. of Physics* **71** 761
[16] d'Arrigo G and Teixeira J 1990 *J. Chem. Soc. Faraday Trans.* **86** 1503
[17] Russo D, Ollivier J and Teixeira J 2008 *Phys. Chem. Chem. Phys.* **10** 4968
[18] Bosio L, Teixeira J, Dore JC, Steytler D and Chieux P 1983 *Mol. Phys.* **50** 733; Bellissent-Funel M-C, Bosio L, Dore JC, Teixeira J and Chieux P 1986 *Europhys. Lett.* **2** 241
[19] Swenson J and Teixeira J 2010 *J. Chem. Phys.* **132** 014508; Teixeira J, *Water* 2010 **2** 702
[20] Luzar A 2000 *J. Chem. Phys.* **113** 10663

Exchange interactions in heavy rare-earths RCo_2 compounds

E.Burzo

Faculty of Physics, Babes-Bolyai University, 400084 Cluj-Napoca, Romania

emil.burzo@phys.ubbcluj.ro

Abstract. The magnetic behaviour of RCo_2 (R = Tm, Er, Ho, Tb) compounds, above the Curie temperatures, T_c, has been analysed. The interdependent exchange interactions at the level of unit cell are discussed in correlation with R5d band polarizations. The clusters evidenced by SANS experiments in paramagnetic region, having correlation length of ≅ 7Å, were associated with the exchange coupled atoms at the level of the unit cell. The exchange interactions are not so high to induce a magnetic ordered phase.

1. Introduction

The RCo_2 compounds, where R is a rare-earth or yttrium crystallize, at normal conditions in a cubic-type structure having Fd3m space group [1]. Below the Curie temperatures, T_c, their crystal structures are distorted due to magnetostrictive effects.

The RCo_2 compounds, where R is a magnetic heavy rare-earth (R = Gd to Tm) are ferrimagnetically ordered, the rare earth and cobalt moments being antiparallelly oriented. The magnetic transitions are of first order when R = Dy, Ho, Er and of second order for R = Gd, Tb, Tm. Above the Curie points, the reciprocal susceptibilities, χ^{-1}, follow non-linear temperature dependences as expected for ferrimagnetic systems [2, 3]. The paramagnetic data were analysed considering that cobalt has an exchange enhanced magnetic susceptibility [4] or an effective intrinsic moment [2, 3].

The polarized neutron diffraction studies performed on RCo_2 compounds with R = Tb [5], Ho [6,7] or Tm [7,8], above the Curie points, in the presence of external fields, evidenced an antiparallel orientation of rare-earth and cobalt moments. A change in the direction of cobalt moment from antiparallelly to parallelly oriented to Tm one was shown in $TmCo_2$, as temperature increased [7, 8]. The paramagnetic data were analysed assuming that cobalt has an exchange enhanced magnetic susceptibility, the cobalt moment being induced by the total field acting on the cobalt atoms [5-8]. The exchange field acting on cobalt, at $T > T_c$, is smaller than the value characteristic for metamagnetic transitions, $H_c \cong 75$ T [9-12]. Thus, only an intrinsic effective cobalt moment can be present. This statement has been confirmed particularly by analysing the thermal dependences of magnetic susceptibilities for RCo_2 compounds with non-magnetic [13, 14] and magnetic rare-earths [15, 16].

The following studies, performed on $ErCo_2$ compound, evidenced also an antiparallel orientation of Er and Co moments at $T > T_c$ [17-19]. The presence of intrinsic cobalt moment was also shown [17]. In addition, the SANS measurements evidenced short-range correlations, having correlation length of ≅ 7 Å [20, 21]. These were associated with the presence of Griffiths–like phase above Curie point. The same antiparallel orientation of rare-earth and cobalt moments was further reported in RCo_2 (R = Ho, Tm) compounds [22-24], in agreement with previous studies [6-8]. The Griffits phase was described as "*establishing a short–range order a kind of low temperature remnant magnetic order of undiluted system*" [23]. The antiparallel orientation of rare-earth and cobalt moments, in the presence of external field, at $T > T_c$, was called parimagnetism. As function of temperature and external field, different arrangements of cobalt and rare-earths moments, above T_c, were also shown [22-24].

In the present paper, starting from the analysis of exchange interactions as well as of cobalt magnetic behaviour at $T > T_c$, the short range order previously reported [20, 21], was associated with the exchange coupled magnetic atoms at the level of the unit cell, having dimensions of ≅ 7.2Å. No magnetic ordered phase, of the Griffits-type, can be shown, at $T > T_c$, in RCo_2 (R= Tb, Ho, Er, Tm) compounds.

2. Exchange interactions in RCo_2 compounds

The exchange interactions in RCo_2 compounds, at $T < T_c$, are rather complex, essentially determining the cobalt moments. Those between R and Co atoms are described as being of 4f-5d-3d type [25, 26]. The exchange interactions between cobalt atoms are of short range. Since of high extension of R5d orbitals, short range interactions are expected to take place between R atoms [27], in addition to those by means of conduction electrons.

The band structure calculations were performed on RM_2 (M = Fe, Co, Ni) compounds with heavy rare-earths [26]. The R5d band polarizations, M_{5d}, in RCo_2 compounds, as function of De Gennes factor, G, are plotted in Fig.1. These are antiparallelly oriented to cobalt moments and can be described by the relation:

$$M_{5d} = M_{5d}(d) + \alpha G$$

where $M_{5d}(d)$ is the R5d band polarization extrapolated at G = 0 and $\alpha = 2.1 \cdot 10^{-2}\ \mu_B$.

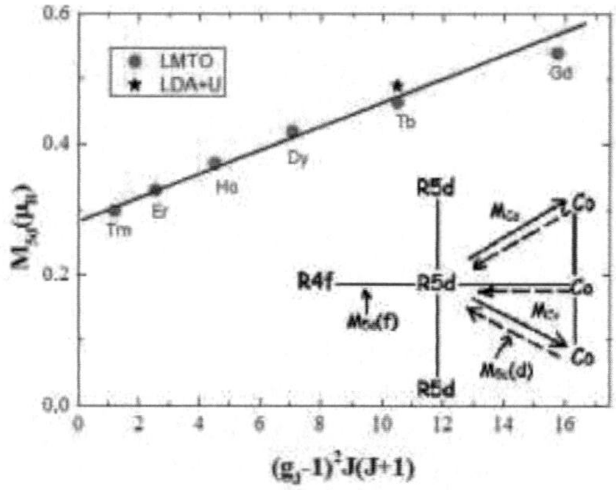

Figure 1. The R5d band polarizations in heavy rare-earths RCo_2 compounds. In the inset, the exchange interactions at the level of unit cell are schematically presented.

Two different contributions are present. The first one $M_{5d}(f) = \alpha G$ is due to local 4f-5d exchange, while the second one, $M_{5d}(d)$, is the result of R5d-Co3d short range exchange interactions, the corresponding hybridizations effects, respectively. The $M_{5d}(d)$ contributions to R5d band polarization were shown to be proportional to $\sum_i z_i M_i$ [26]. By z_i is denoted the number of cobalt atoms situated in the first coordination shell to an R one and M_i are their magnetic moments. This relation is followed in RCo_2 based compounds as evidenced in Fig.2. The slope of this dependence, $|M_{5d}(d)|/\sum_i z_i M_i \cong 2 \cdot 10^{-2}$, is the same as that evidenced in RM_2 (M = Fe, Ni) – based compounds, having similar crystal structures as RCo_2 ones. These data show a strong correlation between the R5d band polarizations and cobalt magnetic moments. As already mentioned, the part played by R5d band polarization in analysing the interactions in RCo_2 compounds is a fundamental question [26].

IOP Conf. Series: Journal of Physics: Conf. Series **848** (2017) 012004 doi:10.1088/1742-6596/848/1/012004

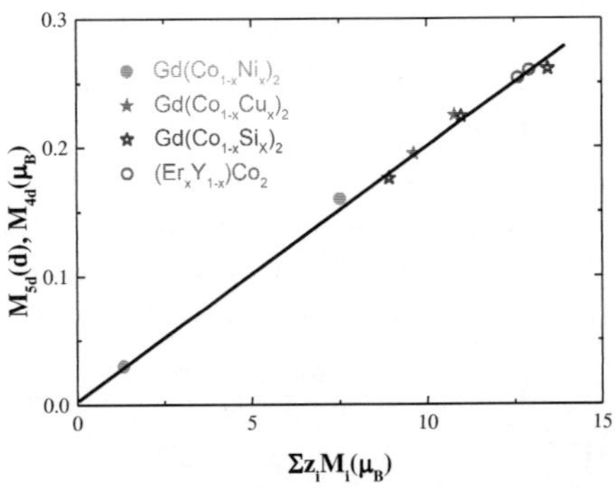

Figure 2. The $M_{5d}(d)$ components of R5d band polarizations, in RCo_2-based compounds, as function of $\sum_i z_i M_i$.

The above data suggest also that the exchange interactions, at the level of the unit cell, are interdependent. The induced cobalt moment by 4f-5d-3d exchange path is stabilized by the Co3d-Co3d short range exchange interactions. Concomitantly with the appearance of a cobalt moment, an additional polarization, $M_{5d}(d)$ is induced on R5d band by a reverse path, as depicted in Fig.1 inset. The cobalt moments, parallelly with R5d band polarizations, are strongly influenced by magnetic dilution effects, both at R and Co sites, as evidenced for example, in $Gd(Co_{1-x}Ni_x)_2$ [28, 29] or $(Gd_xY_{1-x})Co_2$ [30] pseudobinary compounds. Taking into account the above complex and interdependent magnetic interactions, at the level of unit cell, this can be viewed as a "cluster" of magnetic interacting atoms. Above the Curie temperatures, the thermal energy is not enough high to compensate the exchange energy and thus the magnetic coupling evidenced in magnetic ordered state still exist. The magnetic interactions are significantly diminished as compared to those present in ordered phase and thus no magnetic ordering can appear at the level of unit cell; this keeps only a "cluster" feature.

3. Magnetic behaviour of cobalt in RCo_2 compounds at $T > T_c$

The magnetic properties of RCo_2 compounds above the Curie points are also determined by those of cobalt atoms. As mentioned in introduction, either the presence of an exchange enhanced paramagnetism [4] or an intrinsic effective moment [2, 3] were initially considered, at $T > T_c$. Latter studies, evidenced that the cobalt magnetic behaviour, in RCo_2 compounds, can be described by the spin fluctuations model [13-16]. The model [31], takes into account the balance between the frequencies of longitudinal spin fluctuations, which are determined by their lifetime and of transverse fluctuations that are of thermal origin. These phenomena lead to the concept of temperature induced moment. For a weakly or nearly ferromagnet, as cobalt in RCo_2 compounds, the wave number dependent susceptibility, χ_q, has a large enhancement due to electron-electron interactions for small q values. The average amplitude of spin fluctuations $\langle S_{loc}^2 \rangle = k_B T \sum \chi_q$ increases with temperature and reaches an upper limit determined by the charge neutrality condition at a temperature T*. At T > T*, a Curie-Weiss behaviour is predicted, similar as in systems having local moments. The moments are localized in q-space.

By using the calculated density of states and taking into account the effect of spin fluctuations, the temperature dependences of $\langle S_{loc}^2 \rangle$ were calculated in YCo_2 [32] and $LuCo_2$ [33] compounds, by using the procedure previously reported [34, 35]. As can be seen in Fig.3, there is a tendency to saturate $\langle S_{loc}^2 \rangle$ values, at temperatures T > 550 K, where a linear χ^{-1} vs T dependence has been experimentally observed. The same behaviour was already reported in YCo_2 compound [35].

The presence of an intrinsic cobalt moment was shown in RCo_2 compounds with R = Ho, Er and Tm by magnetic circular dichroism, at $T > T_c$. Values of 0.2 μ_B ($ErCo_2$) [17], 0.3 μ_B ($HoCo_2$) and 0.4 μ_B ($TmCo_2$) [24] were reported. The cobalt moment was also shown to increase with temperature [17], as predicted by spin fluctuations model. Taking into account that $M_{Co} = gS$, the corresponding effective cobalt moments in the above compounds, $M_{eff} = g\sqrt{S(S+1)}$, are of 0.66, 0.83 and 1.00 μ_B, respectively. The effective moments increase with temperature, in a similar way as evidenced in Fig.3, up to saturation values. In the asymptotic region of RCo_2 compounds (T > 550 K), the χ^{-1} vs T shows linear dependences. The determined effective cobalt moments decrease little with increasing Curie temperatures, from 3.85 μ_B ($TmCo_2$) to 3.65 μ_B ($ErCo_2$) and 3.25 μ_B ($HoCo_2$). These variations were attributed to partial quenching of spin fluctuations by the internal field [13]. The ratio $r = S_p/S_o$ between the number of spins determined from the effective cobalt moment, S_p, and the saturation one, S_o, follows in RCo_2 compounds, a $T_c^{-2/3}$ dependence [36]. This type of behaviour is really predicted by the spin fluctuations model [37].

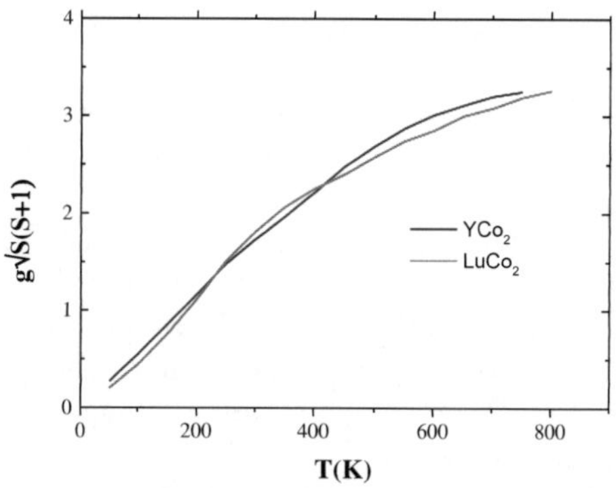

Figure 3. The temperature dependence of the mean square root amplitude $g\sqrt{S(S+1)}$, of the fluctuating cobalt moments, in RCo_2 (R = Lu, Y) compounds.

4. Magnetic behaviour of heavy rare earth RCo_2 compounds in the paramagnetic range

Magnetic measurements were made on $ErCo_2$ compound at temperatures where the Griffits phase is supposed to be present. As seen in Fig. 4, linear dependences of the magnetic moments on the external field are shown. These suggest that no magnetic ordered phase exist at T > T_c.

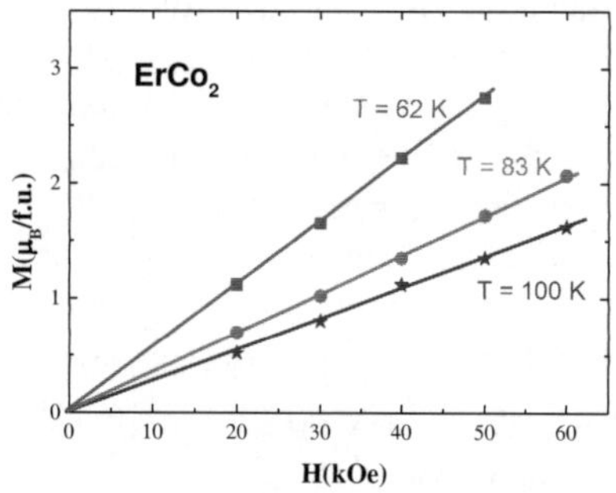

Figure 4. The field dependences of magnetizations in $ErCo_2$,at some temperatures above, T_c.

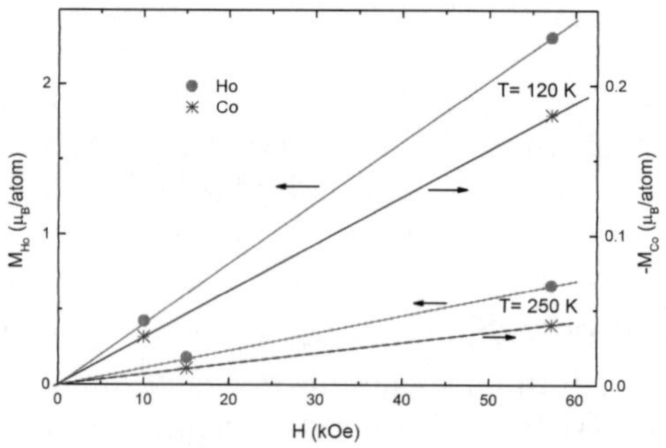

Figure 5. The field dependences of cobalt and holmium moments in $HoCo_2$ as determined by a neutron diffraction study [6].

The magnetic ordered impurities, if exist, are below 0.1 % and influence very little the magnetic data. The same behaviour can be shown in $HoCo_2$, where both Ho and Co magnetic moments are linearly dependent on the external field as evidenced by neutron diffraction studies [6, 7] – Fig.5. These data suggest that there are no magnetic ordered cobalt clusters, of a Griffits phase type. This statement is confirmed also by the evolution with temperature of the rare-earths and cobalt moments, at $T > T_c$, as determined by polarized neutron diffraction, in various external fields [5-8] – Fig.6. There is a linear dependence of cobalt moments on the rare earth ones in the studied temperature range ($40\ K \leq T \leq 300\ K$) and external fields ($10\ kOe \leq H \leq 57.2\ kOe$). The slopes, $a = (M_{Co}/M_R)$, are dependent on the rare-earth partner, decreasing in the same way as the M_{5d} band polarizations as evidenced in Fig.7. On the same figure, the $a \cong 0.041$ value, as determined by polarized neutron on $ErCo_2$ single crystal at $T = 40$ K [23] is also given. Since the small cobalt moments, as well as due to partial quenching of Tm moment, as determined by neutron diffraction, the errors in estimating the a value in $TmCo_2$ are higher than in others RCo_2 compounds.

As already mentioned [26, 38], the R5d-Co3d short range exchange interactions are important in describing the magnetic properties of rare-earth transition metal compounds which is the essential idea of the Campbell model [25]. The 4f-5d-3d exchange interactions are still present above the Curie temperatures, in a relative large temperature range, parallelly with those between cobalt atoms. The exchange couplings are not enough high to induce a magnetic ordering. The presence of Lu5d-Co3d magnetic coupling was also shown in paramagnetic $LuCo_2$ single crystal, at T = 100 K and field H = 57.2 kOe, by a polarized neutron diffraction study [39]. A cobalt moment of 0.016 μ_B has been determined, the corresponding form factor being similar to that of 3d electrons in cobalt metal. The Lu5d band is negatively polarized and of = 0.007(5) μ_B, the ratio $|M_{5d}/\sum_i z_i M_i| \cong (3.5 \pm 2.5)\cdot 10^{-2}$ being, in the limit of experimental errors, the same as that determined in magnetic ordered RCo_2 compounds, in fundamental state ($2\cdot 10^{-2}$).

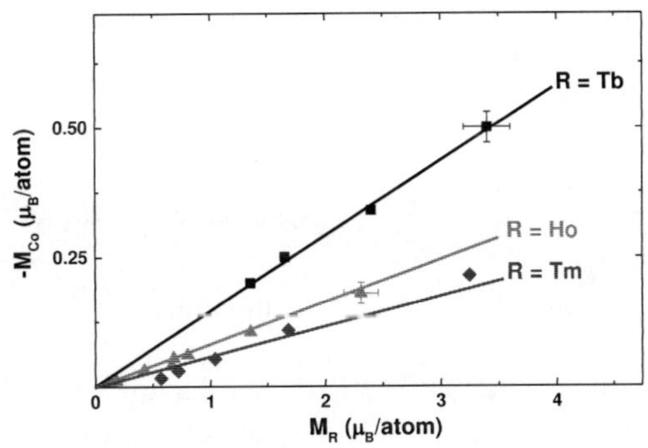

Figure 6. The cobalt magnetic moments in RCo_2 (R = Tb, Ho, Tm) compounds, at $T > T_c$, as function of rare-earths ones.

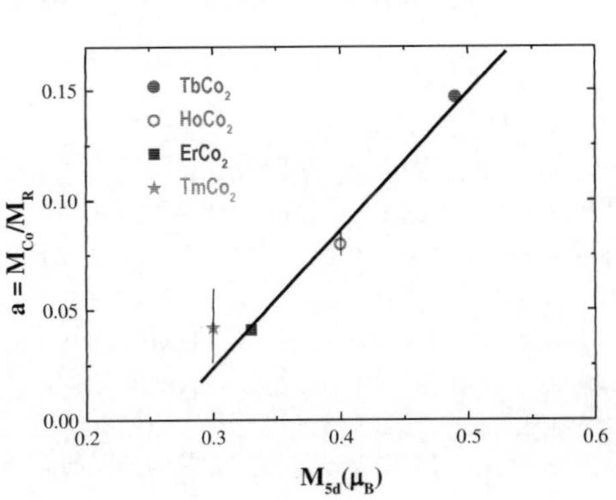

Figure 7. The relation between $a = M_{Co}/M_R$ parameters and the M_{5d} band polarizations.

5. Discussions

The present analysis suggests that the "ferrimagnetic correlations", at $T > T_c$, as evidenced by SANS experiments, having correlation length of ≅ 7-8 Å can be associated with those in the RCo_2 unit cells, where the characteristic exchange interactions between all atoms in ordered phase, are still present in paramagnetic range. Their intensities are not enough to induce a magnetic ordered state. The antiparallel coupling of R and Co moments can be evidenced only in the presence of external field. From energy considerations, the higher rare-earth moment than the cobalt one, will be aligned parallelly to the external field, the 4f-5d-3d coupling, imposing an antiparallel orientation for cobalt moments, up to a characteristic temperature, T_1. The thermal energy is not sufficient to induce a full disordering, in the considered temperature range, as in normal ferrimagnets. Parallel orientations of magnetic moments in external fields can be seen only at higher temperatures, their values depending on the strength of R5d-Co3d interactions characteristic for a given RCo_2 compound. The evolutions with temperature of cobalt moment and of the anisotropy field can determine a more complex magnetic arrangement of rare-earth and cobalt atoms.

Analyzing the magnetic properties of the clusters evidenced by SANS measurements in $ErCo_2$, an effective moment M_{eff}(cell) ≅ 20 μ_B has been reported [18]. The cluster was considered to be constituted from 60 to 100 cobalt and 30 to 50 erbium atoms. The above estimation is unrealistic and can be the result of using the values of the magnetic moments M = gS instead of the effective ones, M_{eff}, in estimating the cluster composition. Really the effective moment of the unit cell containing 16 cobalt and 8 erbium atoms can fit well the above M_{eff}(cell) value. Assuming that the effective moment of erbium is given by its free ion value (9.59 μ_B) and that of cobalt of 0.66 μ_B, as already mentioned, the corresponding effective moment of the unit cell is M_{eff}(cell) ≅ 27 μ_B, supporting our conclusion. Generally, the effective moments in characterizing the systems containing atoms with different magnetic contributions has not a physical significance. The corresponding magnetic behaviour can be described better by the addition law of the magnetic susceptibilities, Curie constants, respectively than of the squares of effective moments.

In high external field, the T_c value of $ErCo_2$ shifts to higher values [40] and as result, deviations from the linear field dependences of magnetizations can be seen at temperatures higher than the Curie point, as evidenced in smaller fields [40]

A more detailed analysis concerning the magnetic behaviour of RCo_2 above the Curie temperatures will be further presented [41].

This work was supported by the Romanian Ministry of Education and Research (UEFISCDI), grant no. PN-II-ID-PCE-2012-4-0028.

References

[1]. E.Burzo, A.Chelkowoski, H.R. Kirchmayr, Landolt Börnstein Handbook, Springer Verlag, Vol.19d2, 1990
[2]. E.Burzo, Phys. Rev. B6, 2882 (1972)
[3]. E.Burzo, Int. J. Magn. 3, 161 (1972)
[4]. D.Bloch, R.Lemaire, Phys. Rev. B2, 2648 (1970)
[5]. D.Gignoux, F.Givord, J.Phys. F: Metal Phys. 9, 1409 (1979)
[6]. D.Gignoux, F.Givord, J.Schweizer, J.Phys. F: Metal Phys. 7, 1823 (1977)
[7]. D.Gignoux, F.Givord, W.C.Koehler, Physica, B86-88, 165 (1977)
[8]. D.Gignoux, D.Givord, F.Givord, W.C. Koehler, R.M.Moon, Phys. Rev. B14, 162 (1976)
[9]. E.Burzo, Rev. Roum. Phys. 23, 693 (1978): J. Less Common Met. 77, 251 (1981)
[10]. E.Burzo, N. Plugaru, I. Creanga, I. Ursu, J. Less Common Met. 155, 281(1989)
[11]. T.Goto, A.H.Katori, T.Sakakibara, H.Mitamura, K.Fukamichi, K.Murata, J. Appl. Phys. 76, 6682 (1991)
[12]. N.N.Duc, T.D.Hien, P.E.Brommer, J.J.Franse, J. Magn. Magn. Mater. 104-107, 1252 (1992)
[13]. E.Burzo. R. Lemaire, Solid State Commun. 84, 1145 (1992)

[14]. E.Burzo, E.Graz, V.Pop, J.Magn.Magn. Mater. 123, 159 (1993)
[15]. E.Burzo, P.Vlaic, D.P.Kozlenko, S.E.Kichanov, N.T.Dang, A.V.Rutkauskas, B.N.Savenko, J. Alloys Comp. 584, 393 (2014)
[16]. D.P.Kozlenko, E.Burzo, P.Vlaic, S.E. Kichanov, A.V.Rutkauskas, B.N.Savenko, Sci. Reports, 5, 8620 (2015)
[17]. J. Herrero-Albillos, L.M.Garcia, F.Bartolomé, A.T.Young, T.Funk, J. Magn. Magn. Mater. 316, e442 (2007)
[18]. J.Herrero-Albillos, F.Bartolomé, L.M.Garcia, A.T. Young, T.Funk, J.Campo, G.J.Cuello, Phys. Rev.B76, 094409 (2007)
[19]. J.Herrero-Albillos, F.Bartolomé, L.M.Garcia, A.T. Young, T.Funk, G.J.Cuello, J.Magn. Magn. Mater. 310, 1645 (2007)
[20]. J.Herrero-Albillos, L.M.Garcia, F.Bartolomé, J.Phys. : Condens. Matter, 21, 216004 (2009)
[21]. C.M.Bonilla, N.Marcano, J.Herrero-Albillos, A.Maisuradze, L.M.Garcia, F.Bartolome, Phys. Rev. B 84, 184425 (2011)
[22]. C.M.Bonilla, I.Calvo, J.Herrero-Albillos, A.I.Figueroa, C.Castan-Guerrero, J.Bartolome, J.A.Rogriguez-Velamazan, D.Schmitz, E.Weschke, D.Paudyal, V.K.Pecharsky, K.A.Gschneidner, F.Bartolomé, L.M.Garcia, J.Appl. Phys. 111, 07E315 (2012)
[23]. F.Bartolomé, C.M. Bonilla, J.Herrero-Albillos, I.Calvo-Almazan, C.Castan, E.Weschke, D.Schmitz, D.Paudyal, Y.Mudryk, V.Pecharsky, K.A.Gschneidner, A.Stunault, L.M.Garcia, Eur. Phys. J. B84, 489 (2013)
[24]. C.M. Bonilla, J.Herrero-Albillos, A.I.Figueroa, C.Castan-Guerrero, J.Bartolome, F.Bartolomé, J.Calvo-Almazan, D.Schmitz, E.Weschke, L.M.Garcia, J.Phys. : Condens. Matter 26, 156001 (2014)
[25]. I.A.Campbell, J. Phys. F.: Metal. Phys. 2, L47 (1972)
[26]. E.Burzo, L.Chioncel, R.Tetean, O.Isnard, J. Phys.:Condens. Matter, 23, 026001 (2011) and references
[27]. E.Burzo, L.Chioncel, I.Costina, S.G. Chiuzbaian, J. Phys.: Condens. Matter, 18, 4861 (2006)
[28]. E.Burzo, D.P. Lazar, M. Ciorascu, Phys. Stat. Solidi (b), 65, K145 (1974)
[29]. E.Burzo, L. Chioncel, J. Opt. Adv. Mater. 6, 917 (2004)
[30]. E.Burzo, D.P.Lazar, J. Solid State Chem. 16, 257 (1976)
[31]. T.Moriya, J. Magn. Magn. Mater. 14, 1(1979); 100, 261 (1991)
[32]. L.Chioncel, E.Burzo, R.Tetean, V.Pop, Mol. Cryst. Liquid Cryst. 417, 513 (2004)
[33]. E.Burzo, AIP Conf. Proc. 1694, 030001 (2015)
[34]. M. Shimizu, Phys. Letters 81A, 87 (1981); Report Progr. Phys. 44, 21 (1981)
[35]. H.Yamada, J. Inoue, K.Terao, S.Kanda, M. Shimizu, J. Phys. F: Metal Phys. 14, 1543 (1984)
[36]. E. Burzo, P. Vlaic, D.P.Kozlenko, S.E.Kichanov, N.T.Dang, E.V.Lukin, B.N.Savenko, J. Alloys Comp. 551, 702 (2013)
[37]. E.P.Wohlfarth, J. Magn. Magn. Mater., 7, 113 (1978)
[38]. H.H.Li, Y.P.Li, J.M.D.Coey, J. Magn. Magn. Mater. 104-107, 1444 (1992)
[39]. D.Gignoux, F.Givord, W.C.Koehler, R.M.Moon, J. Magn. Magn. Mater. 5, 172 (1977)
[40]. J. Woo, Y.Jo, H.C.Kim, A.Pirogov, J.G.Park, H.C.Ri, A.Podlesnyak, J.Schefer, Th. Strassle, A.Teplykh, Physica B, 329, 653 (2003)
[41]. E. Burzo, L. Chioncel, Cogent Physics (sent for publication)

IOP Conf. Series: Journal of Physics: Conf. Series **848** (2017) 012005 doi:10.1088/1742-6596/848/1/012005

Small angle neutron and X-ray studies of carbon structures with metal atoms

V T Lebedev[1], A A Szhogina[1], V Yu Bairamukov[1]

[1] Petersburg Nuclear Physics Institute, NRC Kurchatov Institute, 188300 Gatchina, Leningrad distr., Russia

vlebedev@pnpi.spb.ru

Abstract. Encapsulation of metal atoms inside carbon single-wall cages or within multi-layer cells has been realized using molecular precursors and high temperature processes transforming them into desirable structures. Endohedral fullerenols $Fe@C_{60}(OH)_X$ with 3d-metal (iron) have been studied by SANS in aqueous solutions where they form stable globular clusters with radii $R_C \sim$ 10-12 nm and aggregation numbers $N_C \sim 10^4$. This self-assembly is a crucial feature of paramagnetic fullerenols as perspective contrast agents for Magneto-Resonance Imaging in medicine. Cellular carbon-metal structures have been created by the pyrolysis of diphthalocyanines of lanthanides and actinides. It was established that these ultra porous matrices consist of globular cells of molecular precursor size (~ 1 nm) which are aggregated into superstructures. This provides retain of metal atoms inside matrices which may serve for safety storage of spent fuel of nuclear power plants.

1. Introduction

Carbon structures have a great potential for various applications in biomedicine, electronics, alternative energy sources, ecology etc. that stimulates a progress in their synthesis and chemical modification [1-5]. Among them it should be mentioned a variety of endometallofullerenes (EMF), $M@C_{2n}$ ($2n$ = 60; 70; 76; 80; 82 ...), with rare earth elements (M) and especially new structures encapsulating paramagnetic 3d-elements [6-10]. Presently the EMFs are considered as advanced agents for Magneto-Resonance Imaging in medical diagnostics to improve its efficiency and spatial resolution without overdosing and side effects [11]. On the other hand, fullerenes and EMFs promise a progress in therapy for various diseases including neurodegenerative pathologies also since their water-soluble forms are able to overcome blood-brain barrier [12,13]. However, there is a substantial limitation for biomedical applications of carbon structures. It concerns a preparation of very pure fullerenes C_{60} and C_{70}. Recently basic principles and the ways of preparation of high quality fullerenes C_{60} and C_{70} there were developed [14,15]. Along with it, these authors pursued the aim to obtain new EMFs with 3d-metals. Finally, it was synthesized a new complex $Fe@C_{60}$ with paramagnetic atom [16]. For a long time a creation of such molecules was not successful until a special precursor (iron phthalocyanine) has been found. Further it was embedded into carbon electrodes used for electric arc generating plasma of ionized carbon and metal which create desirable new compound of $Fe@C_{60}$. In hydroxylation process it was transformed into water-soluble form of $Fe@C_{60}(OH)_X$ ($X \sim 30$) [16].

Advances in synthesis of water-soluble forms of fullerenes and EMFs make possible their integration into biomedicine. For this purpose it is necessary a fundamental knowledge of structural features and behaviours of these objects in aqueous solutions by means of neutron and X-ray scattering [17-19].

Water-soluble fullerenes and EMFs exhibit different kinds of assembly into fractal structures (linear or branched molecular chains, globular clusters). The features of such supramolecular organization at scales of 10^0 -10^2 nm depend on molecular concentrations and pH-factor in solutions [10, 17-21]. Suspected mechanisms of carbon structures ordering in solutions involve the contributions of hydrogen bonds, van-der-Waals' and hydrophobic interactions as well as dipole electric and magnetic forces if paramagnetic atoms are confined in carbon cages. Since the behavior of $Fe@C_{60}(OH)_X$ fullerenols governed by various interactions is not predictable, this became a subject of present study.

There was carried out the X-ray analysis of atomic structure of $Fe@C_{60}(OH)_X$ powders [16]. Further by SANS we have studied self-assembly peculiarities for fullerenols in aqueous media. In a series of measurements we increased the concentration towards the threshold of solubility and then decreased to get initial solution. This enabled us to check a reversibility of fullerenols' assembly.

The second part of experiments is devoted to the structural analysis of carbon matrices built of tiny cells with metal atoms captured during pyrolysis of diphthalocyanines (DPC). In these molecules metal atoms (Y, La, Ce, Nd, Tb, U) are coordinated between plane ligands [22]. In neutral atmosphere at high temperatures (800-1700 °C) the molecules in DPC-powders due to destruction have lost light elements (H, N, O). Such samples are ultra porous matrices where metal atoms are localized inside carbon cells which are aggregated. The X-ray study was aimed to determine cells' structural features (dimensions, atoms aggregation numbers) inside matrices at various conditions of pyrolysis [23].

2. Experimental

2.1. Endofullerenes, synthesis and attestation

The synthesis of endofullerenes was carried out in electric-arc (current of 130 A, helium atmosphere, pressure of 380 mm Hg) by the evaporation of carbon electrodes filled with a mixture of graphite and iron phthalocyanine pyrolysate (iron content ~ 10% wt.). To remove residual iron not involved into endohedral complexes, fullerene-containing soot was etched in dilute hydrochloric acid. First, empty fullerenes (C_{60}, C_{70} and higher homologues) were extracted from the soot in o-xylene. Second, the EMFs were isolated from the soot by extraction in N,N-dimethylformamide containing reducing agent to enhance the process efficiency.

After solvent evaporation and drying, solid EMF-extract was examined by X-ray fluorescence analysis (XRF) which showed iron amount (2.4 – 4.8 % wt.). Further this product was hydroxylated in dilute aqueous solution of hydrogen peroxide (65° C) and mixed with a tenfold volume of precipitating mixture (isopropyl alcohol and hexane, ratio 9:1). Finally the precipitate separated by centrifugation was dried under vacuum (80°C, 3 hours). The concentration of iron in resulting product was of 3% wt. (XRF data).

The amount of hydroxyl groups attached to a fullerene cage, $X = 30 \pm 2$, was found by thermo-gravimetric analysis (TGA)[24]. The prepared sample contained 70 % of $Fe@C_{60}(OH)_{30}$ and empty fullerenols $C_{60}(OH)_{30}$ in the remainder and had a solubility of 46 mg/ml.

The molecular structure of the powders of $C_{60}(OH)_{30}$ (reference sample) and $Fe@C_{60}(OH)_{30}$ has been studied by X-Ray diffraction [16]. Chemical bonding hydroxyl groups was confirmed by Fourier Transform Infrared Spectroscopy (FTIR). The FTIR-spectrum of carbon-metal structure showed characteristic frequencies for chemical bonds in molecules of $Fe@C_{60}(OH)_{30}$ (figure 1), i.e. broad band at 3412 cm^{-1} for stretching mode of OH groups and other peaks at 1620 cm^{-1} (C=C), 1390 cm^{-1} (C-OH) and 1150 cm^{-1} (C-OH).

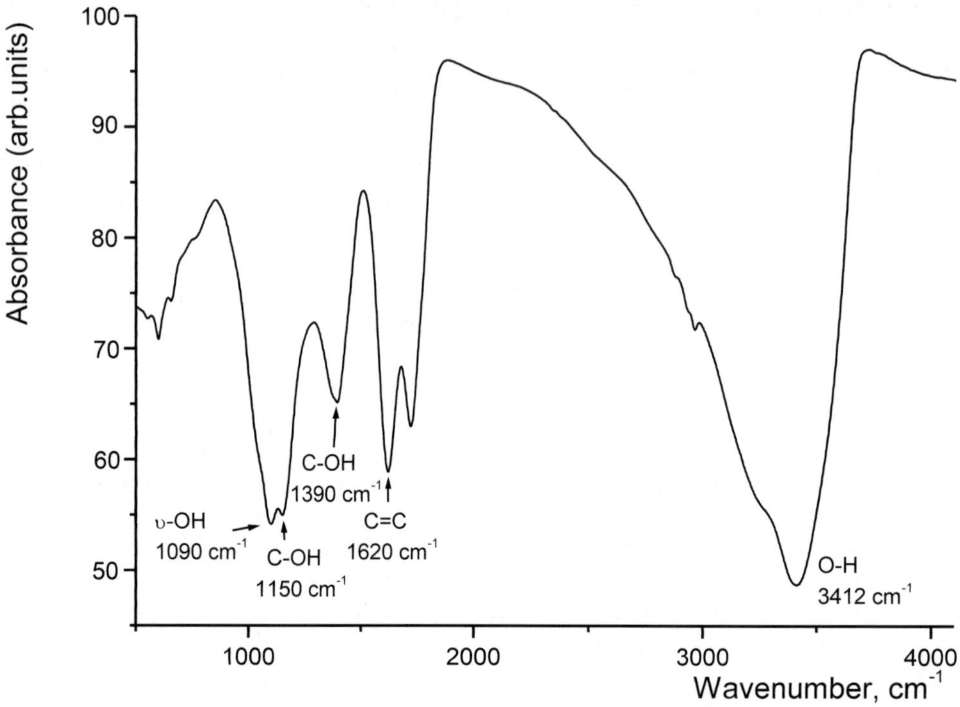

Figure 1. FTIR absorption spectrum for Fe@$C_{60}(OH)_{30}$.

Eventually the detailed analysis enabled us to identify the endohedral structure of new carbon-iron compound [16]. In following SANS experiments fullerenols' ordering in aqueous media has been studied taking into account actual scientific and practical aspects for biomedical applications.

2.2. SANS on fullerenols in aqueous solutions

In light water Fe@$C_{60}(OH)_{30}$ molecule has a good contrast in neutron scattering. This provides the SANS-measurements even at low concentrations. Varying the concentrations (C = 0.24 – 3.9 % wt.), we started from dilute solutions and moved to the threshold of solubility (46 mg/ml). In a series of experiments we collected the SANS-data at ambient temperature in the range of momentum transfer $q = (4\pi/\lambda)\sin(\theta) = 0.04\text{-}0.8\ \mathrm{nm}^{-1}$ (scattering angles 2θ, neutron wavelength λ = 0.3 nm, $\Delta\lambda/\lambda$ = 0.25, "SANS-Membrane", PNPI). Next measurements we carried out using dilution from maximum to minimum concentration to test reversibility in fullerenols' ordering (figure 2).

In both series, the scattering intensities $I(q)$ demonstrate a huge grow (figure 2) at low momentum transfer, $q < 0.4\ \mathrm{nm}^{-1}$. This testifies a strong molecular clustering at the scales ~ $2\pi/q \geq 10^1$ nm. The data in the first series showed a similarity of SANS-profiles shifted in logarithmic scale with concentration rise. The experiments by a dilution exhibited qualitatively the same results. In the first approximation all the data obey the scattering law

$$I(q) = I_0\left[1+(qR_C)^2\right]^{-2} . \quad (1)$$

The parameters I_o and R_C denote forward intensity and correlation radius of molecular clusters. Taking into account a proportionality of intensity parameter $I_o \sim CN_C$ to concentration and aggregation number (N_C), we used normalization $I_o/C \sim N_C$ to reveal the concentration dependence of fullerenols' aggregation degree (figure 3).

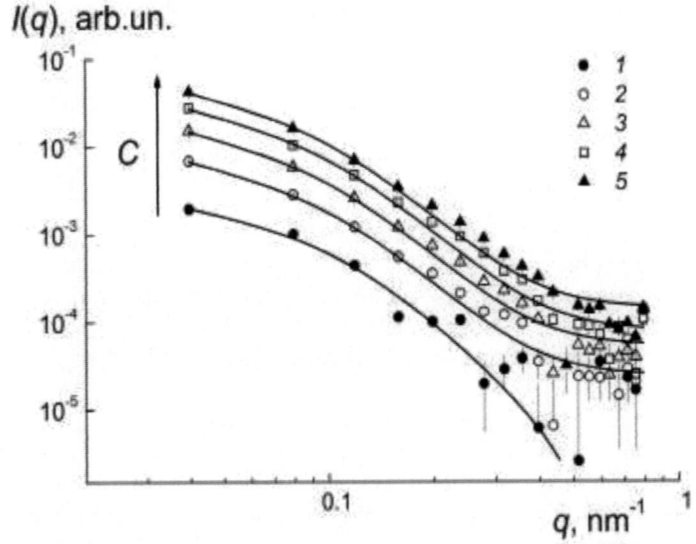

Figure 2. SANS from Fe@$C_{60}(OH)_X$ aqueous solutions (20°C) by the increase of concentration (C, wt.%):1 - 0.24; 2 – 0.67; 3 – 1.4; 4 – 2.5; 5 – 3.9. Lines are fitting functions (1).

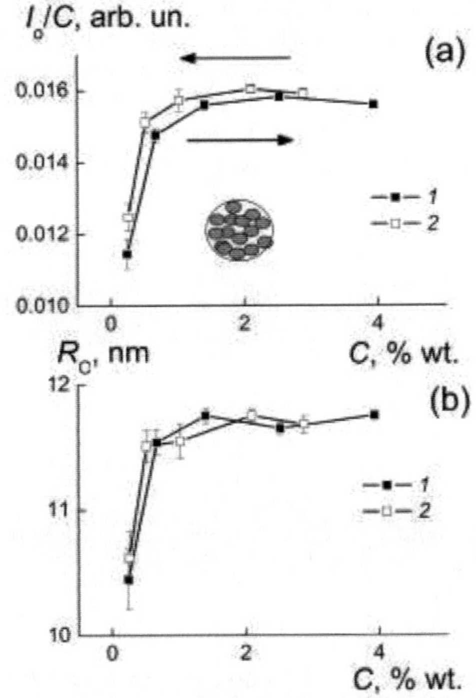

Figure 3. Concentration dependencies of scattering function (1) parameters in solutions with increasing (1) and decreasing (2) fullerenols' content: (a) - forward intensities normalized to concentrations, $I_o(C)/C$; (b) - correlation radii $R_C(C)$ vs. fullerenols' content.

The parameters I_o/C and R_C come upward when concentration increase and achieve saturation at $C \geq 1$ % wt. A reverse dilution of concentrated system does not provide a complete repeat of the parameter $I_o/C \sim N_C$ being a measure of aggregation. The aggregation numbers still remain greater (~ 3 %) by dilution comparatively to concentrating process. Hence, a weak hysteresis in aggregation numbers' behavior exists (figure 3). The N_C magnitudes were computed using the parameters I_o and taking into account the contrast of fullerenols in water and their volume fractions. These high aggregation numbers cover the range of $N_C = (1.1\text{-}1.6)10^4$. Meanwhile the correlation radii of clusters demonstrate a moderate growth, $R_C \sim$ 10-12 nm, without a hysteresis.

Summarizing the results it is worth to note that iron containing fullerenols in aqueous solutions are ordered into stable clusters. Their dimensions and aggregation degrees are not sensitive to strong changes in molecular concentrations. Therefore, these stable small-sized structures with paramagnetic metal atoms seems to be attractive for biomedical applications with dosing variations in living organisms without substantial risks of irreversible sedimentation.

Another kind of carbon structures with metal has been created by the pyrolysis of molecular compounds that provides a formation of porous matrices which strongly hold heavy metals and nuclides [26].

2.3. Pyrolysates of diphthalocyanines of metals, synthesis and structure by X-Ray scattering

Pursuing the aim of reliable isolation of nuclides inside solid matrices, there were developed chemical methods to bond radioactive elements [27]. For preparation of carbon matrices to encapsulate metal atoms, the precursors on the base of diphthalocyanines of metals (M = Y, La, Ce, Nd, Tb, Sm, U) have been synthesized. The methods of their preparation, structure and properties are described in detail [27,28].

A molecule of diphtalocyanine Pc_2M is composed of two ligands (Pc) connected via a metal atom (Figure 4). Pyrolysis of diphtalocyanine (argon atmosphere; temperature 800–900°C) causes molecular destruction with a release of hydrogen, nitrogen and other gases. As a result, in a molecule the carbon atoms at the edges of ligands can be linked via free bonds into a cell similar to fullerene cage. In such a cell (size ~ 1 nm) even ten metal atoms can be easily placed. Upon pyrolysis of diphtalocyanine, monomolecular cells create ultra porous metal–carbon structure, MC_X (X ~ 30–35), with molecular-sized and larger intermolecular voids (~10^0–10^3 nm) [23]. The incorporation of nuclides into the voids of matrix can provide their safe storage. Because of small-sized size holes, a carbon matrix is penetratable only for atoms with radius less than 0.12–0.13 nm. The most of nuclides can leave carbon cells only under their destruction or strong deformation at the temperatures above 1400°C. Below 1000°C a confinement for almost all studied radioactive elements in matrices is close to 100% [27].

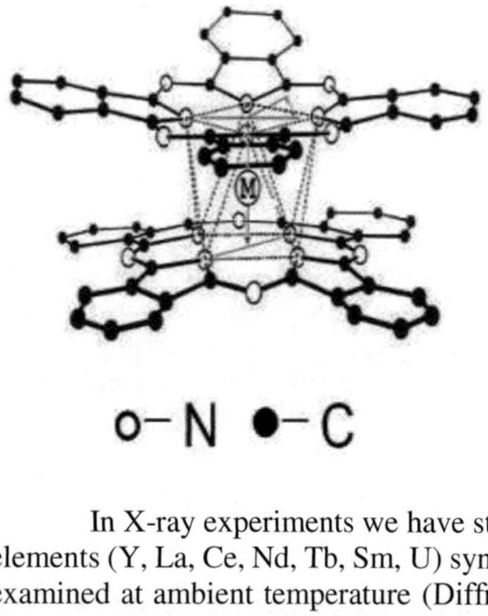

Figure 4. Structure of diphtalocyanine molecule with metal atom (M) between ligands. Nitrogen and carbon atoms are shown.

In X-ray experiments we have studied the structures of carbon cells in pyrolysates with various elements (Y, La, Ce, Nd, Tb, Sm, U) synthesized at 800–1700°C [27]. The powders of pyrolysates were examined at ambient temperature (Diffractometer "Dron-3M", CuK_α radiation, wavelength λ = 0.154 nm) in the interval of scattering vector modules $q = (4\pi/\lambda)\sin(\theta)$ = 7-80 nm^{-1} at the degrees 2θ = 10-150°.

Characteristic data for La-pyrolysates (figure 5) illustrate scattering patterns for the specimens completely amorphous (low-temperature synthesis, T < 1000 °C) and for high temperature materials (T > 1000 °C) containing La-carbide precipitates and crystalline graphite domains coexisting with amorphous phase [28]. The scattering intensities $I(q)$ for low-temperature pyrolysates (800-900 °C) demonstrate broad peaks at q_1 ~ 15-20 nm^{-1}, q_2 ~ 25-40 nm^{-1} and q_3 ~ 45-60 nm^{-1}. In the case of high-temperature pyrolysis (1020 °C), peaks' positions are defined more exactly, $q_{1m} \approx$ 19 nm^{-1}, $q_{2m} \approx$ 30 nm^{-}

[1], $q_{3m} \approx 50$ nm^{-1}. This enables us to estimate the periods $L_1 = 2\pi/q_{1m} \approx 0.33$ nm, $L_2 \approx 0.21$ nm and $L_3 \approx 0.13$ nm.

The smallest parameter L_3 corresponds to the length of carbon-carbon bond, while the others should be attributed to the distances between atomic planes. The L_2 is close to the period of graphite hexagonal lattice $a = 0.246$ nm, and the $L_1 \approx d_{002} = c/2 = 0.335$ nm is a half of c-period [29]. Thus, the data for high temperatures ($T > 1000$ °C) testify atomic order in pyrolysates like in crystalline graphite (figure 5).

In this study we tried to detect subtle structural features of amorphous phase which is not amenable to ordinary X-ray diffraction analysis. Therefore we subtracted the contributions $I_C(q)$ of narrow peaks related to crystal order and approximated by Gaussians. The data were corrected also for background, $I_S(q) = I(q) - I_C(q) - Bg$. Further we used final data to evaluate short range correlations of atoms in these ultra porous structures.

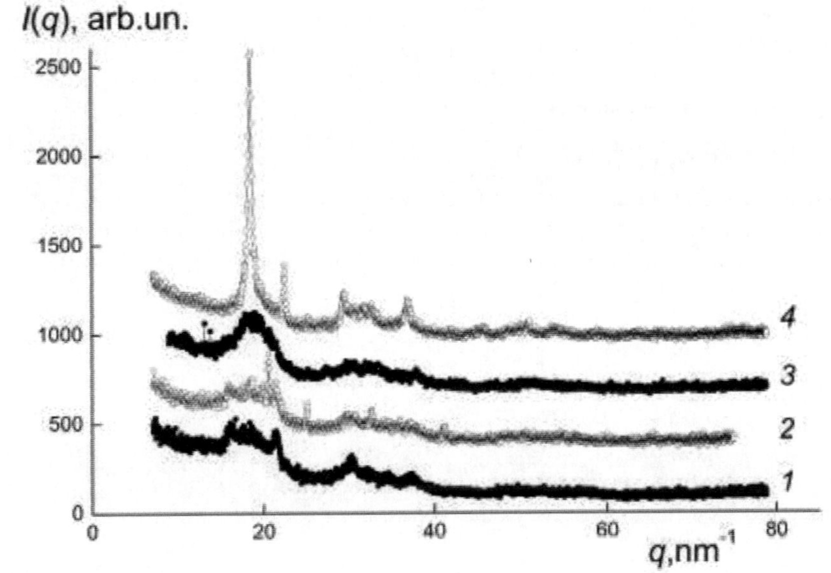

Figure 5. Diffraction patterns for La-pyrolysates synthesized at 800; 900; 1020; 1600 °C (1-4). Curves are shifted for clear presentation.

2.4. Atomic correlations in pyrolysates

For pyrolysates the spectra of atomic correlations [25] are given by Fourier-transform of scattering intensities $I_S(q)$,

$$\gamma(R) \sim \langle \Delta\rho_e(0) \cdot \Delta\rho_e(R) \rangle \sim \left[\frac{1}{2\pi}\right]^3 \int I_S(q)\left[\frac{\sin(qR)}{(qR)}\right] 4\pi q^2 dq \quad . \tag{2}$$

As usual, the $\gamma(R)$ is defined by local deviations of electron density, $\Delta\rho_e(0)$ and $\Delta\rho_e(R)$, from a middle level in the sample when two points are separated by a certain distance R and there is averaging over all possible positions of pairs in sample's volume. For our isotropic samples the spectra are presented in the form $G(R) = R^2\gamma(R)$ which shows spatial correlations between a scattering center and surrounding centers at spherical surface (radius R) (figure 6).

For low-temperature structures (figure 6a) it is visible that the edge of the intense peak quite exactly corresponds to the distance d_{002} between graphene planes in graphite crystal. Indeed, if the atoms belong to neighboring planes with spacing d_{002}, their cross-correlations start at $R \geq d_{002}$. In frames of a simple model [30] for uniform thin parallel planes (discrete positions of atoms neglected) the $G(R)$ increases up to the maximum at $R_{max} = (4/3)^{1/2} d_{002} \approx 0.39$ nm. Really, the position of maxima (R_{max} ~

0.4 nm) in spectra for low-temperatures pyrolysates are in agreement with this model. Meanwhile, only a rough accordance with characteristic periods in graphite crystal (d_{100}, $2d_{002}$) is observed since the peaks are shifted to lower radii (figure 6a). This can be interpreted as a result of curvature of plane fragments. Such fragments are formed of diphthalocyanine ligands (figure 4) which lost light atoms (H, N) by destruction. After partial destruction a molecule of diphthalocyanine probably is a structure like a carbon cage when ligands' edges are linked via free bonds. It means a transformation of plane ligands into curved structures which can be combined in multi-layer stacks. The high-temperature treatment stimulates a growth of compact (globular) structures when single-layer fragments become cross linked. As a result, in spectra neighboring partial peaks overlap and form a broad common hump (figure 6b). Its center position $R \sim 0.6$ nm indicates a characteristic radius of globules which are comparable in size to precursor molecules.

Along with it, the spectra are still modulated that indicates a multilayered structure of globules (figure 6b). In order to find the amount of atoms m (aggregation number) in globular entities we have taken into account the scattering lengths of atoms. Such a length is proportional to atomic number of element (Z). However, in dense matrix a length for each atom should be corrected for scattering length density of matrix. Thus, we used effective lengths (Z_{EC}, Z_{ELa}) for carbon and lanthanum to estimate atomic incoherent scattering intensities for the samples,

$$I_{01} = 4\pi S_1 \sim [\nu Z_{ELa}{}^2 + (1-\nu) Z_{EC}{}^2] \; .$$

Here parameter S_1 is the integral area under the first peak in $G(R)$-spectrum. The integral represents the contributions of lanthanum and carbon atoms with atomic share ν and (1-ν) to the incoherent scattering cross section. Such a peak (maximum at $R \sim 0.1$ nm, figure 6) describes the scattering from electrons in single atoms.

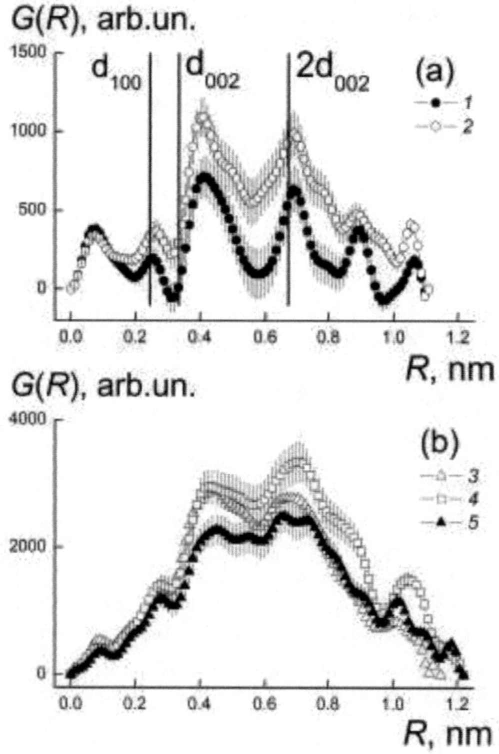

Figure 6. Spectra $G(R)$ for pyrolysates prepared at different temperatures: (a) 800 and 900 °C (1,2); (b) 1020; 1600 and 1800 °C (3,4,5). Straight lines denote the distances between atomic planes in graphite.

Meanwhile, it is overlapped with a peak arisen due to chemical bonds with neighboring atoms. The contribution of this incoherent scattering was found using approximation of $G(R)$-spectra by a sum of Gaussians in the limited range of radii ($0 \le R \le 0.3$ nm).

For each spectrum the incoherent intensity I_{01} was compared to coherent one defined by the total area under the profile of *G(R)*-function,

$$I_0 \sim m[\nu Z_{ELa} + (1-\nu)Z_{EC}]^2 \;.$$

Here the parameter *m* is a total atomic aggregation number for the structures observed in the experimental range of radii $R \leq 1$ nm. For matrices prepared at different temperatures we estimated this parameter from the ratio

$$m = \left[\frac{I_0}{I_{01}}\right] \cdot \frac{[\nu Z_{ELa}{}^2 + (1-\nu)Z_{EC}{}^2]}{[\nu Z_{ELa} + (1-\nu)Z_{EC}]^2}$$

Additionally, the integrals over *G*(*R*)-spectra give the gyration radii of aggregates (R_g) by formula [25]

$$R_g{}^2 = \frac{\frac{1}{2}\int G(R)R^2 dR}{\int G(R)dR}$$

The gyration radii and aggregations numbers for La-carbon structures are presented in figure 7 as dependent on synthesis temperature. The rise of pyrolysis temperature above 1000 °C induces a huge growth in aggregation degree, *m(T)* ~ 100. At the same time, the gyration radii of structures demonstrate only a weak increase, R_g ~ 0.4-0.5 nm. It means a formation of dense globules at high temperatures. Indeed, comparison of their masses ~ *m* and volumes ~ $R_g{}^3$ has shown that the density of high-temperature globules approaches to this parameter for graphite but yet remains lower by 16 %.

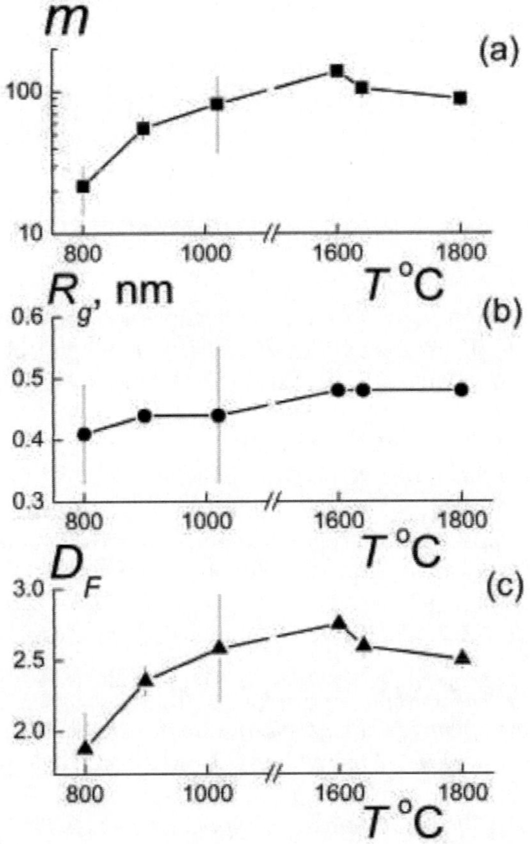

Figure 7. Structural parameters for La-pyrolysates vs. the temperature of treatment:
(a) – aggregation number *m*(*T*);
(b) – gyration radius $R_g(T)$;
(c) – fractal dimension $D_F(T)$.

In general, these structures can be considered as fractal clusters with dimension

$$D_F(T) = \frac{\ln(m)}{\ln\left[\frac{R_g}{r_a}\right]}$$

where $r_A \approx 0.08$ nm is carbon atom radius. The obtained parameter $D_F(T) \sim 1.9$-2.8 strives upward in the temperature range 800-1600°C although in the upper limit (1800°C) it shows a downturn due to increased destruction (figure 7 c). Such structures may be formed via diffusion limited aggregation which leads to a growth of mass fractals when it is possible a transition from plane structures, $D_F \sim 2$, to branched 3D-aggregates, $D_F \sim 2.5$-$2.8 > 2$. These X-Ray studies have revealed small-sized (~ 1 nm) structures which serve as the building blocks. Their association creates multi-level aggregates (10^0-10^2 nm) forming solid matrices. The structural features of various matrices have been analyzed in neutron scattering experiments [23].

3. Conclusions

SANS-studies of new fullerenols $Fe@C_{60}(OH)_X$ have shown that these paramagnetic endohedral objects create stable self-assembled globular structures with moderate and reversible changes of parameters in aqueous solutions in a wide range of concentrations that is crucial for biomedical applications.

Carbon-metal materials prepared of diphthalocyanines and studied by X-ray scattering are characterized as ultra porous matrices composed of molecular-sized structures. By the increase of pyrolysis temperature (800-1800 °C) they undergo a transformation when molecular skeletons lose light atoms and again begin to integrate carbons to form multi-layer globules with high aggregation numbers ~ 100 but their gyration radii ~ 0.5 nm do not exceed the dimensions of diphthalocyanine molecule.

Thus, a combination of X-Ray and neutron scattering together with data treatment complementary in reciprocal and real space has provided a comprehensive analysis of the mechanisms of formation of different kinds of carbon-metal primary structures and understanding subtle features of their assembly into secondary clusters.

Acknowledgements

The work was supported by Russian Foundation for Basic Research (grant N 14-23-01015 ofi-m).

References

[1] *Nanomaterials for application in medicine and biology.* 2008 Ed M Giersig ang G B Khomutov. *Proc. of the NATO Advanced Research Workshop on Nanomaterials for Application in Medicine and Biology*, Bohn, Germany, 4-6 October 2006. (Springer, P.O.Box 17, 3300 AA Dordrecht, The Netherlands. p186)
[2] Avouris P, Chen Z, Perebeinos V 2007 *Nature Nanotechnology* **2** 605
[3] Nicholas R J, Mainwood A, Eaves L 2008 *Phil. Trans. R. Soc. A* **366** 189
[4] Nelson J 2011 *Materials Today* 14. N 10. P. 462-470.
[5] *Carbon nanomaterials sourcebook. Vol. 1. Graphene, Fullerenes, Nanotubes and Nanodiamonds.* 2016 Ed K D Sattler (University of Hawaii at Manoa, Honoluly, USA. CRC Press. Taylor&Francis Group, London, New York. p 591)
[6] Heath J R, O'Brien S C, Zhang Q, Liu Y, Curl R F, Tittel F K, Smalley R E 1985 *J. Am. Chem. Soc.* **107** 7779
[7] Shinohara H 2000 *Reports on Progress in Physics* **63** 843
[8] Koltover V K 2007 *Endohedral fullerenes: from chemical physics to nanotechnology and nanomedicine. In: "Progress in Fullerene Research"* (ed M Lang, New York: Nova Science Publishers, pp. 199-233)
[9] Kozlov V S, Suyasova M V, Lebedev V T 2014 *Russian Journal of Applied Chemistry* **87** 121

[10] Szhogina A A, Kulvelis Yu V, Lebedev V T, Sedov V P 2015 *Russian Journal of Applied Chemistry* **88** 2009
[11] Shevtsov M A, Nikolaev B P, Yakovleva L Y, Dobrodumov A V, Torok Gy, Lebedev V T, Pitkin E 2014 *Appl. Magn. Res.* **45** 339
[12] Harutyunyan A R, Pradhan B K, Sumanasekera G U, Korobko E Yu, Kuznetsov A A 2002
[13] *European Cells and Materials* **3** (Suppl. 2) 84
[14] Grushko Yu S, Sedov V P, Kolesnik S G 2012 *Method of producing fullerene* C_{60}
[15] (RF Patent N 2456233, Publ. 20.07.2012, Bull. №20, priority of the invention of 13.08.2010)
[16] Sedov V P, Szhogina A A, Lebedev V T, Chernenkov Yu P, Aksenov V L , Kovalchuk M V 2014 New endometallofullerenes incapsulating iron atoms. *Reprint of PNPI* N 2963, Gatchina p11
[17] Kyzyma E A, Tomchuk A A, Bulavin LA, Petrenko V I, Almasy L, Korobov M V, Volkov D S, Mikheev I V, Koshlan I V, Bláha P, Avdeev M V, Aksenov V L 2015 *Surface Investigation X-Ray, Synchrotron and Neutron Techniques* (Gordon and Breach Science Publishers, Switzerland) **9** 1
[18] Nikolaev I V, Lebedev V T, Grushko Yu S, Sedov V P, Shilin V A, Török Gy, Melenevskaya E Yu 2012 *Fullerenes, Nanotubes, and Carbon Nanostructures* **20** 345
[19] Lebedev V T, Kulvelis Yu V, Runov V V, Sedov V P, Szhogina A A 2015 *Journal of Optoelectronics and Advanced Materials* **17** 1193
[20] Lebedev V T, Kulvelis Yu V, Runov V V, Sedov V P, Szhogina A A 2014 *Journal of Surface Investigation. X-ray, Synchrotron and Neutron Techniques* (Pleiades Publishing, Ltd.) **8** 1044
[21] Suyasova M V, Kulvelis Yu V, Lebedev V T, Sedov V P 2015 *Russian Journal of Applied Chemistry* **88** 1839
[22] *The Phthalocyanines: properties and applications.* 1989 eds C C Leznoff, A B P Lever (New York: VCH Publ., p 436)
[23] Lebedev V M, Lebedev V T, Orlova D N, Sovestnov A E, Tikhonov V I 2014 *Journal of Surface Investigation. X-ray, Synchrotron and Neutron Techniques* (Pleiades Publishing, Ltd.) **8** 1002
[24] Goswami T, Kumar R., Alam S, Mathur G N 2004 *Thermochimica Acta* **419** 97
[25] Feigin L A, Svergun D I 1987 *Structure Analysis by Small-Angle X-Ray and Neutron Scattering* (Springer Science + Business Media New York) p 333
[26] Tikhonov V I, Moskalev P N, Kapustin V K 2008 *The carbon matrices made of pyrolised phthalocyanines as a base for encapsulation of the long-lived nuclides of iodine, technetium and minor actinides. Proc. 11th Int. Conf. on Environmental Remediation and Radioactive Waste Management ICEM-2007, Belgium. Report* №7084
[27] Tikhonov V I, Moskalev P N, Kapustin V K 2009 *Method of fixation of radionuclides for their stirage and transmutation* (RF Patent. N 2343575. Publ. 10.01.2009. Bull. N 1. Priority of the invention of 27.02.2007)
[28] Sovestnov A E, Kapustin V K, Tikhonov V I, Fomin E V, Chernenkov Yu P 2014 *Physics of the Solid State* **56** 1673
[29] Chung D D L 2002 *J. Material Sciences* **37** 1
[30] Lebedev V T, Sovestnov A E, Kapustin V K, Tikhonov V I, Chernenkov Yu P *Structure of Amorphous Phase of Pyrolysates of Lanthanun Diphyhalocyanines by X-Ray scattering* 2016 *Surface Investigation X-Ray, Synchrotron and Neutron Techniques, Gordon and Breach Science Publishers (Switzerland)* in press.

Effect of polymer on the arrangement of mixed anionic/cationic wormlike surfactant micelles revealed by SANS

A V Shibaev[1], A I Kuklin,[2,3] and O E Philippova[1]

[1] Physics Department, Moscow State University, Moscow 119991, Russia
[2] Joint Institute for Nuclear Research, Dubna 141980, Russia
[3] Moscow Institute of Physics and Technology, Dolgoprudniy 141700, Russia

E-mail: phil@polly.phys.msu.ru

Abstract. We report the results of small-angle neutron scattering (SANS) study of the effect of neutral water-soluble polymer poly(vinyl alcohol) on the structure of salt-free aqueous solutions of charged cylindrical micelles formed by the mixture of anionic (potassium oleate) and cationic (*n*-octyltrimethylammonium bromide (C_8TAB)) surfactants at a fixed molar ratio [C_8TAB]/[potassium oleate] = 0.46. The dependences of the scattering on surfactant concentration were examined both in the presence and in the absence of polymer. It was observed that at high q range, the scattering function is independent of surfactant concentration and added polymer indicating that the local cylindrical structure of the micelles remains unchanged. At intermediate q range, with growing surfactant concentration the scattering function becomes increasingly dominated by a structure peak that reveals enhanced interactions between neighboring micelles. The peak position q* scales with the surfactant concentration C_{surf} as $C_{surf}^{0.6}$ and $C_{surf}^{0.53}$ in the absence and in the presence of polymer, respectively. The most important observation concerns the fact that added polymer decreases significantly the average intermicellar distance, which may be due to microsegregation of polymer and surfactant components typical for microphase separation.

1. Introduction

Mixing of two oppositely charged surfactants can lead to the formation of extremely long wormlike micelles even in the absence of low molecular weight salt [1]. In such system, the electrostatic repulsion between the ionic surfactant head groups can be strongly screened even without added salt because of alternation of oppositely charged surfactant head groups and the release of counterions, which increase the ionic strength of the solution. Reduced repulsion allows tighter packing of surfactant heads thereby favoring growth of cylindrical micelles in length permitting to decrease the number of thermodynamically unfavorable end-caps of spherical shape. The length of mixed micelles can reach micrometers [1]. Such micelles can interlace thus providing viscoelastic properties to the solutions. These properties of wormlike micelles found numerous applications in oil recovery, nanotemplating, drag reduction, health care products and so on [2,3].

The viscoelastic properties of wormlike micellar solutions can be significantly enhanced upon addition of polymeric chains [4-8]. For this aim, hydrophobically modified water-soluble polymers are usually used [4-7]. Such polymers intercalate their hydrophobic groups into wormlike micelles thus leading to the formation of a common network, in which some subchains are formed by polymer, whereas the other ones – by surfactant micelles. The structure of such polymer/surfactant networks was studied by several methods including SANS [4,6,7]. However, such polymers are poorly available commercially, which hinders their wide application.

Recently, it was shown [8] that a considerable enhancement of the viscosity of wormlike micellar solutions can be achieved by using widely available water-soluble polymer poly(vinyl alcohol). The experiments were performed with mixed micelles of anionic and cationic surfactants (potassium oleate and *n*-octyltrimethylammonium bromide (C_8TAB), respectively) without added salt. In this case, the polymer does not interact with micelles as was evidenced by fluorescence spectroscopy with pyrene as a probe, and therefore the addition of such polymer would expect to cause the phase separation in the system. Nevertheless, the obtained polymer/surfactant solutions remain homogeneous. Most probably, the stabilization of these solutions is due to the presence of large amount of surfactant counterions, which can lose their translational entropy, if the phase separation occurs. However, the segregation of polymer and micellar components can be realized at a microscale without macrophase demixing. If the microdomain structure will propagate through the whole system, it will not impede counterions to travel in the most of the volume of the solution thus ensuring the gain in the translational entropy. At the same time, it would permit to realize the local segregation of incompatible polymer and surfactant components resulting in the gain in energy. However, a particular structure of these polymer/surfactant systems remains unclear. Also, the reasons for the observed enhancement of the viscosity of wormlike micellar solutions upon addition of polymer still need to be understood. A suitable method to clarify these points is SANS, as it allows to reveal the changes in the micellar structure and in the mutual arrangement of micelles, when they are mixed with polymer.

Thus, the aim of this paper is to consider the effect of non-interacting polymer PVA on the structure of solutions of mixed micelles of anionic (potassium oleate) and cationic (C_8TAB) surfactants by using SANS technique.

2. Experimental section

2.1. Materials

Surfactants potassium oleate (TCI, purity > 98%) and C_8TAB (ABCR, purity > 98%), polymer PVA Mowiol 4-98 with molecular weight of 27000 g/mol (Aldrich) as well as D_2O (Deutero GmbH, 99.9% isotopic purity) were used as received.

2.2. Samples preparation

First, stock solutions of potassium oleate / C_8TAB and PVA in D_2O were prepared and then they were mixed in appropriate quantities. The obtained samples were stirred for 2-3 days and then left for few days to remove air bubbles.

2.3. Small-angle neutron scattering

SANS experiments were performed on the YuMO spectrometer at the IBR-2 reactor facility (Frank Laboratory of Neutron Physics, Joint Institute for Nuclear Research, Dubna, Russia). The samples were put in dismountable cells with parallel disk-like quartz plates and beam path of 2 mm. The data were recorded at $20.0 \pm 0.3\ ^0C$ by two detectors covering the range of scattering vectors q of 0.007 - 0.5 $Å^{-1}$.

The details of the experiments are reported elsewhere [9]. The intensity profiles were corrected for sample transmission, beam path length and electronic noise, and calibrated at the absolute scale by a vanadium standard. Incoherent scattering from blank solvent (D_2O) was subtracted from the scattering curves. The scattering length densities of the surfactant tails and D_2O are equal to ca. $-0.30 \cdot 10^{-6}$ $Å^{-2}$ and $6.36 \cdot 10^{-6}$ $Å^{-2}$, respectively.

The program SasView was used to fit the scattering curves by form-factor of a cylinder. Only a part of the scattering curve at q higher than the structure peak position q* (q > 0.06-0.07 $Å^{-1}$) was used for fitting, and then the fitted form-factor was calculated in the whole q-range. One fitting parameter (radius of a cylinder) was used.

3. Results and discussion

3.1. Surfactant solutions without polymer

In this work, we studied the structure of solutions of mixed micelles composed of anionic surfactant potassium oleate and cationic co-surfactant C_8TAB at fixed molar ratio [C_8TAB]/[potassium oleate] equal to 0.46. So, under these conditions, the micelles were strongly negatively charged.

Scattering curves of the solutions of mixed micelles in semi-logarithmic format are presented in Figure 1. When normalized with respect to surfactant concentration (Figure 2), all curves coincide perfectly at high q region indicating that the local structure of the micelles in all the samples is the same. This part of the scattering curve can be well fitted by a form-factor of a cylinder (Figure 2) suggesting that the surfactant solutions under study contain cylindrical micelles. From the fit, the radius of micelles is obtained. It equals to 18.7 Å, which is quite close to the length of fully extended oleate chain (19 Å) [10]. Therefore, the radius of mixed micelles is determined by the length of potassium oleate tails, which are in the excess in the surfactant mixture. Previously, similar results were obtained at lower content of C_8TAB co-surfactant [8]. Here we show that further increase of the content of short *n*-octyltrimethylammonium chains up to 0.315 mol % does not affect the radius of mixed micelles.

From Figs. 1 and 2 it is seen that at the intermediate q region the scattering curves have a structure peak, which becomes more pronounced with increasing surfactant concentration. Most probably this peak arises from electrostatic repulsion between similarly charged micelles [12,13]. Fig. 3 shows that the peak position q* scales with surfactant concentration C_{surf} as $q^* \sim C_{surf}^{0.6}$. This scaling behavior is rather close to that expected [12] for strongly interacting cylindrical objects: $q^* \sim C^{0.5}$. From the peak position the average distance between cylindrical micelles d can be estimated. If we assume a hexagonal packing of the micelles, the value of d can be calculated from the following formula (as it is evident from the picture in the inset to Fig.4): $d = d_b / \cos 60^0$, where $d_b = 2\pi/q^*$ [11].

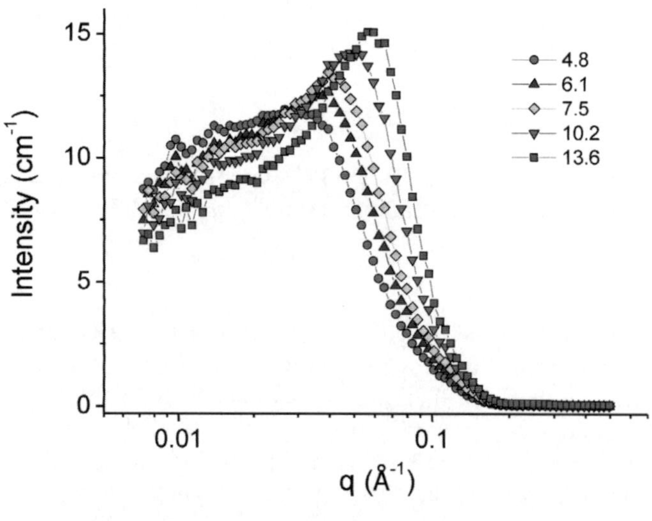

Figure 1. Scattering curves I(q) in semi-logarithmic format for solutions of potassium oleate and C_8TAB in D_2O at a fixed molar ratio [C_8TAB]/[potassium oleate] = 0.46 at 20^0C. Total concentrations of the both surfactants are indicated in the figure.

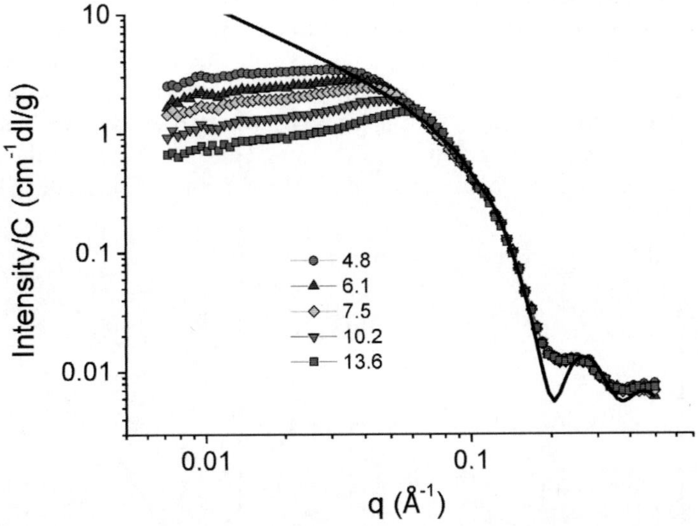

Figure 2. Normalized scattering curves I(q) in double logarithmic format for solutions of potassium oleate and C_8TAB in D_2O at a fixed molar ratio [C_8TAB]/[potassium oleate] = 0.46 at 20^{0}C. Total concentrations of the both surfactants are indicated in the figure. The solid line is a fit of the scattering curves by a form-factor of cylinder with radius R = 18.7 $\pm$ 0.2 Å.

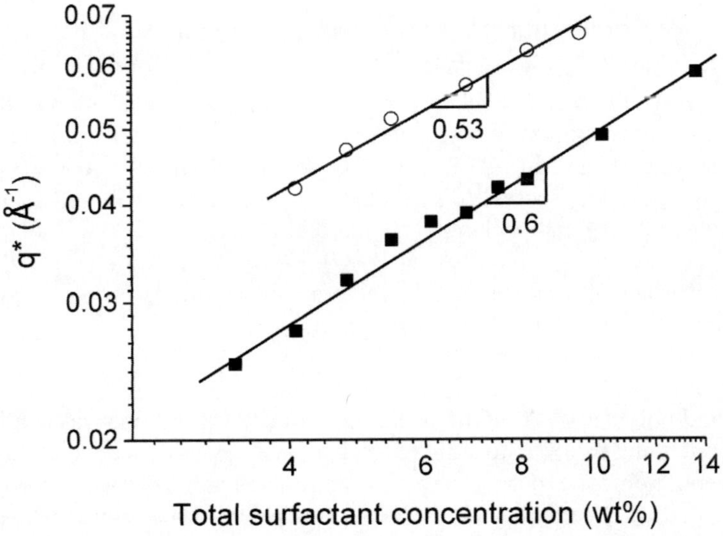

Figure 3. Position of the structure peak as a function of the total surfactant concentration in potassium oleate/C8TAB solutions in D_2O in the presence (circles) and in the absence (squares) of 4 wt% PVA at 20^{0}C. Molar ratio [C_8TAB]/[potassium oleate] = 0.46.

From Fig. 4 it is seen that the micelles come closer to each other with increasing surfactant concentration suggesting that the total number of micelles becomes higher. Simultaneously, in low q region ($q < q^*$), at the increase in the surfactant concentration the scattered intensity decreases, which can be attributed to high osmotic penalty for large spatial scale fluctuations in charge density [6]. This behavior can be related to the strengthening of the intermicellar repulsion at higher content of surfactant in the solution. It should be noted that quite similar scattering curves were previously reported for cationic [6,14] and anionic [15] surfactants at low ionic strength. Here we show that the character of the curves does not change when passing to the mixture of cationic and anionic surfactant.

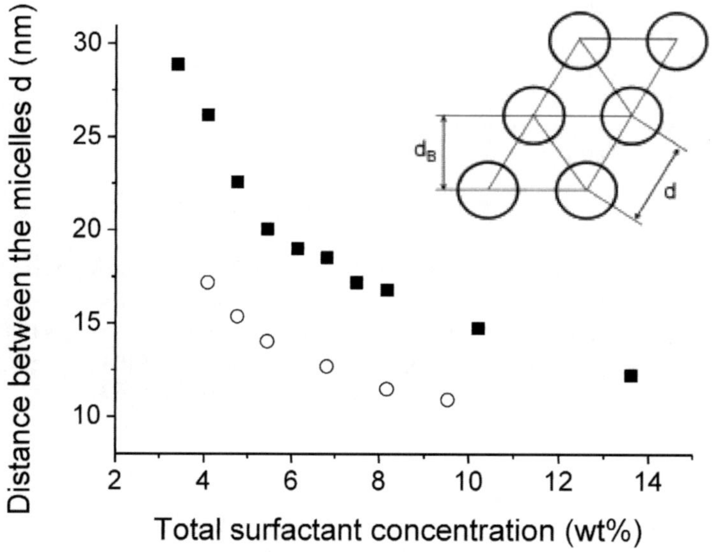

Figure 4. Average distance between cylindrical micelles d as a function of the total surfactant concentration in potassium oleate/C8TAB solutions in D_2O in the presence (circles) and in the absence (squares) of 4 wt% PVA at 20^{0}C. Molar ratio [C_8TAB]/[potassium oleate] = 0.46. Inset: schematic representation of hexagonal packing of cylindrical micelles.

Thus, charged mixed surfactant micelles in salt-free aqueous solutions form cylindrical micelles, which are mutually arranged as a result of electrostatic repulsion.

3.2. Surfactant/polymer solutions

The effect of added neutral polymer PVA on the structure of micellar solutions was studied at different surfactant concentrations ranging from 2.7 to 9.5 wt %, whereas the polymer concentration was fixed at 4 wt %. The PVA concentration corresponds to semidilute regime, since C* value for this polymer is equal to 3.2 wt %. In the studied system, the polymer scattering is rather weak, and the overall scattering is dominated by the micelles. From Fig.5 it is seen that in the presence of polymer, all the scattering patterns show the typical characteristics similar to those of pure surfactant solutions: correlation peak at intermediate q values and cylindrical form-factor at high q region. At the same time, at low q values the patterns for pure surfactants and polymer/surfactant systems are different. Below we discuss these features in more detail.

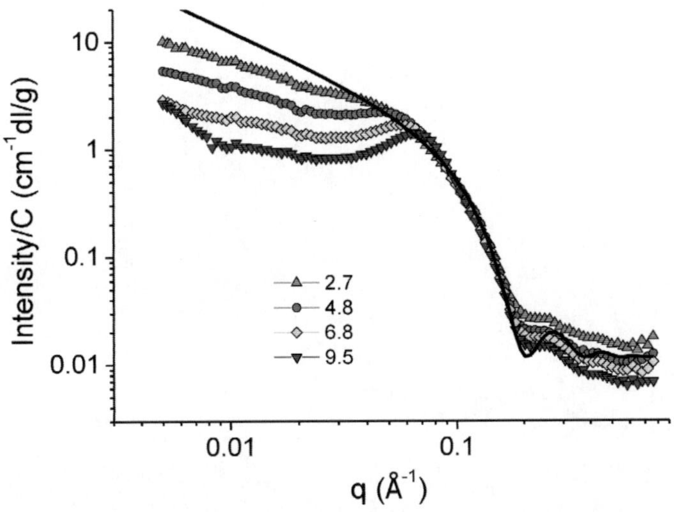

Figure 5. Normalized scattering curves I(q) in double logarithmic format for solutions of potassium oleate and C_8TAB in D_2O at a fixed molar ratio [C_8TAB]/[potassium oleate] = 0.46 in the presence of 4 wt % PVA at 20^{0}C. Total concentrations of the both surfactants are indicated in the figure. Line is a fit of the scattering curves by a form-factor of cylinder with radius R = 18.5 $\pm$ 0.2 Å.

Let us start with high q range. Presentation of the scattering data in the form of Porod plot (Fig.6) emphasizing the high q region evidences that the structure of micelles at this length scale is not affected by the addition of polymer. The same conclusion can be drawn from the fit of the high-q part of the scattering curves by a form-factor of cylinder (Fig.5). Moreover, the fitting results demonstrate that the radius of the cylindrical micelles (18.6 Å) is almost the same independently of the presence of polymer.

As to the intermediate q range, both in the absence and in the presence of polymer we observe a clear structure peak at a characteristic value q* (Figs.5 and 7), and its position exhibits almost the same dependence on the surfactant concentration (Fig.3) as in pure surfactant solution. However, in the presence of polymer the peak becomes narrowly and shifts to higher q values (Fig.7). This means that polymer induces better mutual ordering of cylindrical micelles and their closer packing. For instance, at 6.8 wt % surfactant the average distance between the micelles is equal to 18.5 and 12.7 nm in the absence and in the presence of 4 wt % PVA, respectively. Such behavior can be attributed to a poor miscibility between polymer chains and surfactant micelles, which leads to the local segregation between these components. The segregation pushes the micelles to come closer to each other thus inducing the increase in their local concentration.

At low q range, in the presence of polymer, the scattering intensity is much higher, which may be accounted for by the formation of microphase separated domains enriched either by polymer or by surfactant. The absence of a low q plateau suggests that these domains have the radius much larger than 900 Å ($2\pi/q_{min}$, where q_{min} is the low q limit of the experiment). Thus, polymer induces better mutual ordering and closer packing of similarly charged cylindrical micelles of surfactant.

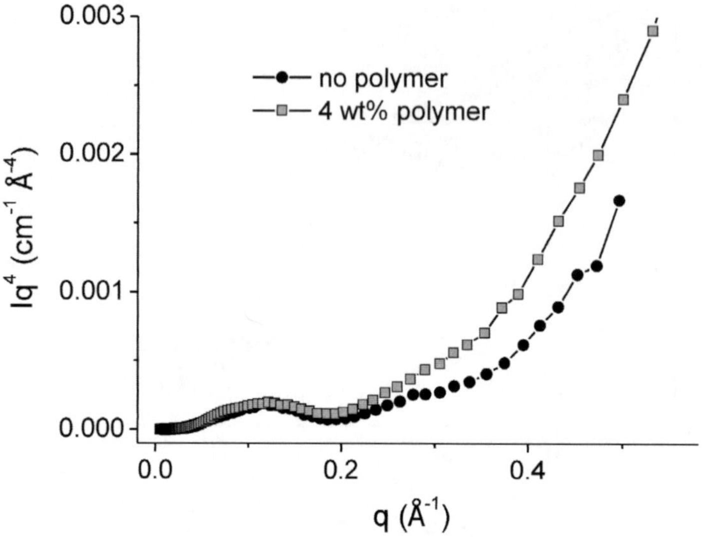

Figure 6. Holtzer plots Iq^4 (q) of the scattering from 4.8 wt % solutions of potassium oleate and C_8TAB in D_2O at a fixed molar ratio [C_8TAB]/[potassium oleate] = 0.46 in the absence (circles) and in the presence (squares) of 4 wt % PVA at 20^{0}C.

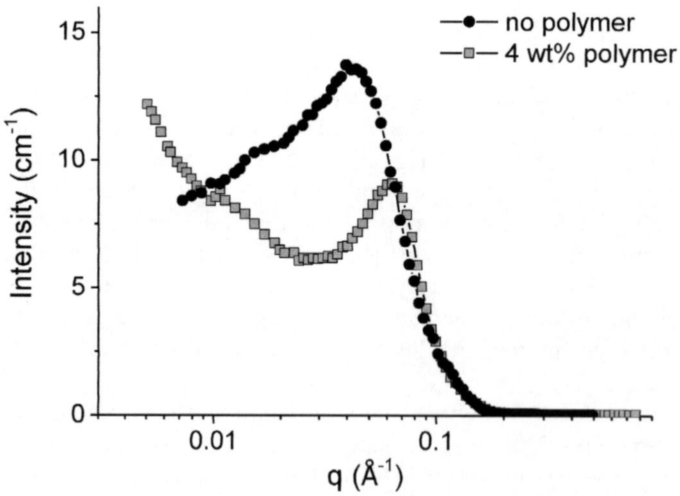

Figure 7. Scattering curves I(q) for 8.2 wt % solutions of potassium oleate and C_8TAB at a fixed molar ratio [C_8TAB]/[potassium oleate] = 0.46 in the absence (black symbols) and in the presence of 4 wt % PVA (grey symbols) in D_2O at 20^{0}C.

Closer packing suggests the increase of the local surfactant concentration, which should induce the growth of micelles in length according to a scaling law $L \sim C_{surf}^{0.5}$ derived theoretically [16,17]. Longer micelles are able to form more entanglements, and thereby enhance the rheological properties, which was indeed observed experimentally upon addition of PVA to the solution of cylindrical micelles formed in the mixture of potassium oleate and C_8TAB.

Thus, the analysis of SANS data allowed not only to reveal the structure of polymer/surfactant system, but also to suggest a reason for the enhancement of rheological properties.

Conclusions

In this article, SANS was used to study the structural changes in mixed surfactant solutions induced by the addition of neutral water-soluble polymer PVA. It was shown that polymer does not affect the cylindrical form of the micelles and their radius. At the same time, it induces better mutual ordering of neighboring micelles and pushes the micelles closer to each other thus increasing their local concentration and therefore triggering the increase in their length. The later fact seems to be crucial for the enhancement of the rheological properties observed upon addition of polymer to surfactant solution.

Acknowledgements

This work was supported by the Russian Science Foundation (project № 15-13-00114).

References

[1] Raghavan S J, Fritz G and Kaler E W 2002 *Langmuir* **18** 3797
[2] Feng Y, Chu Z and Dreiss C A 2015 *Smart Wormlike Micelles: Design, Characteristics and Applications* (Berlin: Springer-Verlag)
[3] Philippova O E and Khokhlov A R 2010 *Petroleum Chemistry* **50** 266
[4] Shashkina J A, Philippova O E, Zaroslov Y D, Khokhlov A R, Pryakhina T A and Blagodatskikh I V 2005 *Langmuir* **21** 1524
[5] Couillet I, Hughes T, Maitland G and Candau F 2005 *Macromolecules* **38** 5271
[6] Lodge T P, Taribagil R, Yoshida T and Hillmyer M A 2007 *Macromolecules* **40** 4728
[7] Ramos L and Ligoure C 2007 *Macromolecules* **40** 1248
[8] Philippova O E and Shibaev A V 2016 *Abstr. World Polymer Congress Macro 2016 (Istanbul)* p 97
[9] Philippova O E, Andreeva A S, Khokhlov A R, Islamov A K, Kuklin A I and Gordeliy V I 2003 *Langmuir* **19** 7240
[10] Shibaev A V, Molchanov V S and Philippova O E 2015 *J. Phys. Chem. B* **119** 15938
[11] Rulkens R, Wegner G and Thurn-Albrecht T 1999 *Langmuir* **15** 4022
[12] Gamez-Corrales R, Berret J F, Walker L M and Oberdisse J 1999 *Langmuir* **15** 6755
[13] Flood C, Dreiss C A, Croce V, Cosgrove T and Karlsson G 2005 *Langmuir* **21** 7646
[14] Schubert B A, Kaler E W and Wagner N J 2003 *Langmuir* **19** 4079
[15] Dreiss C A 2007 *Soft Matter* **3** 956
[16] Cates M E 1987 *Macromolecules* **20** 2289
[17] Magid L J 1998 *J. Phys. Chem. B* **102** 4064

IOP Conf. Series: Journal of Physics: Conf. Series **848** (2017) 012007 doi:10.1088/1742-6596/848/1/012007

Lipid based drug delivery systems: Kinetics by SANS

D Uhríková[1], J Teixeira[2], L Hubčík[1], A Búcsi[1], T Kondela[1], T Murugova[3] and O I Ivankov[3]

[1]Faculty of Pharmacy, Comenius University in Bratislava, Odbojárov 10, 832 32 Bratislava, Slovakia
[2] Laboratoire Léon Brillouin (CEA/CNRS), CEA Saclay, 91191 Gif-sur-Yvette Cedex, France
[3] Frank Laboratory of Neutron Physics, JINR, Joliot-Curie 6, 141980 Dubna, Moscow region, Russia

uhrikova@fpharm.uniba.sk

Abstract. N,N-dimethyldodecylamine-N-oxide ($C_{12}NO$) is a surfactant that may exist either in a neutral or protonated form depending on the pH of aqueous solutions. Using small angle X-ray diffraction (SAXD) we demonstrate structural responsivity of $C_{12}NO$/dioleoylphosphatidylethanolamine (DOPE)/DNA complexes designed as pH sensitive gene delivery vectors. Small angle neutron scattering (SANS) was employed to follow kinetics of $C_{12}NO$ protonization and DNA binding into $C_{12}NO$/DOPE/DNA complexes in solution of 150 mM NaCl at acidic condition. SANS data analyzed using paracrystal lamellar model show the formation of complexes with stacking up to ~32 bilayers, spacing ~ 62 Å, and lipid bilayer thickness ~37 Å in 3 minutes after changing pH from 7 to 4. Subsequent structural reorganization of the complexes was observed along 90 minutes of SANS mesurements.

1. Introduction

Since almost thirty years are complexes of DNA with cationic liposomes studied as potential non-viral vectors for gene therapy [1, 2]. Relationship between the structure of gene delivery vectors and their transfection efficiency have been studied widely, although conclusions are not all consistent. Significant efforts converged in the past years towards the design and development of on-demand delivery systems.

We focus our attention on pH responsive lipid based DNA delivery systems (lipoplexes). Cellular components such as cytoplasm, endosomes, lysosomes, endoplasmic reticulum, etc. are known to keep their own characteristic pH values. For example, the pathway of the gene-vector complex is accompanied by the drop in pH from physiological (pH 7.4) to acidic in lysosome (pH 4.5). Hence, pH-responsive carriers are aimed at increasing the intracellular drug bioavailability by rapidly releasing their payload in the endosomes after cellular uptake [3], and/or facilitating the drug transit to the cytoplasm [4]. pH responsive liposomes are, generally, constructed as a mixture of dioleoylphosphatidylethanolamine (DOPE) and titrable amphiphiles [3, 5, 6]. It is postulated that effective pH responsive formulations should be able to change a property in a narrow pH window.

N,N-dimethyldodecylamine-N-oxide ($C_{12}NO$) (Figure 1) is a surfactant that may exist either in neutral or cationic protonated form depending on the pH of aqueous solutions. A strong polar N – O

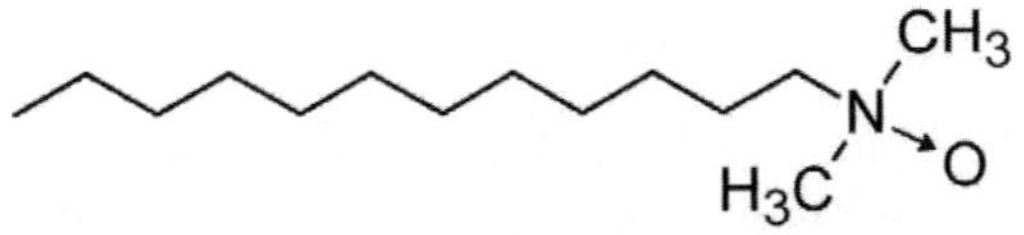

Figure 1. Structure of N,N-dimethyldodecylamine-N-oxide ($C_{12}NO$)

bond with a high electron density on oxygen yields the protonation of molecules $C_{12}N^{+}OH$ at acidic pH (pK ~ 5) [7, 8]. $C_{12}NO$ is widely used in pharmaceutical and cosmetic formulations, as detergent in household dish washing liquids and surface cleaners and in various areas of industry [9, 10]. An amphiphile $C_{12}NO$ incorporates into the biological membrane and can induce changes of fluidity [11], and thickness [12-15] of lipid bilayers. At high concentration $C_{12}NO$ destabilizes lipid bilayers forming non-lamellar phases or mixed micelles [16]. N,N-dimethylalkylamine-N-oxides display antimicrobial [17], immunomodulatory [18] or antiphotosynthetic activity [11]. However, their toxicity is reported as low-to moderate level [9, 19]. Interacting with DNA, $C_{12}NO$ induces a coil – globule phase transition responsive to changes in pH as documented by fluorescence microscopy [20]. The strong influence of pH on DNA/$C_{12}NO$ interaction was confirmed also by others experimental methods including turbidimetry and dynamic light scattering [21], dielectric spectroscopy and circular dichroism [22], fluorescence with pH sensitive probe [23], and viscosity measurements [24].

Our small angle X-ray diffraction (SAXD) experiments revealed the rich structural polymorphism of lipoplexes due to DNA interaction with $C_{12}NO$/ DOPE vesicles [25]. In neutral solutions, the DNA uptake is low, and a lamellar L_{α} phase formed by $C_{12}NO$/DOPE prewails in the complexes at $0.2 \leq C_{12}NO/DOPE \leq 0.6$ mol/mol. In acidic conditions, a condensed inverted hexagonal phase H_{II}^{C} is observed at $C_{12}NO/DOPE = 0.2$ mol/mol. Higher content of $C_{12}NO$ induces hexagonal to lamellar phase transition. Both phases $L\alpha^{C}$ and H_{II}^{C} coexist in complexes for compositions $0.3 \leq C_{12}NO/DOPE \leq 0.4$ mol/mol and pH = 4.9–6.4. Their commensurate lattice parameters, $a_{H}^{C} \approx d_{L}^{C}$, suggest that $L\alpha^{C}$ and H_{II}^{C} phases are epitaxially related. Instead, at the same composition but pH ~ 7, the mixture forms a cubic phase (*Pn3m*) when the complexes are heated to 80 °C and cooled down to 20 °C. A large portion of the surfactant ($C_{12}NO/DOPE > 0.5$) stabilizes the L_{α}^{C} phase in $C_{12}NO$/DOPE/DNA complexes and the distance between DNA strands (d_{DNA}) is modulated by the pH value. Both composition and pH affect the DNA binding in the complexes reaching up to ~95% of the DNA total amount at acidic conditions.

Consequently, $C_{12}NO$ shows low toxicity and good pH responsivity in both aspects: DNA binding and structural polymorphism of the complexes. In the present study we focus our attention on kinetics of $C_{12}NO$ responsivity to changes of pH, i.e. the protonation of the surfactant molecules and DNA binding. Following our knowledge about polymorphic behavior of the system [25], we have chosen vesicles with a composition $C_{12}NO/DOPE=1$ mol/mol, at which a lamellar phase was the only detected structure. SAXD and SANS were used. SANS experiments were performed with the aim to follow structural changes of the vesicles due to DNA binding in the process of $C_{12}NO$ protonation, profiting of the scattering contrast between protonated system in a deuterated solvent. SAXD shows structural changes in complexes as a function of pH of solutions.

2. Material and Methods

2.1. Sample preparation

Neutral phospholipid DOPE (1,2-dioleoyl-sn-glycero-3-phosphoethanolamine) was purchased from Avanti Polar Lipids, Inc.,USA. Deuterium oxide (D_2O, 99.9 %), N,N-dimethyldodecylamine-N-oxide ($C_{12}NO$) and highly polymerized DNA (sodium salt) type XIV from herring testes (average M_r of nucleotide = 308) were purchased from Sigma-Aldrich, USA. NaCl was obtained from Lachema, Brno, Czech Republic. The chemicals were of the analytical grade and were used without further purification. The aqueous solutions were prepared either with redistilled water or D_2O. 150 mM NaCl

was used as buffer. The solution of DNA was prepared by dissolving in a 150 mM NaCl solution. The precise value of DNA concentration was determined by spectrophotometer [25].
DOPE and $C_{12}NO$ were dissolved in chloroform and mixed to molar ratio $C_{12}NO$/DOPE=1. Lipid mixture was dried under a stream of gaseous nitrogen and the residue of chloroform was removed under vacuum. For SAXD experiment, the dry mixture was hydrated by the solution of DNA and the final volume of samples was adjusted by 150 mM NaCl to 2 ml. The pH of samples was adjusted by stepwise addition of 50 mM HCl solution. Afterward samples were homogenized (by vortexing and freezing–thawing). The samples were stored for 1week at 6 °C. The pH of the samples was checked again prior to SAXD measurements. The samples of the $C_{12}NO$/DOPE/DNA complexes were shortly centrifuged. The precipitate with a few drops of bulk solution was enclosed between two Kapton (Dupont, France) windows of a sample holder for X-ray diffraction.
SANS experiment: the dry $C_{12}NO$/DOPE mixture was hydrated by 150 mM NaCl in D_2O to get 1 wt% dispersion (pH ~ 7.3). Dispersions were homogenized by vortexing and freezing-thawing, and finally extruded through a polycarbonate membrane filter with 50 nm pores mounted in a mini-extruder (Avestin, Canada). DNA was added at the ratio DNA:lipid = 1:5 base/mol. 40 mM DCl was used to modulate pH of the sample. Samples were measured in 2 mm quartz cells (Hellma, Germany).

2.2. Experimental methods

SAXD experiment: SAXD measurements were performed at the soft condensed matter beamline A2 at HASYLAB at the Deutsches Elektronen Synchrotron (DESY) in Hamburg (Germany), using a monochromatic radiation with a wavelength λ = 0.15 nm. The scattering pattern was recorded using a 2D Mar CCD detector after tempering the samples at 20 °C. The raw data were calibrated using rat tail collagen [26] and normalized against the incident beam intensity using the purposed-written software A2TOOL. Each diffraction peak of the SAXD region was fitted with a Lorentzian above a linear background. Lattice parameters were determined with uncertainty ±0.1 Å or less.

SANS experiment: SANS measurements were performed at the small-angle time-of-flight spectrometer YuMO at the IBR-2 fast pulsed reactor of FLNP JINR Dubna (Russia). The necessary requirement for the study of kinetics is an experimental setup allowing mixing two substances in the measuring cell during the process of neutron exposure, or to switch the exposure immediately after mixing. A specific device was required to be used under the conditions of YuMO environment. We constructed such device which allows the injection of the adjusted volume of sample into the measuring cell, under an impulse generated by switching a gas compressor. The switch was terminated out of the experimental hutch. Testing measurements have shown that 3 minutes was a sufficient acquisition time to get a scattering curve of satisfactory quality. To follow the process of structural changes, SANS curves were accumulated in 3 minutes cycles, along a total collecting time equal to ~ 90 minutes. The scattered intensity was normalized by using vanadium standard and corrected for the background effect by the blank sample (150 mM NaCl in D_2O). Temperature was fixed at 20 °C.

3. Results and Discussion

3.1. The structure of $C_{12}NO$/DOPE/DNA

Figure 2A shows the diffractogram of $C_{12}NO$/DOPE/DNA complexes formed in a solution at neutral pH (~ 7.2). Two peaks ($L(1)$ and $L(2)$) characterize a lamellar phase (L) with repeat distance d = 54 Å. Fully hydrated DOPE at 20 °C forms an inverted hexagonal phase with a lattice parameter a ~ 76 Å (not shown). The increasing content of $C_{12}NO$ in the mixture induces hexagonal to lamellar phase transition; a lamellar arrangement is observed in all complexes at $C_{12}NO$/DOPE>0.3 mol/mol [25]. The repeat distance, $d = d_L+d_W$, is the sum of the thicknesses of lipid bilayer, d_L, and of water layer d_W between bilayers. The packing of the L phase is rather poor: the peaks are wide most likely due to small positional order of the elementary cells arrangement, a consequence of lamellae fluctuations.

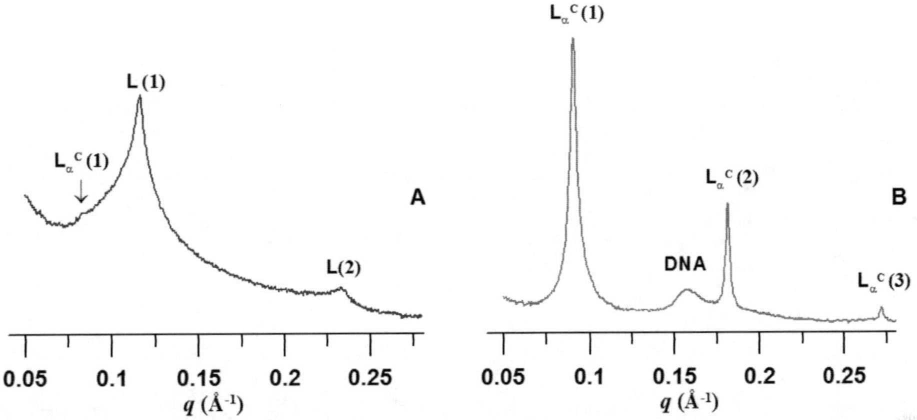

Figure 2. SAXD patterns of $C_{12}NO$/DOPE/DNA complexes prepared in 150 mM NaCl at pH 7.2 (A) and pH 5 (B).

The DNA uptake is low in neutral solutions. A tiny peak observed at q ~ 0.084 Å^{-1} is detected as the first order peak of a condensed lamellar phase L_α^C (d ~ 75 Å) with DNA strands packed in water layers between lipid bilayers. We found that a maximum of ~ 30 % of the total DNA volume in the sample was bound in a condensed lamellar phase L_α^C at $C_{12}NO$/DOPE=1 mol/mol and pH 7.2 (see [25] for details). $C_{12}NO$/DOPE/DNA complexes are sensitive to changes of pH. In acidic solutions, $C_{12}NO$ molecules are protonated, and the surfactant behaves as a cationic agent. The diffractogram of the complex prepared in a solution at pH ~ 5 shows a condensed lamellar phase (Figure 2B). In the acidic solution, the fraction of $C_{12}N^+OH$ is sufficiently high to keep DNA strands arranged regularly between lamellae of L_α^C phase, as manifested by the peak related to DNA – DNA packing. Structural parameters of the complexes at selected pH are summarized in Table 1. SAXD experiment confirmed the very good structural responsivity of the complexes to changes of pH.

Table 1. Structural parameters of $C_{12}NO$/DOPE/DNA complexes

pH	*L* phase *d* (Å)	L_α^C phase *d* (Å)	d_{DNA} (Å)
7.2	54.0	75.0	-
6.1	58.2	70.0	42.2
5.0	-	67.7	36.5
4.0	-	64.3	31.4

3.2. Kinetics of $C_{12}NO$/DOPE/DNA complex formation followed by SANS

In addition to the elastic properties of $C_{12}NO$/DOPE membrane, electrostatic interactions with DNA polyanion also play an important role in the aggregation process, and thus the complexation geometry is dictated by nontrivial interplay between electrostatic and elastic contributions to the free energy of the complexes [27]. In our system, DNA binding is conditioned by the protonation of molecules of the surfactant. We performed small – angle neutron scattering experiment with the aim to follow kinetics of $C_{12}NO$ protonation and DNA binding.

Figure 3 shows SANS curves of $C_{12}NO$/DOPE vesicles and a mixture of DNA – $C_{12}NO$/DOPE, at neutral pH. In spite the fact that $C_{12}NO$/DOPE dispersion was extruded prior to measurements, the curves do not show the typical profile of scattering of unilamellar vesicles like we observed for

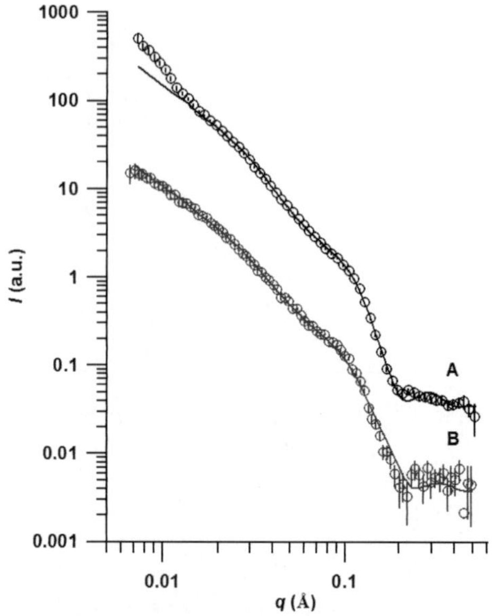

Figure 3. SANS curves of $C_{12}NO/DOPE$ vesicles (A) and a mixture of DNA – $C_{12}NO/DOPE$ (B) in a solution at neutral pH. Full lines show fits using a paracrystal lamellar model.

example in [15]. Due to rapid aggregation, the dispersions are formed by oligolamellar up to multilamellar vesicles.

The SANS data were analyzed using a paracrystal lamellar model with respect to the layer distance distribution and the number of layers in a single cluster [28]. In this model, scattering intensity is given by

$$I(q) = 2\pi(\Delta\rho)^2 \Gamma_m \frac{P_{bil}(q)}{q^2} Z_N(q)$$

where $\Delta\rho$ is the difference in scattering length per unit mass of solute between particles and solvent and Γ_m is the mass per area of the bilayer. The form factor

$$P_{bil}(q) = \left(\frac{\sin(q\xi)}{q\xi}\right)^2$$

where ξ is half the bilayer thickness, and a factor proportional to q^{-2} is typical for an infinitively large two-dimensional sheet. $Z_N(q)$ describes the interference effects for aggregates formed by N layers. Full lines in Figure 3 represent the best fits of experimental data applying this paracrystal lamellar model using SasView fitting software [29]. The average excess of scattering length density per unit mass ($\Delta\rho$) of the solute was calculated from known $C_{12}NO/DOPE$ molar ratio using the appropriate molecular volumes and molecular weights [30, 31]. The ideal model scattering curves were smeared by the instrument resolution ~10 %. According to the model, the best fit of the scattering curve of $C_{12}NO/DOPE$ (Figure 3A, full line) represents a mixture of uni- and ~ 30 % of oligo-lamellar vesicles consisting of more than 1 bilayer separated with a spacing ~ 62.6 Å with polydispersity of the spacing ~10 %. We find the lipid bilayer thickness $d_L = 2\xi = 30.3\pm0.1$ Å at the bilayer composition $C_{12}NO/DOPE$ = 1 mol/mol. To our knowledge, structural data of $C_{12}NO/DOPE$ bilayer are not reported in literature yet. The lipid bilayer consists of a hydrophobic region (d_C) framed by two polar headgroup regions exposed to the water phase. The intercalation of $C_{12}NO$ between DOPE molecules results in a lateral expansion of the interface, and a mismatch between the length of $C_{12}NO$ alkyl substituent and the length of hydrocarbon acyl chains of DOPE (18 carbons) creating defects in the hydrophobic membrane core. These defects, eliminated by chain bending, lead to a decrease of the thickness (d_C) of the hydrophobic region [14, 32]. In our previous work [15] we found that the thickness of the hydrophobic region of dioleoylphosphatidylcholine (DOPC) bilayer (d_C=27.4±0.1 Å) decreases to d_C = 24.9 Å, and the area per 1 molecule increases from A = 70.4 Å^2 (for DOPC) up to A= 106.6 Å^2 at C_{12}NO:DOPC=1 mol/mol. The polar head group of phosphatidylethanolamines is smaller in comparison to that bulky in phosphatidylcholines and also the headgroup region is thinner. For example, at the lipid bilayer thickness d_L=40.5 Å of POPE (palmitoyl-oleoylphosphatidyl-ethanolamine) in liquid-crystalline phase (at 35 °C) the thickness of hydrophobic region d_C = 32.1 Å was determined by [31], meaning that, the obtained lipid bilayer thickness d_L = 30.3±0.1 Å for vesicles of C_{12}NO/DOPE=1 mol/mol sounds reasonable. At low q values (Figure 3A), we see a clear discrepancy between experimental data and the full line represented by the model. Likely, it is explained by the curvature of vesicles, not taken into account by the model. Figure 3B shows the scattering curve of the C_{12}NO/DOPE dispersion with DNA at pH ~ 7. The molecules of C_{12}NO are not charged at neutral pH, and thus DNA binding is weak. The full line representing the best fit (using the same model) indicates neither significant change in population of oligolamellar vesicles (~ 30 %) nor in the lipid bilayer thickness. However, due to intercalation of DNA polyanion between bilayers without efficient screening of its negative charges, the repeat distance increases to ~ 78 Å with polydispersity ~ 20 %. The obtained repeat distance is close to ~ 75 Å detected by SAXD in L_α^C phase at similar conditions.

We injected adjusted volume of diluted DCl in the measuring cell containing a mixture of DNA and C_{12}NO/DOPE vesicles (pH 7) with the aim to follow kinetics of DNA binding due to protonation of C_{12}NO molecules. SANS data were collected in 3 min cycles. Dependences of SANS intensity $I(q)$ on scattering vector q for DNA – C_{12}NO/DOPE dispersion along the measuring time are depicted in Figure 4. The initial q dependence (red circles) shows the scattering of DNA – C_{12}NO/DOPE dispersion at pH ~ 7, prior to DCl injection. In a solution at neutral pH, the interaction between DNA and C_{12}NO/DOPE vesicles is weak, and we observed uni – or oligolamellar structures, as discussed above. The injected DCl changes pH in the measuring cell from 7 to 4. The interaction of DNA with C_{12}NO/DOPE vesicles under acidic conditions results in the disintegration of vesicles and the formation of multilamellar structures, similarly to DNA - cationic liposome complexes. All curves were fitted by the same paracrystal lamellar model using SasView software. Excellent accord is obtained. An example of the distribution of residuals is shown in Figure 4, inset. The system was followed along 90 minutes. Figures 5A - C show the time evolution of the obtained structural parameters: the number of layers (Figure 5A), the lipid bilayer thickness (Figure 5B) and the spacing (Figure 5C). Our experiment indicates that both processes the protonation of C_{12}NO molecules and DNA binding are rapid. We detected substantial growth of clusters due to DNA complexation reaching up to ~32 layers already in three minutes after the DCl injection (Figure 5A). Beyond this rapid aggregation, we detected further growth of aggregates. The number of layers increases gradually, and about ~ 4 additional layers were attached during the followed time interval. Cryoelectron microscopy observations [33] report that most likely the DNA – cationic liposomes complex growth begins with one DNA – coated vesicle. In this way DNA acts as a molecular "glue" enforcing close apposition of neighbour vesicle. Multilamellar lipid – DNA complexes appear to form by mechanism that involves the rupture of an approaching vesicle and subsequent adsorption of its membrane to a "template" vesicle or a lipid-DNA complex. The authors [33] do not exclude also another mechanism, when the attached vesicle is flattened, and is adsorbed to a "template" vesicle. However, in this way, the complex grows always with even number of layers. In such mechanism one can not exclude alternation of layers consisting of the lipid only, and those with DNA. The complexes frequently exhibit partially open bilayer segments on their outside surfaces. DNA accumulates near the edges of such unusually terminated membranes. The full line in Figure 5A shows the fit of experimental points by a Boltzmann function indicating the growth of complexes by cooperative process. Our experiment shows that DNA complexation is rapid and the "ground" of complexes was formed in time domain before three minutes after the acidification. The observed growth can be interpreted as an adsorption

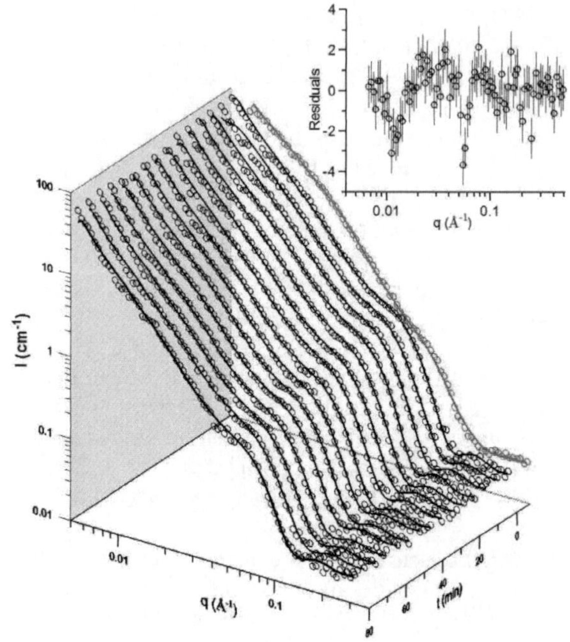

Figure 4. Dependences of SANS intensity $I(q)$ on scattering vector q for DNA – $C_{12}NO$/DOPE dispersion as a function of time; prior (red points) and after DCl injection. Full lines show fits using a paracrystal lamellar model.
Inset: An example of the distribution of residuals.

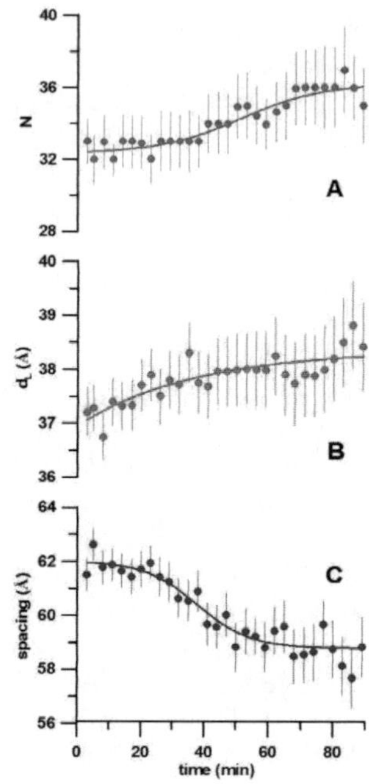

Figure 5. Time dependence of the structural parameters: the number of layers (A), the lipid bilayer thickness (B) and the spacing (C).

of further oligo-lamellar vesicles to the complex at any of its rims (with DNA accumulated) that might exist.

Figure 5B shows changes in the lipid bilayer thickness. Both $C_{12}NO$ protonation and structural changes in the polar region of the bilayer due to DNA binding result in the increase of the lipid bilayer thickness up to d_L ~ 37 Å, in three minutes after the DCl injection. Next, a slight increase in d_L (~ 1Å) manifests internal re-organization of the complexes. Modelling [34, 35] and experimental studies [36] confirms a direct involvement of the neutral lipids in the electrostatic interaction with DNA. Due to the cationic additive, the P^--N^+ dipole of the phospholipid headgroup changes its orientation [37], exposing the N^+ end towards the aqueous environment and interacts with the complexed DNA. Accordingly, DNA can induce additional changes of the orientation of P^--N^+ dipole due to Coulombic attraction. This results in a reduction average area occupied by PE headgroup. Reorientation of P^--N^+ dipole, and H^+(or D^+) binding due to the $C_{12}NO$ protonation can yield electrostatic interactions at the headgroup level. As a consequence, chain-chain interactions within the hydrophobic core increases balancing forces in the membrane lateral pressure profile. We suppose that these molecular changes are responsible for the lipid bilayer increase, as shown in Figure 5B.

Figure 5C shows time evolution of the spacing. Both experiments (SAXD and SANS) confirm a weak binding of DNA between the lamellae of the $C_{12}NO$/DOPE vesicles at neutral pH. After DCl injection, H^+ (D^+) diffusion can be considered as the driving force of aggregation. $C_{12}NO^+H$ molecules act as a cationic surfactant. The drop of the spacing from ~ 78 Å to ~ 62 Å in three minutes proves the efficient screening of negative charges of DNA condensed between the lipid bilayer. It is worth mentioning that the model for data treatment includes polydispersity of the spacing, which is ~ 43 %.

The resolution of SANS instrument is lower in comparison to SAXD. Indeed, SANS curves (Figure 4) show a broad peak at q ~ 0.1 $Å^{-1}$ smeared because of instrumental resolution. The obtained value, ~ 62 Å, is in good accord with the repeat distance d = 64.3 Å of a condensed lamellar phase of $C_{12}NO$/DOPE/DNA complexes at pH 4 derived from SAXD (Table 1). The thickness of $C_{12}NO$/DOPE bilayer forming the complexes is d_L ~ 37-38 Å (Figure 5B). Thus the water gap of thickness d-d_L ~ 25 Å fits exactly for an intake of hydrated DNA strands with a diameter ~ 25 Å [38]. The contrast between DNA and buffer is small at the used DNA concentration and thus its contribution to the SANS scattering intensity is negligible. Figure 5C shows a marked drop in the spacing, from ~ 62 Å to ~ 59 Å, proving an evolution of the system during the time followed by SANS experiments. In spite of high polydispersity (~ 43 %), to accommodate DNA with a diameter ~ 25 Å into a lamellar phase with spacing ~ 59 Å would suppose thinning of the lipid bilayer. As mentioned above, one of mechanisms of the complexes growth supposes an adsorption of flattened vesicles to a "template" vesicle [33]. In this way, the complex grows always with even number of layers. Moreover bilayers that include DNA can alternate with bilayers without DNA. Such structures are frequently formed at DNA aggregation with neutral phosphatidylcholines in presence of divalent cations, and were detected by SAXD [39 - 41]. The coexistence of two lamellar phases within one structure was observed also for $C_{12}NO$/DOPE/DNA complexes at pH > 6 (see Table 1) [25]. A careful inspection of structural data tells, that the repeat distance of L_α^C phase decreases with pH, while L phase swells due to excess of non-compensated positively charged $C_{12}NO^+H$ molecules. Note that, at acidic conditions the repeat distances of both phases are close, and likely can-not be resolved by SANS technique. In respect to high polydispersity of the spacing, we suppose that DNA is condensed in each of both structures; however bilayers depleted of DNA are slightly prevailing in clusters with smaller spacing (~59 Å). The observed drop of the spacing can be attributed to internal reorganization, and likely related to redistribution of DNA, actually its release. Lipoplexes show best stability at the isoelectric point, when the numbers of positive and negative charges are equal [27]. The overcharging of the complex away from its isoelectric point is caused by changes of the bulk structure with absorption of excess DNA or cationic lipid. Importantly, overcharged complexes are observed to move toward their isoelectric charge-neutral point at higher concentration of salt co-ions, with positively overcharged complexes expelling cationic lipid and negatively overcharged complexes expelling DNA [42]. Moreover, high ionic strength reduces the fraction of bound DNA in the complexes, and the isoelectric point is attained at lower ratio DNA/cationic additive than the one that can be estimated by calculations based on their nominal charges [43]. Our lipoplexes were formed in aqueous medium at high ionic strength (150 mM NaCl). The fit of experimental points in Figure 5C by Boltzmann function shows t =38.0±2.9 min as inflection point for DNA release. However, the sigmoid in Figure 5A, related to the growth of complexes, shows the inflection point later, at t = 51.3±4.9 min. This time succession offers an explanation why DNA released from the complexes can mediate the adsorption of the next approaching vesicle and form additional layers in this lamellar structure.

4. Conclusion

We used N,N-dimethyldodecylamine-N-oxide ($C_{12}NO$) for designing of a pH sensitive DNA delivery vector. SAXD experiments show the necessary structural responsivity of $C_{12}NO$/DOPE/DNA complexes to changes of pH in aqueous medium at high ionic strength. SANS data in kinetic experiments were analysed using a paracrystal lamellar model in order to follow time evolution of the complexes, their growth, changes in the lipid bilayer thickness and spacing. The protonation of the surfactant molecules in $C_{12}NO$/DOPE mixture is rapid, and multilamellar $C_{12}NO$/DOPE/DNA complexes with ~ 32 layers are formed before 3 minutes. In the time domain above 3 minutes, there is further reorganization of the complexes likely due to moving the structure towards its higher stability at isoelectric point.

Our experiments confirm that N,N-dimethylalkylamine-N-oxides show excellent pH responsi-vity changing their property from nonionic to cationic in a narrow window of pH. This ability is preserved also when molecules of the surfactant are mixed with a neutral phospholipid, what suggests their utilization for designing genetic material delivery vectors, also for nanocarriers on targeted drug delivery.

Acknowledgements
The authors thank S. S. Funari for his assistance on SAXD measurements. This work benefitted from SasView software, originally developed by the DANSE project under NSF award DMR-0520547. The research has received funding from the European Community's Seventh Framework Programme (FP7/2007-2013) under grant agreement n° 226716 (HASYLAB project II-20100372 EC); JINR project 04-4-1121-2015/2017 and grant VEGA 1/0916/16.

References

[1] Felgner P L, Gadek T R, Holm M, Roman R, Chan H W, Wenz M, Northrop J P, Ringold G M and Danielsen M 1987 *Proc. Natl. Acad. Sci. U. S. A* **84** 7413
[2] Lasic D D and Templeton N S 1996 *Adv. Drug Deliv. Rev.* **20** 221
[3] Fattal E, Couvreur P and Dubernet C 2004 *Adv. Drug Deliv. Rev.* **56** 931
[4] Sakaguchi N, Kojima C, Harada A, Koiwai K and Kono K 2008 *Biomaterials* **29** 4029
[5] Ganta S, Devalapally H, Shahiwala A and Amiji M 2008 *J. Control. Release* **126** 187
[6] Simoes S, Moreira J N, Fonseca C, Duzgunes N and de Lima M C P 2004 *Adv. Drug Deliv. Rev.* **56** 947-65
[7] Maeda H, Tsunoda M and Ikeda S 1974 *J. Phys. Chem.* **78**1540
[8] Búcsi A, Karlovská J, Chovan M, Devínsky F and Uhríková D 2014 *Chem. Pap.* **68** 842
[9] Singh S K, Bajpai M and Tyagi V K 2006 *J. Oleo Sci.* **55** 99
[10] Devínsky F 1985 *Acta Fac. Pharm. Univ. Comen.* **40** 63
[11] Šeršeň F, Leitmanová A, Devínsky F, Lacko I and Balgavý P 1989 *Gen. Physiol. Biophys.* **8** 133
[12] Dubničková M, Kiselev M, Kutuzov S, Devínsky F, Gordeliy V and Balgavý P 1997 *Gen. Physiol. Biophys.* **16** 175
[13] Barlow D J, Lawrence M J, Zuberi T, Zuberi S and Heenen R K 2000 *Langmuir* **16** 10398
[14] Karlovská J, Degovics G, Lohner K, Lacko I, Devínsky F and Balgavý P 2004 *Chem. Phys. Lipids* **129** 31
[15] Belička M, Kučerka N, Uhríková D, Islamov A Kh, Kuklin A, Devínsky F and Balgavý P 2014 *Eur. Biophys. J.* **43** 179
[16] Uhríková D, Kučerka N, Islamov A, Gordeliy V and Balgavý P 2001 *Gen. Physiol. Biophys.* **20** 183
[17] Devínsky F, Kopecká-Leitmanová A, Šeršeň F and Balgavý P 1990 *J. Pharm. Pharmacol.* **42** 790
[18] Bukovský M, Mlynarčík D and Ondráčková V 1996 *Int. J. Immunopharmacol.* **18** 423
[19] Warisnoicharoen W, Lansley A B and Lawrence M J 2003 *J. Pharm. Sci.* **92** 859
[20] Mel'nikova Y S and Lindman B 2000 *Langmuir* **16** 5871
[21] Wang Y, Dubin P L and Zhang H 2001 *Langmuir* **17** 1670
[22] Bonincontro A, Marchetti S, Onori G and Santucci A 2003 *Chem. Phys. Lett.* **370** 387
[23] Goracci L, Germani R, Savelli G and Bassani D M 2005 *Chembiochem* **6** 197
[24] Marchetti S, Onori G and Cametti C 2005 *J. Phys. Chem. B* **109** 3676
[25] Hubčík L, Funari S S, Pullmannová P, Devínsky F and Uhríková D 2015 *Biochim. Biophys. Acta* **1848** 1127
[26] Roveri N, Bigi A, Castellani P P, Foresti E, Marchini M and Strocchi R 1980 *Boll. Soc. Ital. Biol. Sper.* **56** 953
[27] May S, Harries D and Ben Shaul A 2000 *Biophys. J.* **78** 1681
[28] Bergstrom M, Pedersen J S, Schurtenberger P and Egelhaaf S U 1999 *J. Phys.Chem. B* **103** 9888
[29] SasView, http://www.sasview.org/
[30] Belička M, Klacsová M, Karlovská J, Westh P, Devínsky F and Balgavý P 2014 *Chem. Phys. Lipids* **180** 1

IOP Conf. Series: Journal of Physics: Conf. Series **848** (2017) 012007 doi:10.1088/1742-6596/848/1/012007

[31] Kučerka N, van Oosten B, Pan J, Heberle F A, Harroun T A and Katsaras J 2015 *J. Phys. Chem. B* **119** 1947
[32] Balgavý P and Devínsky F 1996 *Adv. Colloid Interface Sci.* **66** 23
[33] Huebner S, Battersby B J, Grimm R and Cevc G 1999 *Biophys. J.* **76** 3158
[34] Bandyopadhyay S, Tarek M and Klein M L 1999 *J. Phys. Chem. B* **103** 10075
[35] Pink D A, Quinn B, Moeller J and Merkel R 2000 *Phys.Chem.Chem. Phys.* **2** 4529
[36] Ryhanen S J, Saily MJ , Paukku T, Borocci S, Mancini G, Holopainen J M and Kinnunen P K J 2003 *Biophys. J.* **84** 578
[37] Scherer P G and Seelig J 1989 *Biochemistry* **28** 7720
[38] Radler J O, Koltover I, Salditt T and Safinya C R 1997 *Science* **275** 810
[39] McManus J, Radler J O and Dawson K A 2003 *Langmuir* **19** 9630
[40] Francescangeli O, Stanic V, Gobbi L, Bruni P, Iacussi M, Tosi G and Bernstorff S 2003 *Phys. Rev. E* **67** 011904
[41] Uhríková D, Hanulová M, Funari S S, Khusainova RS, Šeršeň F and Balgavý P 2005 *Biochim. Biophys. Acta* **1713** 15
[42] Koltover I, Salditt T and Safinya C R 1999 *Biophys. J.* **77** 915
[43] Pullmannová P, Bastos M, Guangyue B, Funari SS, Teixeira J, Lacko I and Uhríková D 2012 *Biophys. Chem.* **160** 35

Lipid membranes loaded with Ca^{2+} and Zn^{2+} cations

N Kučerka[1,2], E Dushanov [2], K T Kholmurodov[2,3], J Katsaras[4] and D Uhríková[1]

[1] Department of Physical Chemistry of Drugs, Faculty of Pharmacy, Comenius University, 83232 Bratislava, Slovakia
[2] Frank Laboratory of Neutron Physics, Joint Institute for Nuclear Research, 141980 Dubna - Moscow Region, Russia
[3]Dubna International University for Society, Nature and Man, 141980 Dubna - Moscow region, Russia
[4] Joint Institute for Neutron Sciences, Oak Ridge National Laboratory, Oak Ridge, TN 37831-6453, USA

kucerka@fpharm.uniba.sk

Abstract. We have studied the interactions of calcium and zinc with the biomimetic membrane made of dipalmitoyl-phosphatidylcholine (DPPC). The small angle neutron experiment on oriented multilamellar samples showed clearly differences in the effects of the two cations. For both, a bilayer thickness increases due to divalent metal ion (Me^{2+}) binding, reaching the maximum at stoichiometry Me^{2+}:DPPC~1:7 mol/mol. However, while the further increase in Ca^{2+} results in a bilayer thinning down to the level of pure DPPC, the Zn^{2+} binding indicates the behaviour of a typical isotherm, reaching a level of saturation.

1. Introduction

It is well known that divalent metal cations play important role in cell's physiology and biochemistry. In addition to calcium and magnesium, metals such as iron, manganese, copper, zinc, nickel and cobalt are essential at the appropriate concentration, yet toxic beyond normal levels. Among the first-row transition metals, zinc is second only to iron in terms of abundance and importance in biological systems. Zn^{2+} plays a fundamental role in several critical cellular functions such as protein metabolism, gene expression, structural and functional integrity of biomembranes, and in metabolic processes [1]. Compared to other micronutrients, zinc exists in biological systems in high concentrations, particularly in biomembranes. Concentration of zinc in animal organels ranges from $<10^{-9}$ M in cytoplasm to $>10^{-3}$ M in some membrane vesicles [2].

Cell membrane properties such as membrane fluidity, bending and rigidity moduli, electrostatics, and aggregation and fusion are tightly associated with ions that are prevalent in both the cytosol and the exterior of the membrane. Interestingly, the divalent metal cations were found to play a dominant role in affecting bilayer structure. For example, it is well known that Zn^{2+} plays a fundamental role in several critical cellular functions such as protein metabolism, gene expression, structural and functional integrity of biomembranes, and in metabolic processes, while Ca^{2+} was shown to alter a bacterial membrane in a manner limiting its water penetration. The cation binding depends strongly on the property of the cation and the membrane lipid head-group.

Our observations agree well with the notion that the effect of Ca^{2+} and Zn^{2+} cations on the structure of lipid bilayer can most likely be rationalized in terms of electrostatic interactions, rather than that of geometrical constraints due to bilayer curvature, and thus reinforcing the special importance of these cations. The electronic structure of calcium results in its high affinity to

IOP Conf. Series: Journal of Physics: Conf. Series **848** (2017) 012008 doi:10.1088/1742-6596/848/1/012008

electronegative groups PO_4 and CO_2 and therefore, also other electronegative moieties such as ester oxygens and/or carbonyl groups of the lipid headgroup can be directly involved in complex formation. The molecular dynamics simulations have been performed to gain more detailed information.

2. Results and Discussion

We have performed the small angle neutron experiment on oriented multilamellar samples to decouple effects due to electrostatic interactions from those of geometrical constraints found in curved vesicular bilayers. The experiment involved the contrast variation measurements, in which the series of D_2O/H_2O solutions were utilized for the sample hydration. This approach allows to solve the scattering phase problem by requiring the scattering form factors to change linearly as a function of D_2O content [3]. We also employ the procedure suggested by Leonard et al. [4] for unambiguous phase determination as described in details elsewhere [5].

The neutron scattering length density (NSLD) profiles of bilayers were obtained by a successful Fourier transformation of corrected contrast varied diffraction data [5]. While the profile corresponding to the 100% D_2O contrast condition provides the best estimate of bilayer steric thickness, lower contrast profiles reveal more details in the region of lipid head group. This is best seen in the case of 8% D_2O measurements, where the net NSLD of water equals to zero, providing essentially the lipid only NSLD profile of bilayer (Fig. 1a). In addition, the contrast varied diffraction data can be subtracted from each other either, providing thus effectively the probability distributions of water and lipid molecules. Figure 1b shows graphically the example of water probability distributions resulting from the subtraction. This approach does not require any assumptions on the functional form of NSLD profiles nor on the probability distributions resulting from their direct subtractions. The water distribution is obtained from averaging all of the separately subtracted pairs of contrast varied NSLD profiles, providing also an estimate on standard deviation error.

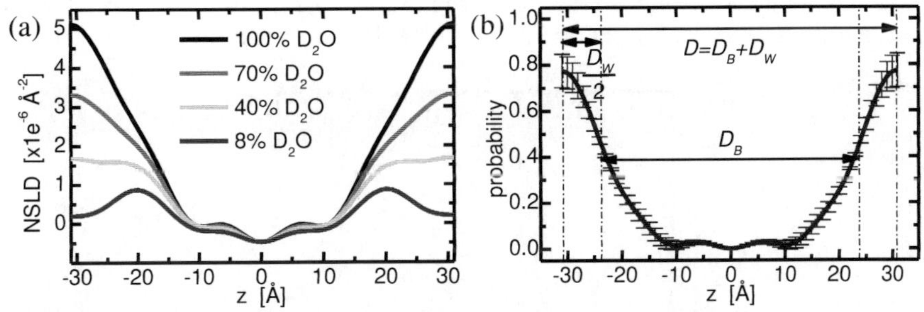

Figure 1. (a) Bilayers prepared of DPPC mixed with Zn^{2+} at 10.4:1 ratio shown through the NSLD profiles calculated from measurements at various contrast conditions. (b) Water probability distribution obtained by the NSLD subtraction procedure.

The obtained results confirm clearly the binding of both cations to the lipid bilayers. It is reflected in the increased lamellar repeat distance D that most likely results from the charge induced repulsion of bilayers. Interestingly, the further scrutiny of parameters reveals these changes in D, that is a sum of D_B and D_W (see Fig. 1b), are not distributed between D_B and D_W in a same manner for different ion concentrations. In addition, there is an apparent difference in the influence of the two ions examined. Let us first explore closer the effect of Zn^{2+} ions. The Fig. 2a shows a monotonic increase of D and D_B over the concentration range up to 0.14 mol/mol (i.e., DPPC:Zn^{2+}=7:1), after which the parameters seem to reach plateau.

IOP Conf. Series: Journal of Physics: Conf. Series **848** (2017) 012008 doi:10.1088/1742-6596/848/1/012008

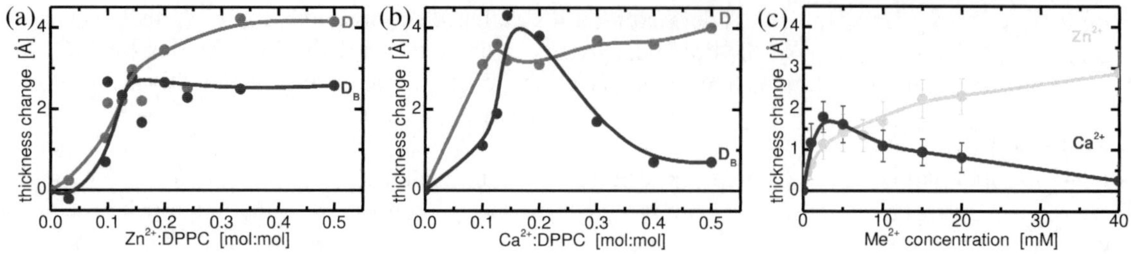

Figure 2. (a) Zn^{2+} and (b) Ca^{2+} cation-induced changes to bilayer structure expressed through lamellar repeat distance D (red) and bilayer thickness D_B (blue). (c) The results of cation-induced changes to the bilayer structure of unilamellar vesicles dispersed in water solution as published previously [6].

Ca^{2+} loaded bilayers clearly response to the increasing amount of cations differently when compared to Zn^{2+}. While the changes to the bilayer structure follow the same trend for both ions in the range up to 0.14 mol/mol, they differ at higher concentrations. The most pronounced difference is observed in the bilayer thickness D_B that decreases with further ionic increase, displaying a maximum at the 7:1 ratio of DPPC:Ca^{2+}. The D_B decreases almost to the level of pure DPPC bilayer at ratios above 2.5:1. This is supported also by the previously published results reproduced in Fig. 2c, in which unilamellar vesicles dispersed in the water solution loaded with either Zn^{2+} or Ca^{2+} ions displayed the same trends as in our results [6]. In addition, the same effect in the case of planar and curved bilayers rules out the involvement of the curvature impact similarly to conclusions for charged lipid bilayers [7].

We utilized molecular dynamics simulations for obtaining more structural details on the submolecular level in the most intriguing concentration point of 7:1 ratio of DPPC:Me^{2+}. Figure 3 focuses on the two major head group components represented by phosphate and carbonyl that are anticipated to be the primary targets for cation bindings due to their electronegativity.

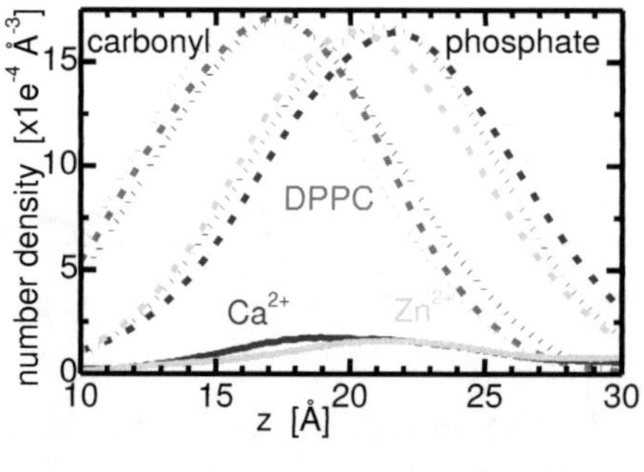

Figure 3. Number density of carbonyl (dotted lines) and phosphate groups (dashed lines) simulated for pure DPPC bilayer (red) and bilayers loaded with Ca^{2+} (blue) or Zn^{2+} (cyan) ions.

The addition of Ca^{2+} or Zn^{2+} ions affects clearly the DPPC head group as is evident by the positional change of its two components. While Ca^{2+} causes a small shifting of both carbonyl and phosphate components outwards, about twice as big shift towards the bilayer center is observed in the case of Zn^{2+}. On the other hand, the cations themselves seems to be localized in opposite directions, where Ca^{2+} penetrates deeper into the lipid bilayer.

3. Conclusions

The effects of Ca^{2+} and Zn^{2+} on the DPPC bilayer are different, as shown in changes of the lipid bilayer thickness. The radial distribution functions of cation-lipid pairs calculated for several lipid hedgroup atoms reveal a strong binding of Ca^{2+} to phosphate group (forming a contact pair Ca^{2+}-P and non contact pair Ca^{2+}-O) at low concentrations, while likely contributing to a screening effect at higher concentrations. The location of Zn^{2+} shows more-less uniform correlation with various lipid headgroup atoms. The weak correlation of zinc with lipid headgroup then results into the saturation of structural changes with its increasing concentration.

Acknowledgments
The authors acknowledge support from VEGA grant 1/0916/16 and JINR project 04-4-1121-2015/2017.

References

[1] Christianson D W, 1991 *Adv. Protein Chem.* **42** 281
[2] Williams R J P 1988 *Zinc in Human Biology* (London: Springer-Verlag) p 15
[3] Worcester D L and Franks N P 1976 *J. Mol. Biol.* **100** 359
[4] Léonard A, Escrive C, Laguerre M, Pebay-Peyroula E, Néri W, Pott T, Katsaras J and Dufourc E J 2001 *Langmuir* **17** 2019
[5] Kučerka N, Nieh M-P, Pencer J, Sachs J N and Katsaras J 2009 *Gen. Physiol. Biophys.* **28** 117
[6] Uhríková D, Kučerka N, Teixeira J, Gordeliy V and Balgavý P 2008 *Chem. Phys. Lipids* **155** 80
[7] Kučerka N, Pencer J, Sachs J, Nagle J F and Katsaras J 2007*Langmuir* **23** 1292

IOP Conf. Series: Journal of Physics: Conf. Series **848** (2017) 012009 doi:10.1088/1742-6596/848/1/012009

Nanoporous active carbons at ambient conditions: a comparative study using X-ray scattering and diffraction, Raman spectroscopy and N_2 adsorption

A A Shiryaev[1,*], A M Voloshchuk[1, †], V V Volkov[2], A A Averin[1], S D. Artamonova[1]

1) Frumkin Institute of Physical Chemistry and Electrochemistry RAS, Leninsky pr. 31, korp. 4, Moscow 119071, Russia

2) Institute of Crystallography RAS, Leninsky pr. 53, Moscow 119333, Russia

*E-mail: shiryaev@phyche.ac.ru

Abstract. Furfural-derived sorbents and activated carbonaceous fibers were studied using Small- and Wide-angle X-ray scattering (SWAXS), X-ray diffraction and multiwavelength Raman spectroscopy after storage at ambient conditions. Correlations between structural features with degree of activation and with sorption parameters are observed for samples obtained from a common precursor and differing in duration of activation. However, the correlations are not necessarily applicable to the carbons obtained from different precursors. Using two independent approaches we show that treatment of SWAXS results should be performed with careful analysis of applicability of the Porod law to the sample under study. In general case of a pore with rough/corrugated surface deviations from the Porod law may became significant and reflect structure of the pore-carbon interface. Ignorance of these features may invalidate extraction of closed porosity values. In most cases the pore-matrix interface in the studied samples is not atomically sharp, but is characterized by 1D or 2D fluctuations of electronic density responsible for deviations from the Porod law. Intensity of the pores-related small-angle scattering correlates positively with S_{BET} values obtained from N_2 adsorption.

1. Introduction

The sorption capacity and efficiency of carbon adsorbents strongly depend on pore size distribution and on structure of pore-matrix interface. Analysis of vapour adsorption isotherms is well established, widespread and important method of investigation of the porous structure. The effective pore size is usually calculated in the framework of Dubinin's volume filling theory (MVFT) [1, 2], density functional theory (DFT) [3], Kawazoe or BJH equations [4]. Complementary techniques, such as Small-Angle X-ray (SAXS) and neutron scattering (SANS) [5–10], pulsed NMR [11] and positron annihilation spectroscopy [12] are used for characterization of porous solids as well. These methods are nondestructive and rather fast compared to sorption experiments but provide information about various structural parameters and on different scales. Application of several complementary

† - deceased

techniques is the only reliable way to investigate the sorbent structure in detail. In this contribution we report results of investigation of activated carbons made from furfural (FAS-type) and of activated carbonaceous fibers (ACC-type) using several complementary techniques: Small- and Wide Angle X-ray Scattering (SWAXS), X-ray diffraction, Multi-Wavelength Raman Scattering (MWRS) and N_2 adsorption. This study attempts to elucidate correlations between results obtained using different analytical methods and to improve understanding of structure of activated carbons on several structural levels.

2. Experimental

Technology of synthesis of FAS-type activated carbons (AC) was developed at SPA "Neorganika" (Electrostal, Russia). Extreme mechanical strength of these materials and high sorption capacity of this type of AC make them extremely important in applications requiring high mechanical and adsorption properties, e.g., biomedicine, radiochemistry etc. Detailed description of these ACs is given in [13], general preparation scheme is the following: 1) liquid moulding of synthetic furfural monomer at 105 °C, 2) carbonization of spheroidal sorbent grains in inert medium at 750 °C, 3) steam activation at 850 °C. The samples differ by duration of the activation process: from 5 to 15 hours. After the activation the spheroidal grains were washed in distilled water and dried. In our work the sorbents were studied in as-received state.

The carbonaceous fibers (ACC) were prepared by thermal treatment of impregnated viscose fibers.

The nitrogen adsorption characteristics were measured using ASAP-2020 Analyzer (Micromeritics). We report results obtained by MP method, since it allows determination of pore size distribution for supermicro- and mesoporous adsorbents. The obtained values can be recalculated to other comparative methods (t-, αS-, n-methods).

The samples used for the structural characterization were shelf-stored at identical conditions for comparable periods of time (at least 12 months), assuring attainment of "equilibrium" degree of moisture saturation, which, of course, is sample-dependent due to variations in pore sizes.

It is well-known that water and other adsorbents may influence structure of sorbents considerably. Kinetics of water desorption may be extremely slow even at elevated temperatures (e.g., [14-16]). This applies even to samples in high vacuum in transmission electron microscopy where physisorbed water is most likely absent, but chemically-bound fraction may well be retained. For example, investigation of loose nanodiamond powder using infra-red spectroscopy showed that even vacuum heating of the powder at 400 °C for few hours does not eliminate surface OH-groups completely [17]. Therefore, frequently employed outgasing of activated carbons at 80-100 °C "in dry air" prior to structural studies may lead to large sample-to-sample variations due to differences in pore structure and relative fraction of pores of different sizes, strongly influencing desorption kinetics of adsorbed moisture. For consistency, in this contribution we report structural peculiarities of the samples stored for prolonged period of time at ambient conditions and thus being in equilibrium with environment. Validity of this approach to the studied samples is confirmed by presented comparative studies at various temperatures and duration of storage in vacuum. The adsorption measurements were performed according to common routine after prolonged vacuum drying.

SWAXS patterns were recorded using monochromatic CuK_α-radiation in broad range (0.1–27 nm^{-1}) of scattering vectors, q, using dedicated diffractometer SAXSess (Anton Paar). Samples of the FAS-type activated carbons were measured in standard X-ray capillaries or between two layers of special low-scattering polymer film; the fibers-based samples were measured without packaging. Kratki collimation scheme was employed; desmearing was performed using standard software. All measurements were performed at room temperature in evacuated chamber (residual pressure 5–10 mbar); therefore, partial desorption of physisorbed atmospheric species might took place. However, due to high brightness of the setup the measurements were generally fast and total exposure of the samples to vacuum was less than 5 minutes. For comparative purposes, for some samples the scattering curves were obtained immediately after reaching the target pressure (less than one minute) and after continuous pumping for up to 24 hours and after vacuum annealing at 50 °C for up to 24 hours. The difference was found to be negligible, see below.

X-ray diffraction patterns were recorded using Empyrean (Panalytical BV) diffractometer in Bragg–Brentano geometry using Ni-filtered CuKα-radiation in the angular range 10–120° (2Θ). No binder was employed to exclude influence of sample humidity on sorbent structure, which is discussed in detail in [11, 14] and all reported patterns were acquired for the samples dried directly in the sample holder in hot air for several hours. However, for comparative purposes several samples were also studied in wet state.

In-depth analysis of XRD data for carbons requires fitting of the whole pattern (total scattering) [18]. In the same time, the total scattering approach is based on models of sample structure and validity of these models requires independent confirmation. For comparative purposes the diffraction patterns were deconvoluted into components assuming Pearson VII peaks shape. Since the deconvolution is always not very robust, in subsequent analysis we operate with relative intensities, rather than with areas.

Raman spectra were acquired in quasi-backscattering geometry in air at room temperature at low power of 532 and 785 nm lasers using microscope (Bruker Senterra). Absence of sample alteration by the laser beam was checked both visually and by controlling persistence of spectral features. Several scans were averaged to improve signal to noise ratio; in addition for every sample several spots were analysed for consistency. Spectral decomposition to individual components is described in corresponding part of the paper.

3. Results and discussion

3.1. Sorption

Parameters of porous structure of the activated carbons (see Table 1) were calculated from the nitrogen adsorption isotherms at 77 K using several methods. The BET method was used for calculation of S_{BET} [4]; b) the comparative MP-method [19–22] gave mesopore surface area (S_{ME}) and total surface area (SMP) for samples with rather wide micropores (supermicropores); c) nanopore surface area calculated as: $S_{NANO} = S_{BET} - S_{ME}$; Dubinin's micropore volume filling theory (MVFT) [1, 2] provided micropore volume (W_0) and characteristic energy of adsorption (E_0); micropore size (x_0) calculated using Stoeckli et. al. equation [23]. Figure 1 shows the N_2 vapour adsorption isotherms for majority of the studied samples. The average isotherm for adsorption on the surface of nonporous carbon black was adopted as the reference isotherm [21, 22].

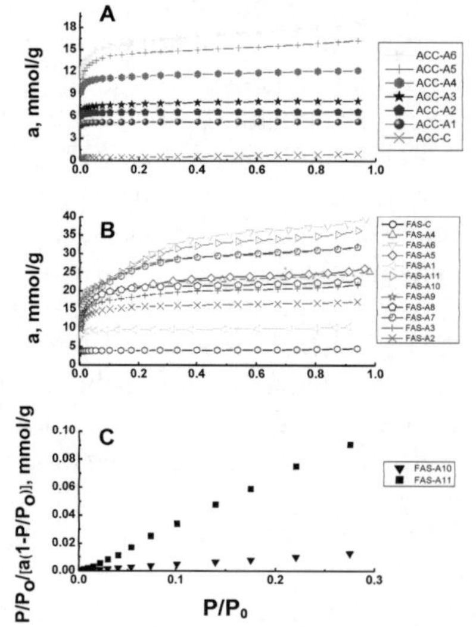

Figure 1. Nitrogen adsorption isotherms of studied samples. A – activated fibers (ACC); B - carbonizate (FAS-C) and activated (FAS-A) carbons; C – BET adsorption isotherms in linear form in the region used for calculation of S_{MP} for supermicro- and mesoporous carbons.

IOP Conf. Series: Journal of Physics: Conf. Series **848** (2017) 012009 doi:10.1088/1742-6596/848/1/012009

Table 1. Parameters of porous structure of the carbonizate (FAS-C), activated carbons of FAS-type and of activated fibers (ACC) with different degrees of activation

Sample	S_{BET}, m²/g	S_{MP}, m²/g	S_{ME}, m²/g	S_{NANO}, m²/g	W_o, cm³/g	E_o, kJ/mol	x_o, nm	x_{BET}, nm
FAS-C	350		20	328	0.13	45	0.18	0.88
FAS-A1	680		30	650	0.32	32.4	0.29	1.06
FAS-A2	1300		70	1230	0.53	20.5	0.59	0.90
FAS-A3	1500		90	1410	0.60	20.3	0.61	1.02
FAS-A4	1660		160	1600	0.67	19.6	0.60	1.05
FAS-A5	1600		120	1480	0.60	19.4	0.64	1.1
FAS-A6	2300		290	2010	0.67	18.2	0.79	1.20
FAS-A7	1900	1800	190	1710	0.73	17.9	0.83	1.16
FAS-A8	1650	1820	80	1570	0.70	17.2	0.94	0.95
FAS-A9	2030	2100	170	1850	0.81	17.1	0.94	1.10
FAS-A10	1600	1700	140	1460	0.65	16.5	1.06	1.04
FAS-A11	1950	1950	220	1730	~1.20	15.0	1.60	~1.30
ACC-C	38		30	8	0.014	42	1.8	0.8
ACC-A1	460		3	457	0.18	39.2	0.19	0.8
ACC-A2	580		4		0.22	45.2	0.16	0.8
ACC-A3	680		50		0.27	28.0	0.33	0.8
ACC-A4	940		35	905	0.39	28.3	0.79	0.9
ACC-A5	1250		80	1170	0.52	18.3	0.79	0.9
ACC-A6	1400	1470	140	1260	0.55	19.9	0.63	0.94
ACC-A7	1459		120		0.56		0.74	0.74
ACC-A8	1530		280		0.623		0.87	0.82
ACC-10	23				Non-porous			
ACC-UT					Non-porous			

* $S_{NANO} = S_{БЭТ} - S_{ME}$, **$x_{BET} = 4V_{tot}/S_{BET}$ (V_{tot} - the total pore volume)

IOP Conf. Series: Journal of Physics: Conf. Series **848** (2017) 012009 doi:10.1088/1742-6596/848/1/012009

3.2. X-ray diffraction

X-ray diffraction patterns of the studied sorbents are typical for nanocrystalline carbons (Fig. 2). In the 2Θ>10° range the patterns are dominated by broad 002 graphite peak. Formally, the corresponding interlayer spacings are between 0.36 and 0.39 nm, which are notably larger than those for ideal (0.334 nm) or even turbostratic graphite. Such spacings are encountered in non-graphitising carbons and in poorly graphitized carbonaceous substances with impurities. Application of the Debye–Scherrer formula gives crystallite sizes in the order of 1–2 nm. However, as shown in [24, 25], when the size of graphitic crystallites becomes smaller than 2–3 nm, the width and position of the 002 diffraction peak cannot be reliably used for determination of size and lattice spacing because of strong influence of detailed structure and morphology of the graphene layer stacks and surface termination.

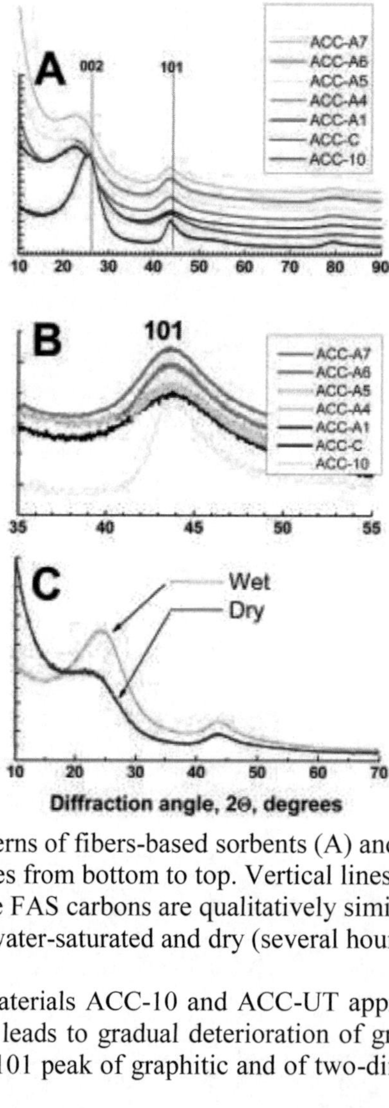

Figure 2. X-ray diffraction patterns of fibers-based sorbents (A) and close view of the 101+10 peak (B). The burn-off degree increases from bottom to top. Vertical lines indicate peak positions for ideal graphite. Diffraction from the FAS carbons are qualitatively similar. C - Comparison of X-ray diffraction patterns recorded in water-saturated and dry (several hours in flow of 40-50 °C air) states.

The patterns of non-porous materials ACC-10 and ACC-UT approach that of hexagonal graphite. Increase of the activation degree leads to gradual deterioration of graphitic structure. The peak at 42 degrees represents an overlap of 101 peak of graphitic and of two-dimensional 10 peak of turbostratic carbon.

Comparison of X-ray diffraction patterns recorded in water-saturated and dry (several hours in flow of 40-50 °C air) states (Fig. 2c) confirms very significant influence of moisture on X-ray diffraction patterns of the FAS ACs, see also [11, 13]. This effect is explained mostly by structural peculiarities of particular sample: depending on wettability and porosity of the material adsorbed water molecules may either align adjacent graphitic crystallites increasing L_a or, in contrast, may introduce disorder.

This effect is currently investigated in detail. Note that this dramatic effect of moisture is not present in SAXS data, see below.

For the activated carbon fiber samples (ACC) a fairly good positive correlation between the intensity of the ratio $I_{(101)+(10)}/I_{(002)}$ peaks and degree of activation (manifested in specific surface area (S_{BET}) and maximal adsorption volume (W_0)) is observed (Fig. 3). Note that existence of the correlation for the ACC-type samples indicates that diffraction halo of adsorbed water plays very minor role if any. In the same time, the FAS-type sorbents form another group with no or little dependence of the sorption parameters from the diffraction data. The correlation observed for ACC-samples could reflect either increase of lateral dimensions of graphitic domains with constant or decreasing thickness, or growing contribution of turbostratic carbon. The differences in morphology of the graphitic domains are partly inherited from precursors. In addition, one cannot exclude differences in gas-transport reactions during activation process which may promote lateral growth of graphene stacks (e.g., [26]).

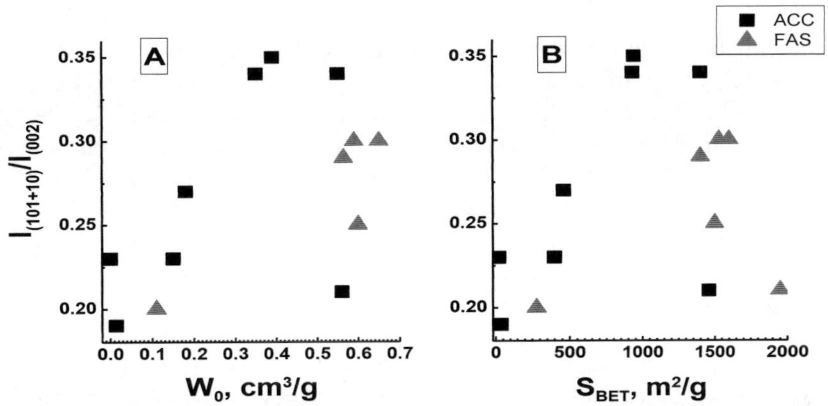

Figure 3. Relationships between sorption parameters and ratio of XRD peaks intensities. For ideal hexagonal graphite the $I_{(101)}/I_{(002)}$=0.17.

3.3. Small-angle scattering

3.3.1. Modeling of scattering curves from pores with diffuse boundaries. According to the Porod law a desmeared SAXS curve for slit-shape scatterers with sharp interfaces should be linear in coordinates $q^4I(q) - q^2$ at high values of the scattering vector. However, for almost all samples deviations from the Porod law are clearly observed (see below), in accordance with many of previous SAS studies of sorbents of different types [5]. Several approaches to describe such deviations could be used: 1) heavily polydisperse distribution of the scatterers; 2) assumption of fractal nature of the interface [27, 28], and 3) various models of the interface boundary [29, 30]. Since Guinier plots of our samples show existence of monodisperse scatterers, the polydispersity should not play an important role. The fractal approach is very popular, but it provides little information for comparison with other techniques. It should be also kept in mind that the fractal dimension derived from experimental scattering curves is sensitive to details of background subtraction [31].

In our view the most useful approach to explain deviation of SAS curves of carbonaceous sorbents should relate the deviations from the Porod law to structural features of the material, such as finite width of the interface or short-range disorder [29, 30]. It is important to emphasise that interpretation of SAS studies of activated carbons is usually performed in assumption of atomically flat pore boundaries of graphene layers. This assumption is largely based on two famous papers devoted to investigation of fibers-based carbons using SAS [32] and TEM [33]. However, recent high resolution TEM studies of various carbonaceous sorbents (e.g., [34, 35]) clearly show that generalization of this assumption to all carbonaceous materials is a gross oversimplification. Some of furfural-based samples discussed in the current work were analysed using TEM protocol from [34] and it is obvious that many pores cannot be approximated by a void formed by atomically flat walls [37]. For example, some of the walls are formed by stacks of ruptured graphene layers for which the formalism of "sharp" interface is not applicable. In the same time, corrugated walls are not uncommon. Below we present results of direct modeling of the scattering patterns from objects of variable roughness and show that the walls roughness may well account for observed deviations from the Porod law. Depending on exact topology of the wall its roughness may influence the scattering curve similar to mathematically-derived effects of finite thickness of the void-matrix interface or fluctuations of electron density.

To demonstrate influence of the scatterer roughness on SAXS patterns two independent computational approaches were employed.

The first approach is based on direct calculation of the SAXS curve from a scatterer with a given shape and consists of calculation of the SAS curve from a given cluster of "quasi-atoms" (QA) and is extremely successful in evaluation of SAXS data for monodisperse particles [37]. The "quasi-atoms" (QA) approach is based on filling a volume of a given shape and size with solid quasi-atoms. The size of QA is determined by user, therefore, the surface of the body can be made very smooth (very small QAs) or rough/corrugated. In order to avoid appearance of Bragg peaks from QA packing and to explicitly create variations is density of the body, the QA can be placed with variable degree of overlap. According to the Babinet principle scattering curves for a cluster of QAs and for a pore with the same shape/size and contrast will be undistinguishable.

The second approach is based on direct calculation of the scattering curve from a body with a surface layer with a gradient of the scattering density. The thickness of the layer and profile of electron density variations are user-determined as a radial density profile ρ(r). The details of calculation of scattering intensity from quasi-atom models are described in [37]. Scattering from a spherically symmetric particle with a radial density profile ρ(r) is computed as $I(q) = 4\pi \int_{s_{\min}}^{s_{\max}} \rho(r) \frac{\sin(qr)}{qr} \cdot r^2 \cdot dr$ [38].

These two kinds of simulation of the surface density gradients differ in essential detail. The quasi-atom modeling may describe both scattering from inhomogeneities in a particle shell and effect of the scattering from an object with gradually changing density. The modeling by a radial density function describes only the latter case which does not influence scattering intensity at high angles in comparison to the scatterers with sharp boundaries, but may describe the scattering behaviour at intermediate angles corresponding to those used in the present study.

Using the first (quasi-atoms) approach we have modeled scattering from ellipsoidal scatterers consisting of quasi-atoms of different radii and embedded into ellipsoidal shell. Variations of the thickness and QA radii in the shell produce objects with virtually identical sizes, but with widely different surface roughness/corrugation. Examples of the modeled bodies are shown in Fig. 4. Porod plots for sets of representative models are shown in Fig. 5.

IOP Conf. Series: Journal of Physics: Conf. Series **848** (2017) 012009 doi:10.1088/1742-6596/848/1/012009

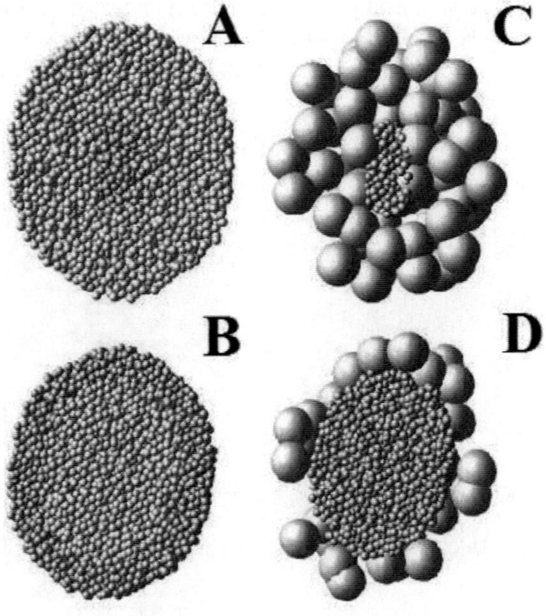

Figure 4. Examples of the bodies (cross-cuts) constructed for modeling of the scattering curves from objects with variable roughness. Different colors depict the "core" and "shell". A, B – small quasi-atoms (smooth surface); C, D – large quasi-atoms (rough/corrugated surface. Various combinations were modeled, see text for details.

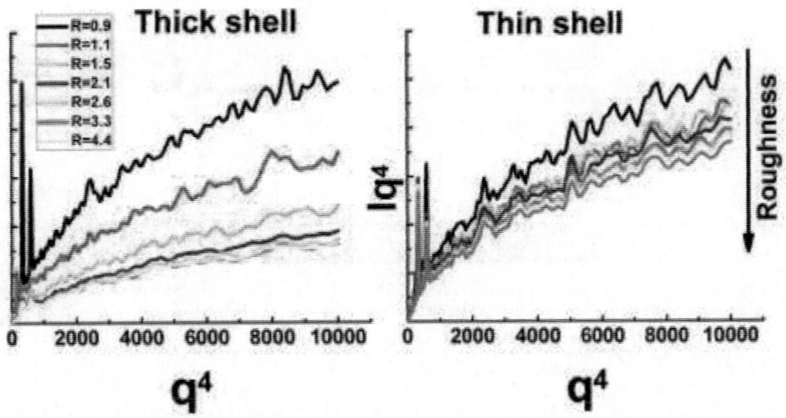

Figure 5. Porod graphs for the bodies consisting of quasiatoms of different sizes (see text and Fig. 4 for details). Radius of the shell QA's is indicated. The curves are normalized to "dry volume" of the bodies [37].

IOP Conf. Series: Journal of Physics: Conf. Series **848** (2017) 012009 doi:10.1088/1742-6596/848/1/012009

The modeling of the scattering curves for bodies with user-defined density profiles was performed for several general cases including very sharp boundary, boundary with a smooth decay in density and a case where the density has a peak, roughly mimicking kind of an adsorbate layer, see Fig. 6c. The power of the fitting exponent of the corresponding part of the scattering curves on log-log plot (the "slope" in our notation) show clear deviations from the ideal -4 value.

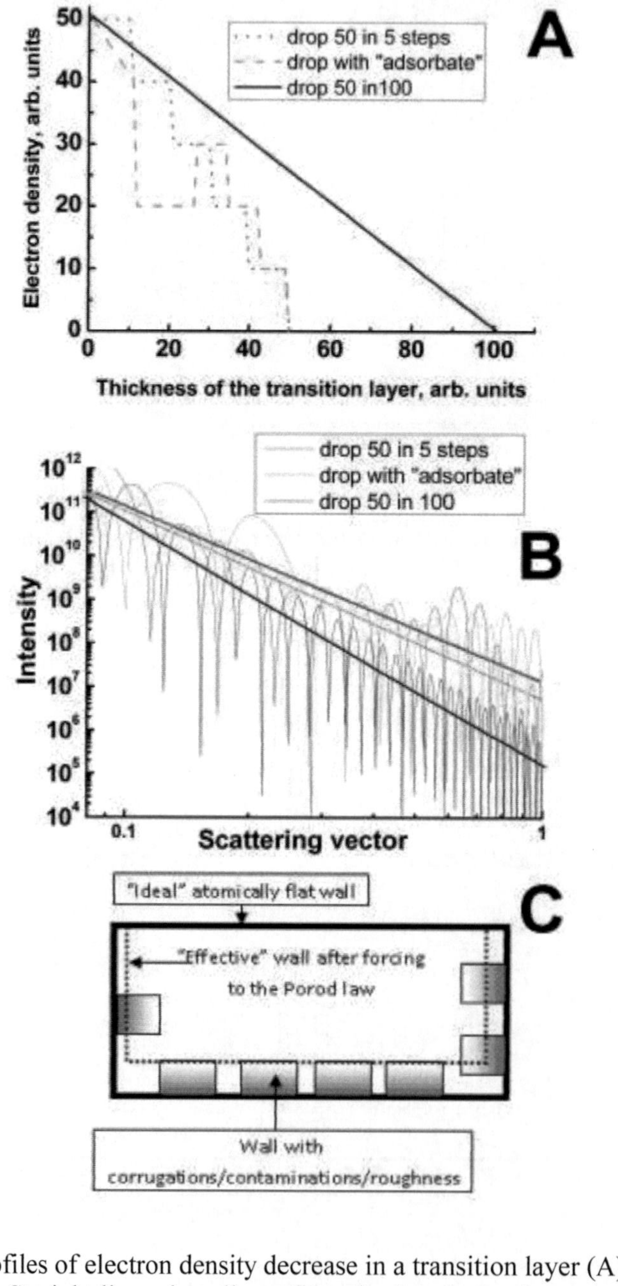

Figure 6. Simulated profiles of electron density decrease in a transition layer (A) and scattering curves from such bodies (B). Straight lines show linear fit to the data in the shown range of the scattering vectors, C – schematic image illustrating weakness of the concept of open/closed porosity evaluation from the SAS data.

IOP Conf. Series: Journal of Physics: Conf. Series **848** (2017) 012009 doi:10.1088/1742-6596/848/1/012009

The principal conclusion of these rather straightforward calculations is that subtraction of a constant background from a SAS curve to forcibly obey the Porod rule explicitly implies that the scattering pores possess atomically sharp walls, the assumption which is not always correct (see TEM results [34, 36]). Such subtraction effectively replaces a pore with rough/corrugated walls by some kind of an "effective" pore with unclear physical meaning in general case (Fig. 6c). Therefore, the combined use of SAS and adsorption results to extract "closed porosity" fraction is meaningful only in cases where the flat wall approximation is valid. Of course, subtraction of atomic scattering due to chemical elements comprising the sample is necessary, but in cases of activated carbons this contribution is typically small. Obviously, the angular region where the described deviations from the Porod plot will be manifested depends on ratio of the scatterers size and wavelength of the incident radiation and in some cases the deviations will occur in range inaccessible for SAXS instruments. However, our simple modeling as well as theoretical works by Ruland [29, 31] shows that one should be very careful prior to "standard" subtraction of a constant. Of course, background signal from atomic scattering by carbon and other important elements present in a sample should be subtracted, but in case of rather strong scattering from active carbons this contribution is usually very small.

3.3.2. Experimental data. SAXS patterns of the sorbents reveal their hierarchical structure (Fig. 7). As discussed in the Introduction, most SAXS studies of activated carbons employ some degree of degassing. The X-ray diffraction patterns of the studied carbons notably depend on amount of moisture (chapter 3.2, fig. 2c). In order to check influence of the moisture on the SAXS patterns several samples were measured in several states: 1) immediately after reaching the required conditions (P=5-10 mbar at 23 °C), 2) after pumping for 0 up to 24 hours at RT or at 50 °C. The measurements themselves were performed both at 25 and at 50 °C. The relatively low T of 50 °C was chosen to minimize eventual modification of the samples with highly developed porosity. Figure 7b shows that the scattering patterns coincide within measurement error except the region corresponding to diffraction from poorly ordered graphite ($10<q<20$ nm^{-1}), where rearrangements of the crystallites (reversible on cooling) are observed. The discrepancy with the XRD results can be tentatively explained by two interrelated factors: 1) as shown in independent study [36 and in preparation], SAXS is sensitive mostly to pores accessible to N_2 molecules and, depending on the AC type, these pores are not necessarily serve as principal water reservoir, 2) the scattering and diffraction phenomena differ in their origin, providing complementary information: diffraction patterns are mostly determined by (relatively) ordered crystallites and their mutual arrangement, whereas the scattering is determined by contrast of electronic density and (in isotropic case) is not sensitive to spatial arrangement of the domains.

For most samples the Debye–Bueche plot ($I^{-1/2}$-q^2) is linear in range of scattering vectors corresponding to pore scattering region (Fig. 7c). This suggests applicability of the random two-phase medium approximation. However, no clear relation between fit parameters of the Debye-Bueche plot and sorption data exists.

The major fraction of the total scattering intensity is due to polydisperse large heterogeneities with sizes exceeding the resolution of the employed setup (65 nm). It is important to emphasise that the Guinier plot for the low q region is usually not linear, or the linearity range is fairly short. This observation indicates that the large scatterers do not represent a monodisperse system. These heterogeneities could be tentatively assigned to graphitic crystallites or to macropores; their role in the sorption processes is minor. The power of the fitting exponent of the corresponding part of the scattering curves on log-log plot (below we use the term "slope") differ for the two studied types of carbons. For the furfural-based carbons the slope values cluster between -3.7 and -4.2 without clear correlation with activation degree. For the fibers-derived carbons the slopes show greater variability between -5 and -2.5. We note here that following standard approach to treat SAXS data from activated carbons one should subtract a constant to force the scattering curve to obey the Porod law. However, as shown above (see paragraph 3.2.1) such procedure implies rather strong assumptions about structure of the material and should not be applied "by default".

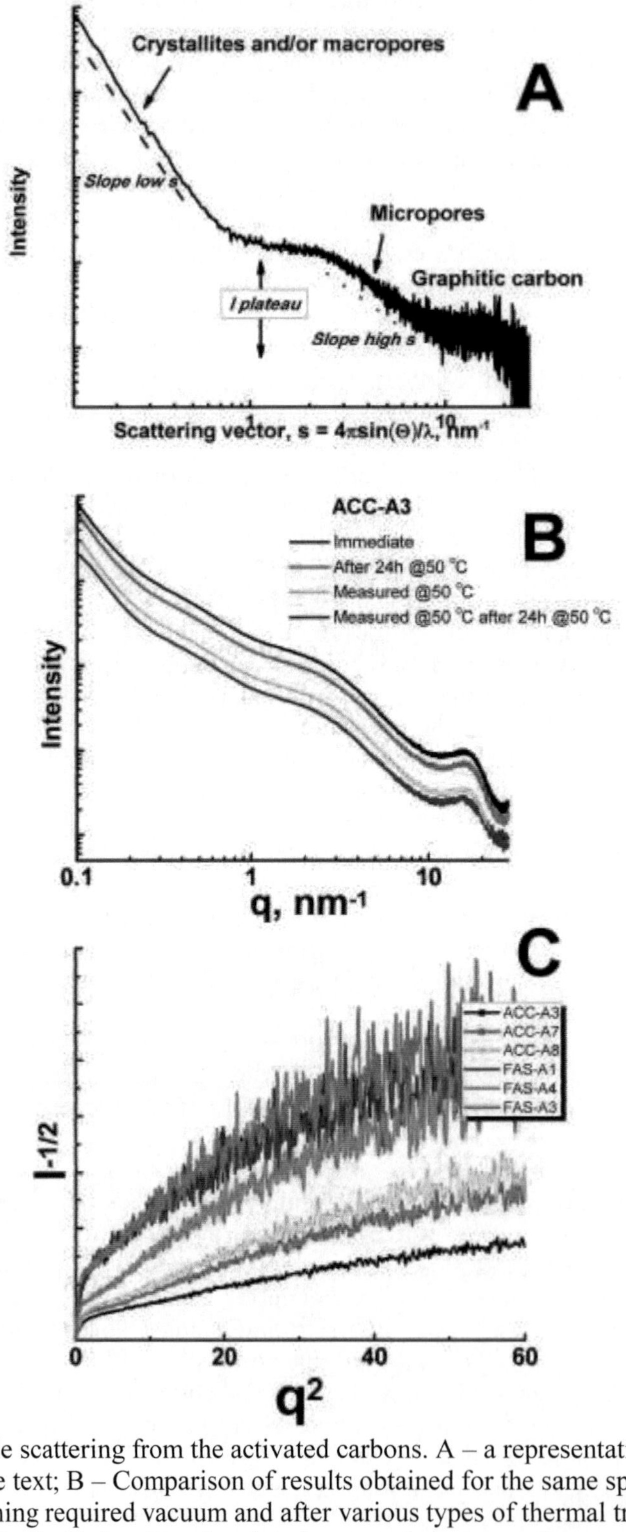

Figure 7. Small-angle scattering from the activated carbons. A – a representative SAXS curve with notations used in the text; B – Comparison of results obtained for the same specimen recorded immediately after reaching required vacuum and after various types of thermal treatment (see legend and text); C – Debye-Bueche plots for several representative samples.

A broad correlation between the activation and the slope is observed: the activation generally pushes the slope towards smaller values (-2.5) (Fig. 8). Since the drop of intensity of these regions is always not less than 1.5 orders of magnitude, with certain caution one may apply the concept of fractals, though this approach is somewhat difficult to compare with other characterization methods. Note also, that as mentioned above the scatterers are polydisperse and in case of their power-law size distribution the scattering curve will be undistinguished from fractal system (e.g. [27]).

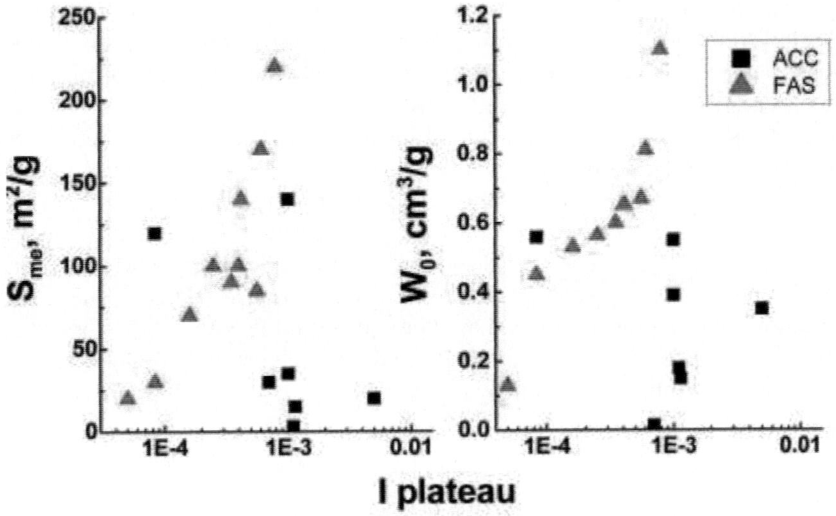

Figure 8. Sorption parameters as function of relative intensity of SAXS curve plateau. Positive correlation is observed for the polymer-based sorbents, but not for the fibers-based ones.

For the furfural-based samples we may assume presence of surface fractal. However, for the fibers-based samples the activation leads to transition from the surface to mass fractal ("perfectly rough" object corresponds to the slope -3). Since a mass fractal is a structure including branching and crosslinking of structural units to form a three-dimensional network, applicability of this concept to fibers-derived samples with rather loose structure is not unexpected. Interestingly, for few samples of both types the observed slopes are ≤4. According to [29] such behavior is due to existence of a transition layer on surfaces of the scatterers. Therefore, large scale heterogeneities (crystallites or macropores) in some of the furfural-derived sorbents as well as fibers-based samples with low degrees of activation possess finite thickness transition layer on surfaces. Below we show that similar phenomenon is also observed for micropores in some samples.

At larger values of the scattering vectors (usually at $q \geq 3$ nm^{-1}, i.e. at scales less than ~2 nm) scattering from pores responsible for the sorption dominates. The Guinier plots ($\log(I)$–q^2) at medium to high q values (~2–8 nm^{-1}) contain linear region, indicating monodisperse character of the scatterers. Gyration radii are in the range 0.5–0.9 nm. The slope of this part of the scattering curve on double logarithmic plot ($\log(I) - \log(q)$) is roughly –2 (see below for detailed consideration). With certain caution this may be interpreted as an indication of lamellar shape of the scattering heterogeneities. Scattering curves from several samples are even more complex, possibly suggesting existence of three populations of the heterogeneities or several types of pore–matrix interfaces [30]. A clear correlation between SAXS and sorption data is observed for furfural-derived sorbents. Tentative explanation of this behaviour is given in "Discussion" section.

Intensity of small-angle scattering is proportional to product $(\Delta\rho)^2V$, where V is the volume of a scatterer and $\Delta\rho$ – contrast of electronic density between the matrix and the scatterer. Therefore, relative intensity of the plateau on the scattering curves may serve as an indication of the relative volume of the pores, weighted with fraction of adsorbed water. For the furfural-derived samples the sorption characteristics show broad positive correlation with the plateau intensity (figure 8). Therefore, assuming that the scatterers (i.e. the pores) in the sorbents of a given type possess roughly similar electronic density, the correlation between S_{BET} and $I_{plateau}$ is logically explained by higher volume of the pores in the samples with higher degrees of activation. In the same time, for the fibers-based sorbents such correlation does not hold. We do not have a unique explanation of such behavior, but presumably the microstructure of these materials on nanometer scale is less well-constrained due to weaker interaction between the structural units. Due to Babinet's principle the SAXS pattern acquired at a single wavelength cannot discriminate between a pore and a solid object of the same size and shape. It is therefore possible that in case of the fiber-based sorbents contributions of heterogeneities of very different nature (e.g., pore and a crystallite) to the scattering pattern are comparable. This smears the effect of activation.

The observed deviations from the Porod law are positive, i.e. the power exceeds 2. Such behaviour indicates presence of 1D and 2D fluctuations of electron density of the pore-bulk interface. Following [29] these types of the fluctuations can be explained by presence of vacancies and by variations of interlayer spacings in graphitic crystallites close to the interface. Few samples of highly activated furfural-based sorbents, but with the lowest S_{BET} values, show power law with an exponent close to 4, suggesting that the pores are approaching three-dimensional quasi-isometric shape. For the studied sorbents a correlation between the sorption characteristics with the SAS curve slope may be observed: e.g. S_{BET} increases with decreasing slope (fig. 9). The observed correlation may be logically explained by higher specific surface in materials with very rough interfaces.

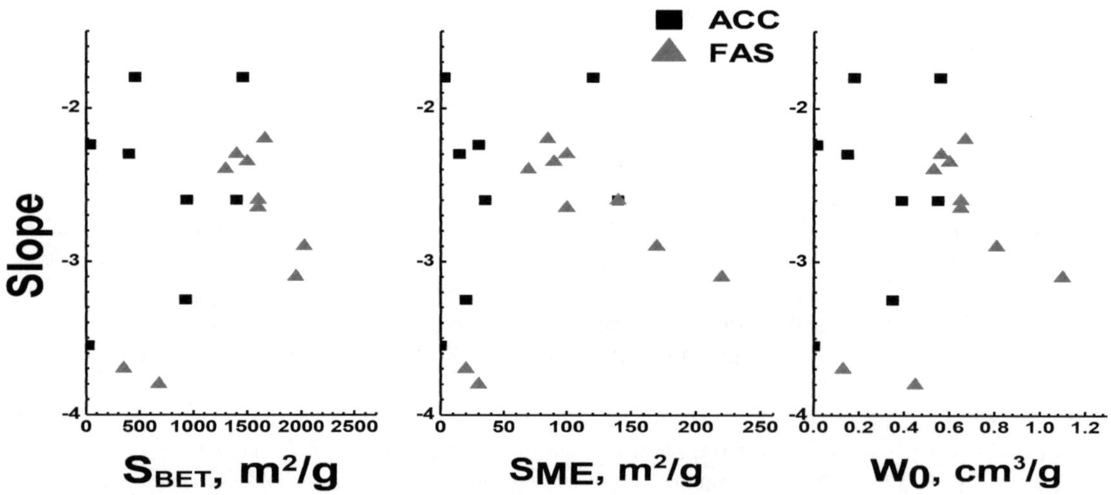

Figure 9. Relationships between parameters derived from sorption experiments and slope of SAXS curves in region related to pores.

Comparison of pores sizes obtained from sorption and SAXS data shows that x_{BET} and x_0 (Fig. 10) are broadly correlated with the Guinier radius (R_g). For the active carbons with higher degree of activation and wider micropores the pore sizes obtained from adsorption and X-ray data broadly correlated. However, no correlation is observed for activated carbons with small degree of activation.

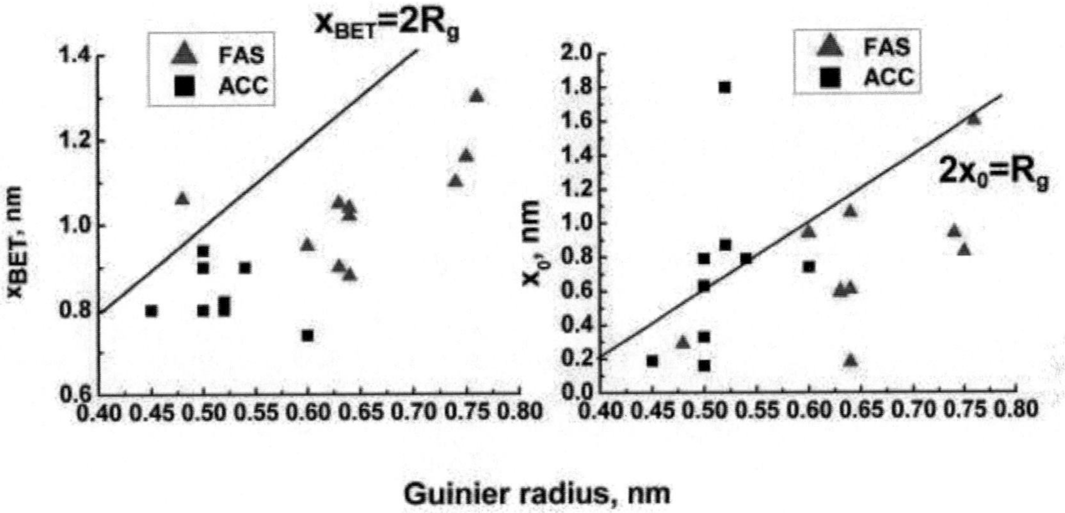

Figure 10. Correspondence between pores sizes obtained from SAXS and sorption.

3.4 Raman scattering

Raman spectra of all samples are dominated by strong broad G and D bands of sp^2-bonded carbon. In most cases weak second order bands are also observed. Experimental spectra were decomposed into several components using Fityk software [39]; the number of potentially existing components was inferred from inflection points. Voigt profile (convolution of Gaussian and Lorentzian profiles) of the lines was assumed in order to account for various types of peak broadening. In most cases 4–5 components were needed to obtain good fit of the data. All spectra possess well-known G and D bands; the D' was poorly pronounced in most cases. Spectral decomposition employing only D and G bands often leaves a noticeable residue in the region around 1500 cm^{-1}. The fit may be partly improved by using Breit–Wigner–Fano shape taking into account extended phonon density of states (PDOS) of graphitic carbon. However, since the PDOS for the studied materials are poorly constrained due to wide variations in sizes and morphologies of graphene stacks, we employ a "phenomenological" approach and introduce an additional broad band underlining the D and G components; an example of such decomposition is shown on Figures 11a and 12. Similar decomposition was earlier used by several authors [39-43] and it was suggested that the "extra" band (G2 following notation from [43]) is related to poorly ordered sp^2-carbon, for example, to defects outside the aromatic plane. Results of our work also support the view that this band is not purely a mathematical trick, but is indeed an important issue in Raman studies of carbon-based sorbents. Following [40] we suggest that it represents a G-band of (semi)amorphous carbons present in the samples. This assignment is based on several independent observations: 1) X-ray diffraction clearly shows presence of considerable amount of poorly ordered carbon in all studied sorbents; 2) the position of the band corresponds well to the G-

IOP Conf. Series: Journal of Physics: Conf. Series **848** (2017) 012009 doi:10.1088/1742-6596/848/1/012009

band of amorphous carbon [44]; 3) in unconstrained fit this band has almost pure Gaussian character, indicating that responsible bonds possess broad distribution of interatomic angles; 4) this band shows pronounced dispersion. Note that recent computational study [45] assign the raman intensity between 1400 and 1550 cm^{-1} to coupling of breathing modes and asymmetric stretching caused by defects within the PAH clusters and impurities.

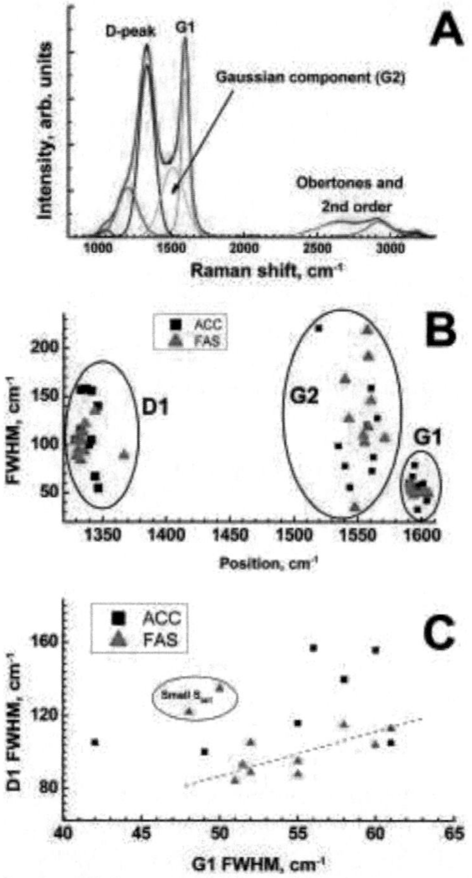

Figure 11. A - Example of spectral decomposition of Raman spectrum acquired at 532 nm excitation. A Gaussian band around 1500 cm^{-1} is clearly observed. B – position of the studied samples at FWHM-position plot, see text. C – relationships between the FWHM values of the G1 and D1 bands.

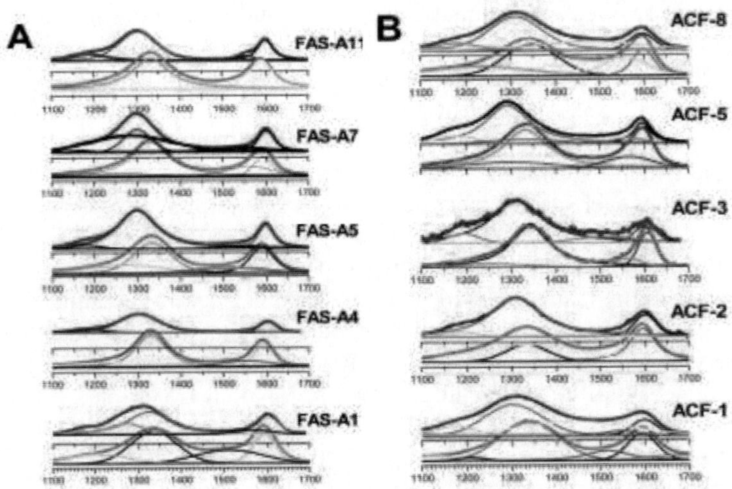

Figure 12. Raman spectra of studied sorbents. A- FAS – furfural-based active carbons; B – Activated Carbon Fibers - based sorbents. For every sample spectra excited at 532 nm (bottom curve) and 785 nm (top curve) are shown. The degree of activation increases from bottom to top.

Decomposition of the D band into several components is also possible, and example of such decomposition is shown, for instance, in [43] (their bands D1 and D2). If the assignment of the broader G2 band to disordered carbon is correct, one may expect presence of the corresponding D2 band. However, the FWHM of the D2 band is always very large (250-400 cm^{-1}) both in [43] and in our own work. In our view bands of such width are not easily ascribed to a single component, reflecting kind of a "continuum". Subsequently, position of the maximum, FWHM and other characteristics of this band often change significantly between independent decomposition runs. Therefore, this band was not included into our evaluation. FWHM and positions of the components are shown on Fig. 11b. Decomposition of spectra recorded at 532 and 785 nm excitation gave qualitatively similar results; therefore, only the former data are shown.

A broad correlation between FWHM of the D1 and G1 peaks is observed for the FAS samples except the specimens with very low S_{BET} (fig. 11c). At present structural grounds for the correlation are not yet clear. Intensity and FWHM of the G2 band broadly correlates with porosity derived from sorption measurements, see fig. 13. For the fibers-derived samples the G2 band becomes less important with increased activation. For the FAS-samples the relation is more difficult to explain.

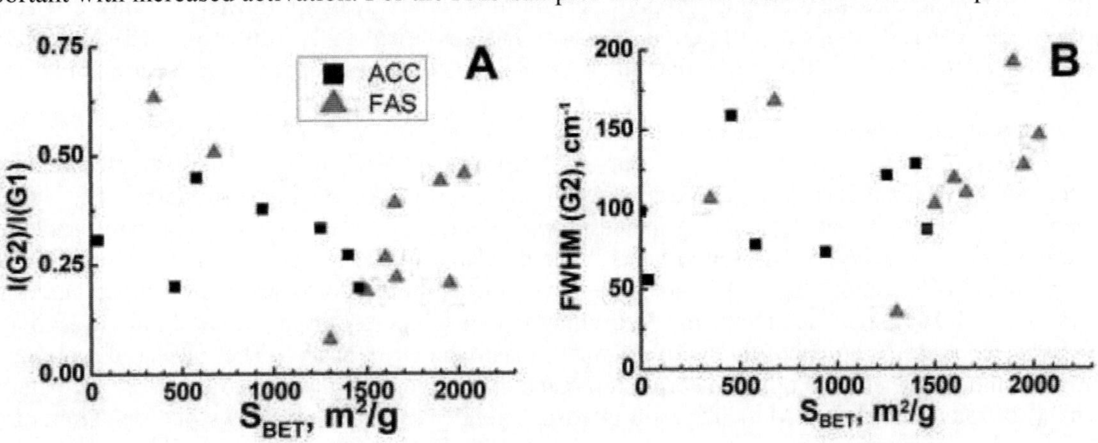

Figure 13. Relationships between intensity and FWHM of the G2 band with porosity.

Virtually all spectra show weak peak around 1180 cm^{-1} and sometimes a second feature at 1030-1060 cm^{-1} (figs. 11a, 12). These peaks might correspond to sp^3-carbon, e.g. in olefinic chains [29], to transpolyacethylene (TPA) – like structures or symmetric breathing modes of PAH rings of different sizes [45]. Some support of the assignment to TPA is given by existence of a feature in pair-distribution function analysis (not shown), but persistence of the feature at 785 nm excitation weakens this assignment. Due to general weakness of the peak it is difficult to correlate it with other structural or sorption parameters. However, its intensity seems to be lower in samples with medium degree of activation.

In order to get better constraints on the structure of the samples excitation at different wavelengths was used. The dispersion of the main G1-band (close to 1600 cm^{-1}) was generally small and appears to be unrelated to the activation degree. The small dispersion implies that the responsible sp^2-carbon is arranged into relatively large crystallites (several nm in size). In the same time, X-ray diffraction data indicates that the average crystallites are smaller than 2 nm. No clear correlation between the FWHM values for the G1 or D1 peaks or of I(D)/I(G) ratio and X-ray-derived size (Tuinstra-Konig relationship) was observed. The discrepancy may be ascribed to several factors including various physical mechanisms of the signal formation in X-ray diffraction and Raman scattering [46]. In addition, these two methods are sensitive to somewhat different populations of the crystallites present in complex material such as carbonaceous sorbent. Presumably, the G-peak at 1600 cm^{-1} in Raman spectra originates mostly from relatively perfect large crystallites; the (semi)amorphous component gives rise to broad G-peak around 1500 cm^{-1}. No clear correlation between the width of the G peak or I(D)/I(G) ratio with sorption characteristics is observed. In the same time, a weak negative correlation is observed for the D peak properties and sorption – its FWHM decreases with increasing S_{BET}, likely reflecting changes in the defects types with progressive activation.

The dispersion of the D band demonstrates high variability from sample to sample (Fig. 12). The studied sorbents can be separated into two main groups with fairly different values of the dispersion. Most of the polymer-based samples are characterized by relatively small dispersion, whereas the fibers-based ones are clustered in the region of high dispersion values.

4. Conclusions

Carbonaceous sorbents produced from furfural and from impregnated viscose fibers were characterized after storage at ambient conditions using complementary techniques permits to establish correlations between sorption parameters and some structural characteristics. The sorbents could be represented as a system comprising nanosized graphite-like crystallites of various degrees of perfection wrapped into amorphous carbon. The amorphous fraction diminishes on activation. In wet state water molecules may arrange graphitic crystallites, dramatically influencing X-ray diffraction patterns. In contrast to X-ray diffraction, moisture has only negligible influence on SAXS patterns from the studied ACs, implying that overall shape and size of pores accessible to N_2 are not strongly affected.

A correlation between the Small-Angle X-ray Scattering results and N_2 sorption data is observed for carbons obtained from the same precursor. According to SAXS the studied samples represent hierarchic structures consisting of large crystallites and fairly monodisperse pores. The relative intensity of the pore-related scattering increases with degree of activation. An important problem in interpretation of the SAXS data for sorbents is the dependence of the scattering curve slope on pore-matrix interface structure. In most samples the deviations from the Porod law are positive, suggesting presence of 1D and 2D fluctuations of electron density of the pore-bulk interface. In highly-activated carbons the pores approach three-dimensional quasi-isometric shape. The observed correlation between the SAXS curve slope and sorption parameters is due to higher specific surface in materials with very rough interfaces. Modeling using two independent approaches shows that deviations from the Porod law should not be ignored in analysis of SAS data for active carbons. For active carbons with higher degree of activation and wider micropores the pore sizes obtained from adsorption and X-ray data broadly correlated. However, no correlation is observed for activated carbons with small degree of activation. X-ray diffraction patterns of the sorbents change considerably with humidity of the powdered sorbents, severely complicating data analysis.

Spectral decomposition of Raman spectra in addition to obvious G and D bands reveals almost pure Gaussian residue in the region around 1500 cm^{-1}. Presumably this band is due to poorly ordered sp^2-carbon, possibly being a G-band of (semi)amorphous carbons present in the samples.

Acknowledgements
This work was partly supported by RFBR grant 15-05-03351 (to AAS). The measurements were performed using equipment of CKP FMI IPCE RAS.

References

[1] Dubinin M M 1975 Physical adsorption of gases and vapors in micropores. In: ed. D A Cadenhead. *Progress in surface and membrane science* (New York: Academic Press) **9**. 1–70
[2] Dubinin M M 1989 Fundamentals of the theory of adsorption in carbon adsorbents: characteristics of their adsorption properties and microporous structures *Carbon* **27** 457–467
[3] Ustinov E A, Do D D 2004 Application of Density Functional Theory to analysis of energetic heterogeneity and pore size distribution of activated carbons *Langmuir* **20** 3791–3797
[4] Gregg S J, Sing K S W 1982 *Adsorption, surface area and porosity* (London: Academic Press)
[5] Hoinkis E. Small angle scattering of neutrons and x-rays from carbons and graphites 1997 In: ed. P A Thrower. *Chemistry and Physics of Carbon.* **25** (Dekker: New York) p.71–241
[6] Findenegg G H, Jahnert S, Muter D, Prass J, Paris O. 2010 Fluid adsorption in ordered mesoporous solids determined by in situ small-angle X-ray scattering *Phys Chem Chem Phys* **12** 7211–7220
[7] Gun'ko V M, Meikle S T, Kozynchenko O P, Tennison S R, Ehrburger-Dolle F, Morfin I, and Mikhalovsky S V. 2011 Comparative characterization of carbon adsorbents and polymer precursors by Small-angle X-ray scattering and nitrogen adsorption methods *J Phys Chem C* **115** 10727–10735
[8] Laslo K, Czakkel O, Dobos G, Lodewyckx, Rochas C, Geissler. 2010 Water vapour adsorption in highly porous carbons as seen by small and wide angle X-ray scattering *Carbon* **48** 1038–1048
[9] Nguen T X, Bhata S K 2012 Characterization of accessible and inaccessible pores in microporous carbons by a combination of adsorption and small angle neutron scattering *Carbon* **50** 3045–3054
[10] Scherdel C, Reichnauer G. 2012 The impact of residual adsorbate on the characterization of microporous carbons with small angle scattering *Carbon* **50** 3074–3082
[11] Vartapetian R Sh, Khozina E V. 2006 Structure of rigid and swirling adsorbents according to pulsed NMR data on the mobility of adsorbed water and benzene molecules *Colloid J* **68**(1) 1–19
[12] Petrov M P, Wang C L, Weber M H, Lynn K G 2003 Positron annihilation techniques suited for porosity characterization of thin films *J Chem Phys B* **107** 2715–2734
[13] Tsivadze A Yu, Gur'yanov V.V and Petukhova G A 2011 Preparation of spherical activated carbon from furfural, its properties and prospective applications in medicine and the national economy *Prot. Metals and Phys. Chem. Surf.* **47**(5) 612–620
[14] Vartapetyan R Sh, Voloshchuk A M 1995 The mechanism of the adsorption of water molecules on carbon adsorbents. *Russ Chem. Rev* **65**(11) 985–1001
[15] Laszlo K, Czakkel O, Dobos G, Lodewyckx P, Rochas C, Geissler E 2010 Water vapour adsorption in highly porous carbons as seen by small and wide angle X-ray scattering *Carbon* **48** 1038 –1048
[16] Scherdel C, Reichenauer G 2012 The impact of residual adsorbate on the characterization of microporous carbons with small angle scattering *Carbon* **50** 3074-3082
[17] Maturilli A, Shiryaev A A, Kulakova I I, Helbert J 2014 Infra-red reflectance and emissivity spectra of nanodiamonds *Spectroscopy Lett.* **47**(6) 446-450

[18] Ruland W, Smarsly B 2002 X-ray scattering of non-graphitic carbons: an improved method of evaluation *J Appl Cryst* **35** 624–633
[19] Mikhail S Sh, Brunauer S, Bodor E E 1968 Investigation of a complete pore structure analysis. I. Analysis of micropore *Colloids & Interface Sci.* **26** 45–53.
[20] Mikhail S Sh, Brunauer S, Bodor E E 1968 Investigation of a complete pore structure analysis. I. Analysis of four silica gels *Colloids & Interface Sci.* **26** 54–61
[21] Voloshchuk A M, Dubinin M M, Moskovskaya T A 1988 Porous structure and surface chemical state of carbon adsorbents. 1. Selection of reference isotherm of nitrogen vapors on surface of carbon adsorbents *Izv. Akad. Nauk SSSR. Ser. Khim.* Iss. 2 277–288
[22] Voloshchuk A M, Dubinin M M, Moskovskaya T A 1988 Porous structure and surface chemical state of carbon adsorbents. 2. Nitrogen adsorption at 77 K on nonporous carbon adsorbents *Izv. Akad. Nauk SSSR. Ser. Khim.* Iss. 10 2203–2208
[23] Stoeckli H F, Rebstein P, Ballerini L 1990 On the assessment of microporosity in active carbons, comparison of theoretical and experimental data *Carbon* **28**(6) 907–909
[24] Rutman A M, Skakov Yu A 1989 Radial distribution functions of atoms and interference functions of partly ordered carbon materials. I. Influence of edge atoms of disperse layers *Sov Phys Cryst* **34**(3) 338–341
[25] Fujimoto H 2003 Theoretical X-ray scattering intensity of carbons with turbostratic stacking and AB stacking structures *Carbon* **41** 1585–1592
[26] Davydov V A, Shiryaev A A, Rakhmanina A V, Filonenko V P, Lyapin S G, Vasiliev A L, Roddatis V V, Autret C, Agafonov V N, Khabashesku V N 2011 Transformations of polyhedral carbon nanoparticles under high pressures and temperatures *Carbon* **49** 2389-2401
[27] Schmidt P W 1982 Small-angle scattering studies of disordered, porous and fractal systems *J. Appl. Cryst.* **15** 567-569
[28] Schmidt P W, Avnir D., Levy D., Hohr A., Steiner M., Roll A 1991 Small angle X-ray Scattering from the surfaces of reversed phase silicas: Power-law scattering exponents of magnitudes greater than four *J. Chem. Phys.* **94** 1474-1479
[29] Ruland W 1971 Small-angle scattering of two-phase systems: determination and significance of systematic deviations from Porod's law *J. Appl. Cryst.* **4** 70-73
[30] Ciccariello S, Goodisman J, Brumberger H 1988 On the Porod law *J Appl Cryst* **21** 117-128
[31] Ruland W 2001 Apparent fractal dimensions obtained from small-angle scattering of carbon materials *Carbon* **39** 323–324
[32] Perret R., Ruland W 1969 Single and multiple X-ray Small-angle scattering of carbon fibers *J. Appl. Cryst.* **2** 209-218
[33] Fourdeux A, Herinckx C, Perret R, and Ruland W 1969 La structure des fibres de carbone *C. R. Seances Acad. Sci., Ser. C,* **269** 1597-1600
[34] Pré P, Huchet G, Jeulin D, Rouzaud J-N, Sennour M, Thorel A 2013 A new approach to characterize the nanostructure of activated carbons from mathematical morphology applied to high resolution transmission electron microscopy images *Carbon* **52** 239-258
[35] Hu C, Liu A C Y, Weyland M, Hadi Madani S, Pendleton P, Rodriguez-Reinoso F, Kaneko M, Biggs M J 2015 A multi-method study of the transformation of the carbonaceous skeleton of a polymer-based nanoporous carbon along the activation pathway *Carbon* **85** 119-134
[36] Pre P, Shiryaev A A, Voloschuk A M, Rouzaud J-N 2013 Characterization of nanoporous carbons from X-Ray scattering, TEM images and (N_2,CO_2) adsorption *Ext. abstr. Carbon 2013 conf.*
[37] Svergun D I 1999 Restoring low resolution structure of biological macromolecules from solution scattering using simulated annealing *Biophys. J.* **76** 2879-2886
[38] Feigin L A and Svergun D I 1989 Structure analysis by Small-Angle X-Ray and Neutron Scattering (New York and London: Plenum Press) pp. 335
[39] Wojdyr M 2010 Fityk: a general-purpose peak fitting program *J Appl Cryst* **43** 1126–1128

[40] Beny-Bassez C, Rouzaud J-N 1985 Characterisation of carbonaceous materials by correlated electron and optical microscope and Raman microscopy In: *Scanning electron microscopy* (Chicago: SEM Inc) 119-132

[41] Sadezky A. Muckenhube H, Grothe H, Niessner R, Poschl U 2005 Raman microspectroscopy of soot and related carbonaceous materials: Spectral analysis and structural information *Carbon* **43** 1731–1742

[42] Schwan J, Ulrich S, Batori V, Ehrhardt H, Silva S R P 1996 Raman spectroscopy on amorphous carbon films *J Appl Phys* **80** 440-448

[43] Shimodaira N and Masui A 2002 Raman spectroscopic investigations of activated carbon materials *J. Appl. Phys.* **92** 902-909

[44] Ferrari A C 2001 A model to interpret the Raman spectra of disordered, amorphous and nanostructured carbons *Mater Res Soc Proc* **675** W11.5.1-W.11.5-12.

[45] Smith M W, Dallmeyer I, Johnson T J, Brauer C S, McEwen J-S, Espinale J F, Garcia-Perez M 2016 Structural analysis of char by Raman spectroscopy: Improving band assignments through computational calculations from first principles *Carbon* **100** 678–692

[46] Zickler G A, Smarsly B, Gierlinger N, Peterlik H, Paris O 2006 A reconsideration of the relationship between the crystallite size L_a of carbons determined by X-ray diffraction and Raman spectroscopy *Carbon* **44** 3239–3246

IOP Conf. Series: Journal of Physics: Conf. Series **848** (2017) 012010 doi:10.1088/1742-6596/848/1/012010

Neutronographic investigations of supramolecular structures on upgraded small-angle spectrometer YuMO

A I Kuklin[1,2*], A V Rogachev[1,2], D V Soloviov [1,2,3], O I Ivankov[1,2,3], Yu S Kovalev[1,2], P K Utrobin[2], S A Kutuzov[1], A G Soloviev[1], M I Rulev[1,4] and V I Gordeliy[1,2,5]

[1] Joint Institute for Nuclear Research, Dubna, Russia
[2] MIPT, Dolgoprudny, Russia
[3] Institute for Safety Problems NPP, Chernobyl, Ukraine
[4] Lomonosov Moscow State University, Moscow, Russia
[5] Institute Biological Structure, Grenoble, France
E-mail: kuklin@nf.jinr.ru

Abstract. The work is a review of neutronographic investigations of supramolecular structures on upgraded small-angle spectrometer YuMO. Here, key parameters of small-angle spectrometers are considered. It is shown that two-detector system is the basis of YuMO upgrade. It allows to widen the dynamic q-range twice. In result, the available q-range is widened and dynamic q-range and data collection rate are doubled. The detailed description of YuMO spectrometer is given.The short review of experimental researches made on the spectrometer in the polymers field, biology, material science and physical chemistry is given. The current investigations also have a methodological aspect. It is shown that upgraded spectrometer provides advanced world level of research of supramolecular structures.

1. Introduction

In 2012 the first work about scientific orientation of small-angle spectrometer (SANS – MURN, JINR, Dubna) celebrated its fortieth anniversary [1]. Authors considered many methodical aspects of small-angle scattering: contrast variation, homogeneous and heterogeneous particles and deuterium exchange. Average scattering densities are shown for different proteins and RNA. Biological orientation of YuMO was obvious for the authors. Despite the success in proteins crystallization and synchrotron science in recent decades defines the trend in structural biology, there are still a number of problems connected with the native state of proteins and the importance of SANS for them is difficult to over-estimate. At the same time, other areas of research arose and some of which are presented in this short review.

In recent years the spectrometer was significantly upgraded and got a name "YuMO" in honor of one who created the spectrometer ideology (Yuri Mechislavovich Ostanevich). In this work, some methodical and scientific aspects of research with the help of small-angle neutron scattering on example of investigations on the small-angle neutron spectrometer on pulsed reactor IBR-2 (after 2011 IBR-2M) are shown.

Investigation of supramolecular objects acquires more relevance with regard to the intensive development of nanotechnologies and corresponding scientific tasks [2, 3]. Also the requirements to the methods of investigations of substances on nanoscales are increasing.

Small-angle scattering of thermal neutrons method is one of the major means of structural characteristics determination for condensed matter in range 2 – 100 nm. Small-angle spectrometer YuMO on the 4th channel of reactor IBR-2, JINR, Dubna is almost the only spectrometer of this type that is systematically and efficiently used for condensed matter research in different organizations all over the Russia and membership countries of JINR.

As a rule in leading world centers there are two or more spectrometers of this type. Such number of spectrometers is connected with big amount of users of small-angle neutron scattering. Subsequently it is defined by the importance of structural information obtained by the method and the wide spectrum of tasks it solves – from material sciences to structural biology [4].

In the review mentioned above [1] the authors have seen biological objects as the main field of research on the small-angle spectrometer. Mentioned in the article the detailed consideration of small-angle scattering is still relevant. In fact, the questions of H_2O/D_2O contrast variation with isotopic exchange taking into account are considered and formula of average scattering lengths calculation for ribosome proteins, RNA and DNA are shown. Authors used research experience in the world and in Laboratory for Neutron Physics of the Joint Institute for Nuclear Research. When considering questions about high-resolution method for protein crystals investigation, SANS is pointed out as non-destructive method. However, as practice showed, the main disadvantage of neutron sources for high-resolution neutronography is their weak flux. High fluxes are also the key features for small-angle spectrometers.

Small-angle spectrometers belong to the class of installations with low-resolution. Because of the small-angles of scattering the main contribution into resolution is usually made not by the time parameter (that is limited for IBR-2 by the neutron pulse width), but by the angular resolution that is defined by collimation and the size of the minimal sensible element of the detector part of spectrometer. This statement is also true for other pulse sources. There is another situation for stationary reactors working practically with the only one wavelength (more precisely – with very narrow neutron wavelengths range) and using mechanical velocities selector for monochromatization of neutrons.

In order to have satisfactory wavelength resolution one has to reduce the neutron flux on the sample. Usually the resolution is lower than 5% (under adequate flux decrease) in case one does not use perfect crystal (usual resolution for wavelength selector is approximately 10% [5]). In result under the same resolution neutron fluxes on the sample on stationary reactors and pulse sources become equal and on IBR-2 with comparably low average power (2 MW) neutron fluxes are comparably high [6]. Subsequently the data collection rate can also be high.

In 2013 small-angle spectrometer on the 4th channel of IBR-2 celebrated its thirtieth anniversary [7]. It was the first time the installation got through the drastic upgrading in 1990, which was connected with personal computers rapid development. This allowed to make significant spectrometer management automatization even in comparison with contemporary situation. The basic progress was the concept of measuring scattering curve in absolute scale at the feasibly shortest time. This allowed obtaining interesting results in different fields of condensed matter physics [8-17]. New paradigm of installation involved flexibility when making an upgrade and preserving the existing instrument advantages. In 2000th was done modernization in basis of it is two-detector system [17-20].

Small-angle spectrometer with direct geometry (the sample "is oriented" to the zone) is more suitable to fast pulsed reactor IBR-2 with current length of starting "pulse" (nearly 300 us). The instrument upgrade (first of all usage of two-detector system) was based on maximization of advantages of method of neutron detection versus their time of flight on pulse neutron source. It was the use of the full spectrum of

IOP Conf. Series: Journal of Physics: Conf. Series **848** (2017) 012010 doi:10.1088/1742-6596/848/1/012010

thermal neutron wavelengths, which ensured the overlap of the spectra of small-angle neutron scattering obtained from two detectors.

2. Upgraded small-angle neutron scattering spectrometer

With a significant variety of small-angle spectrometers all over the world the key parameters are: intensity and density of the flux on a sample, [21], q-range and dynamic q-range (the range of q-vectors that is available simultaneously [22, 23], that is defined by the ratio q_{max}/q_{min}) and resolution. Upgrading of these key parameters is that was done on the spectrometer YuMO. The opportunities of sample investigation were broadened as to allow studying samples under pressure, in external magnetic field and under light illumination [24]. In addition, we did optimization of spectrometer configuration and created user-friendly interface for experiments management and data treatment for spectrometer users.

In [19, 20, 25] there is a description of new software and in [26] – details of nodes of automated spectrometer YuMO. Contemporary methods of smoothing while treating experimental data for obtaining SANS curve with spectrometer resolution taken into account are given in [27]. In [28, 29] the specificities of spectra calibrations for one-detector spectrometer case with vanadium standard, the ratio between time of measurement with typical standard and sample, spectra correction with regard to detector gap time and opportunities of hydrogen detection in mixtures of water and heavy water are given.

Key feature of the upgrading was the usage of two-detector system that is the basic element of YuMO spectrometer at the moment [18, 23]. This system allows not only widen the dynamic q-range and lessen twice the time of experimental data collection, but also it allows conducting measurements for unique samples in wide q-range simultaneously. Consequently, the widened q-range implies the increase of accuracy of obtained results that for model object is usually the only way to recover the form-factor and space distribution unequivocally for investigated objects [30]. The increase of dynamic q-range is also important for short-living samples measurements or when investigating kinetics.

The other feature of YuMO spectrometer in new installation scheme is obtaining SANS curve in absolute scale even during the experiment [31]. This gives an opportunity of obtaining additional information about the structure of the object and its interaction for example with solvent.

It is worth mentioning about further perspectives of YuMO development. New unique position-sensitive detector was created, already passed successfully preliminary tests and was mounted in working position on the spectrometer. It has high spatial resolution and will make it possible to investigate structure and properties of anisotropic objects and provide unique resolution of small-angle spectrometer YuMO [32].

2.1. Upgraded spectrometer description

The scheme of upgraded spectrometer is shown on Figure 1. Behind the shelter after reactor zone with thermal (or cold) moderator there is chopper, partially decreasing background and satellite pulses of neutron power, which are formed by additional reflector.

The flux of thermal neutrons was formed by collimator system (4,6 Figure 1) so that neutrons coming onto the sample were forming a neutron beam with maximum cross-section 22 mm and intensity of $3x10^7$ neutrons/sec [6, 21].

The first collimator (4) together with adjustable collimator (7) that have a retractable “nose” and two discs (5,8) with discrete set of cadmium rings that have inner holes with different sizes, constitute new system of forming a neutron beam. The system is automatically controlled with the help of computer.

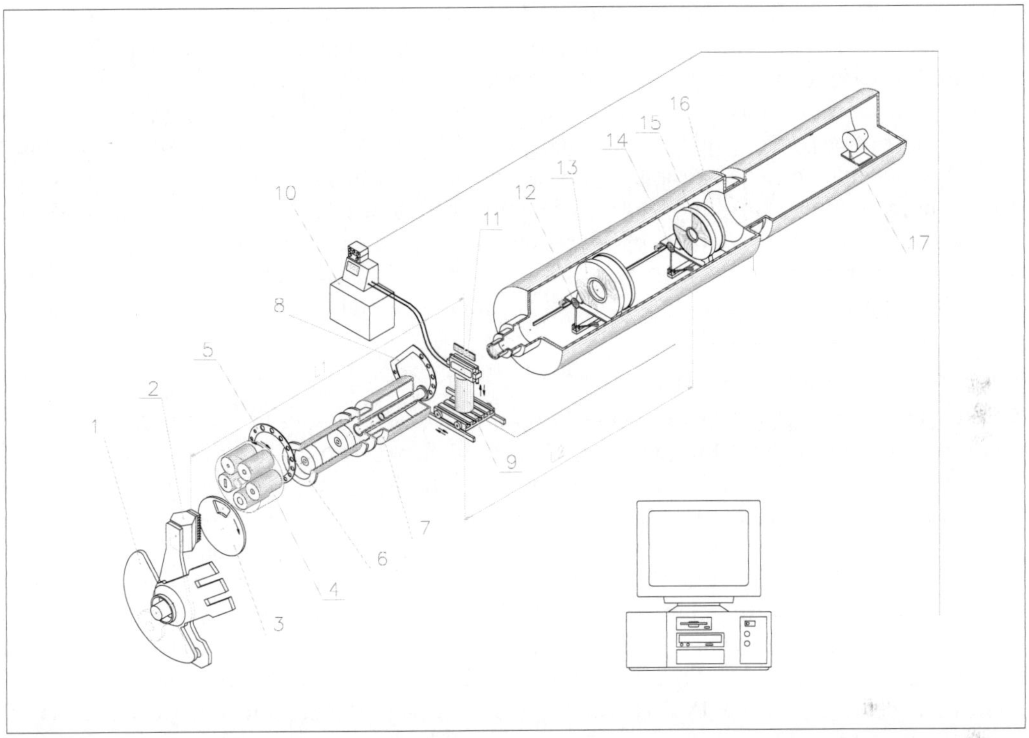

Figure 1. The scheme of main nodes of upgraded small-angle spectrometer YuMO on IBR 2 reactor. 1 is two reflectors, 3 is chopper, 4 and 6 is collimation system, 9 is samples table, 10 is thermostat, 11 – holder for 25 samples, 13, 15 is OLD and New detector correspondently, 12, 14 is vanadium standards, 16 is changeable vacuum tube, 17 is direct beam detector.

Vanadium windows on the retractable part of adjustable collimator and on vacuum tube of detectors replaced with quartz glass “Heraeus”.

Sample table can move in horizontal and vertical directions. The samples are put in the box made of aluminum alloy inside of which there are channels for the thermal liquid. The box is connected with computer controlled thermostat (10). Working range of temperatures is –20°C до +180°C. The box windows are made of aluminum foil (0.1 mm). At the same time, 25 cuvettes “Helma” (24x47x7 mm) could be placed inside the box. The sample table is not under vacuum that makes easy to mount different devices on it, for example, light source, pressure chamber, thermostatic volume (temperature box), magnetic installation etc.

Neutron flux scattered on the sample comes into stage vacuum volume where ring neutron detectors (13, 15) and direct beam detector (17) are placed. Vanadium standards that are placed directly in front of each detector (12, 14) during the experiment are periodically put into and out of the beam that allows avoid unpleasant effects of long-periodical power oscillations and obtain SANS curve in absolute units of scattering cross-section. Besides, such a procedure allows data treatment simultaneously with the experiment conduction.

Vanadium standards are preliminary adjusted relative to the axis of spectrometer with the help of the laser. In front of the second (NEW) detector, the computer-controlled mask (15) is placed for checking the anisotropy of the scattering picture. The main electronic modules made in WINDOWS 10 standard. Both detectors can move with the help of computer controlled stepping motors inside of vacuum tube with length of approximately 10 meters. The software for controlling the installation is a system that controls temperature, motors and data collection [19]. During the experiment conduction, the system controls the main spectrometer parameters. Also there is an option of remote experiment control.

The monitor counter placed in front of the sample can be used for spectra normalization, checking the work of vanadium standards and at the same time tracking the neutron beam and its intensity.

Direct beam detector is used for normalization in case of samples that have strong scattering or strong adsorption when standard procedure works not accurately enough [29].

2.2. Two-detector system

YuMO spectrometer upgrading based on development, creation and usage of two-detector system [22, 23] based on ring scattering detectors [33]. In the vast majority of experiments using slow neutrons requires the widest possible q-range. The standard solution of the problem is detector movement along the base sample-detector and/or creation of a detector with a big surface. The disadvantage of detector movement is the increase of data collection time at least twice and losing the sample state in case it has irreversible phase transition of limited lifetime. Creation of big-surface detector is limited by technical reasons and big costs. Multi-detector system has no such disadvantages. Its working principle is the following. A part of neutrons scattered on the sample is registered by the nearest (first) to the sample detector and the other part (scattered under lower angles) comes through the central hole of the detector and is registered by the next (second) detector (Figure 1) [34]. Having come through the central hole of the second detector the part on neutrons is registered on the third detector etc. The overlapping of the spectra is provided by time-of-flight technique. At the time of this writing the spectrometer had two detectors for tracking scattered neutrons and one direct beam detector.

In case of two-detector system it is necessary to take into account additional scattering on vanadium standard of the first detector in active state (inside the neutron beam) that is registered on the second detector (for the first detector all the formula are as for one-detector installation) [34]:

$$\frac{(d\Sigma/d\Omega)_s}{(d\Sigma/d\Omega)_v} = \frac{I_s^n T_v^o}{I_{s+v}^n - I_s^n} \frac{d_v^n \Omega_v^n T_v^n}{d_s \Omega_s^n} \left(1 + \frac{d_v^o \Omega_v^{on}}{d_v^n \Omega_v^n T_v^n}\right) \qquad (1.1)$$

where I_s^n и I_{s+v}^n – scattering intensity on the sample that is registered on the second detector and scattering intensity on the sample and vanadium standard simultaneously registered on the second detector; Ω_s^n – the solid angle formed from the sample position to the second detector; Ω_v^n – solid angle formed from vanadium standard position to the second detector; Ω_v^{on} – solid angle formed from the position of vanadium standard of the first detector to the second detector; $(d\Sigma/d\Omega)_v$ – differential cross-section of vanadium standards; d_s – sample thickness; d_v^n , d_v^o and T_v^n , T_v^o – thicknesses and transmissions for vanadium standards of the first and the second detectors respectively. In the right part of (1.1) there are constants and measured values. Consequently, it is possible to save the intensity scale (cross-section scattering) in absolute units.

Formula mentioned above are used only in case of low-scattering samples (cross-section is lower than 10 cm^{-1}). For strongly scattering samples (cross-section is higher than 10 cm^{-1}) the expression I^n_{s+v} - I^n_s from (1.1) becomes approximately zero and fluctuate significantly that implies fluctuation of the scattering intensity. In this case it is necessary to use another normalization and instead of expression I^n_{s+v} - I^n_s use $I^n_{\phi+v}$ - I^n_{ϕ}, where I^n_{ϕ} and $I^n_{\phi+v}$ – scattering intensity on background sample (solvent of matrix) and background sample and vanadium standard registered on the second detector simultaneously. In this case formula is the following (for the first detector all the formula are as for one-detector installation):

$$\frac{(d\Sigma/d\Omega)_s}{(d\Sigma/d\Omega)_v} = \frac{I^n_s T^o_v}{I^n_{\phi+v} - I^n_{\phi}} \frac{d^n_v \Omega^n_v T^n_v{}^n T_{\phi}}{d_s \Omega^n_s T_s} \left(1 + \frac{d^o_v \Omega^{on}_v}{d^n_v \Omega^n_v T^n_v}\right) \quad (1.2)$$

where I^n_s и I^n_{s+v} – scattering intensity on the sample that is registered on the second detector and scattering intensity on the sample and vanadium standard simultaneously registered on the second detector; Ω^n_s – the solid angle formed from the sample position to the second detector; Ω^n_v – solid angle formed from vanadium standard position to the second detector; Ω^{on}_v – solid angle formed from the position of vanadium standard of the first detector to the second detector; $(d\Sigma/d\Omega)_v$ – differential cross-section of vanadium standards; d_s – sample thickness; d^n_v , d^o_v and T^n_v , T^o_v – thicknesses and transmissions for vanadium standards of the first and the second detectors respectively. In the right part of (1.2) there are constants and measured values. Consequently, it is possible to save the intensity scale (cross-section scattering) in absolute units.

Experiments in most neutron research centers are conducted with several positions of the detector and then all the data are merged and shown on one plot with preliminary normalization of each spectrum. Such approach requires rationale with taking into consideration boundary conditions and, for example, its application to solutions where the concentration of investigated particles is time-dependent or to solutions with low kinetics is problematic. Such problematic samples could be lipids or big organelles.

For time-of-flight spectrometers, q-range is defined by experimenter with regard to sample cross-section and ratio signal/background. Two-detector system allows solving this problem. When overlapping two spectra from different detectors q-range limitations become experimentally defined.

For ring detectors with central pinhole there is no necessity to make spectrum correction in the near zero angles as in case with beam-stop. These additional advantages of YuMO spectrometer allow controlling scattering curve boundary conditions.

2.3. Spectrometer resolution

It is worth to remind that small-angle spectrometers belong to the class of low-resolution installations. Nowadays, within the development of mathematical methods and for tasks with lattice parameter smaller than 50Å the spectrometer resolution often plays the key role. In terms of obtaining the best resolution, a multi-detector system has important advantages. Indeed, under the same spatial resolution the nearest position of the detector to the sample allows obtaining higher resolution. Consequently using of two detectors gives better resolution if compare them with using one (big) detector with the same q-range, but placed in the between.

Pulse neutron sources have significant advantage not only because the time resolution (and consequently neutron wavelength) is much better than on stationary reactors, but also they have a possibility of widening q-range. In our work [27] it is shown that using of dynamic window in smoothing procedure allows widen q-range in case of bad statistics. Widening of q-range implies increase in accuracy of obtained results that for model object is usually the only way to recover the form-factor and space distribution unequivocally for investigated objects.

Commonly used for calibration of X-ray spectrometers silver behenate can be used and for neutrons. However, nuclear contrast is not so high. We evaluated the possibility of using silver behenate not only for calibration purposes but also for resolution [35].

3. Physical parameters of spectrometer and investigation of supramolecular structures

The scheme of experiment mentioned above allows widening q-range of slow neutrons that are scattered on the sample and registered on detectors. Consequently, it widens the range of transferred momentum at least twice in comparison with moving detector on base sample-detector at the same time of experiment.

On the Figure 2 the comparison of two spectra (before and after upgrade) is shown. For some samples the minimum q-value is 4×10^{-3}Å^{-1}.

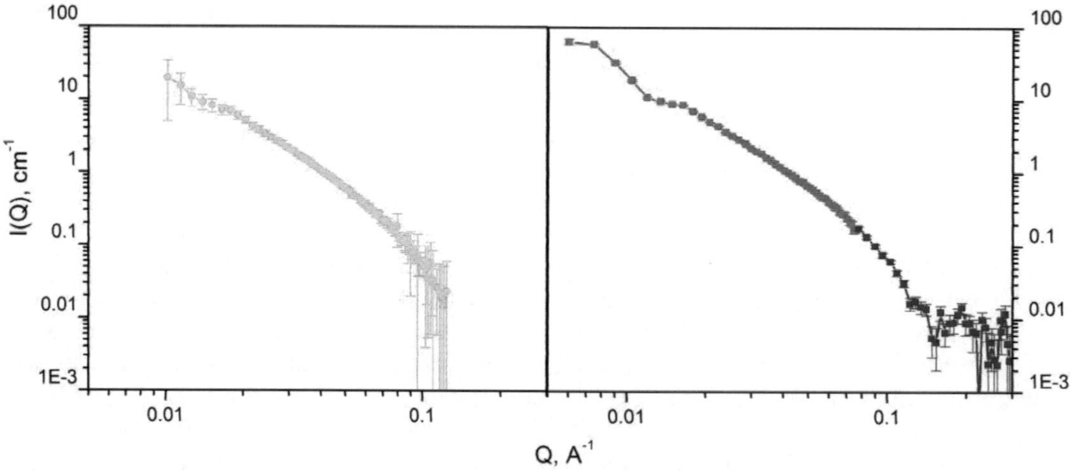

Figure 2. SANS curves for single lipidic membranes measured in 1999 with one detector (left) and two-detector system in 2000 (right: red – NEW detector and blue – OLD detector). Measurements time was 20 min.

Figure 3 demonstrates the possibilities of spectrometer for low-scattering sample with short time of measurement and the value of dynamic q-range that is eventually limited by the sample size. On the Figure 4 there is a curve where intensity of scattered neutrons has a five orders change. The ratio q_{min}/q_{max} (dynamic q-range) in this case is more than 90.

In addition, the optimization of spectrometer parameters was made. The questions of data consistency in the area of overlapping experimental points from two detectors were investigated [22]. Devices allowing studies of the samples under pressure and with external magnetic field are created. Available

IOP Conf. Series: Journal of Physics: Conf. Series **848** (2017) 012010 doi:10.1088/1742-6596/848/1/012010

temperature range on the sample is widened and new device for experiments with high pressure opens new direction for investigation of condensed matter [36-38] on YuMO spectrometer.

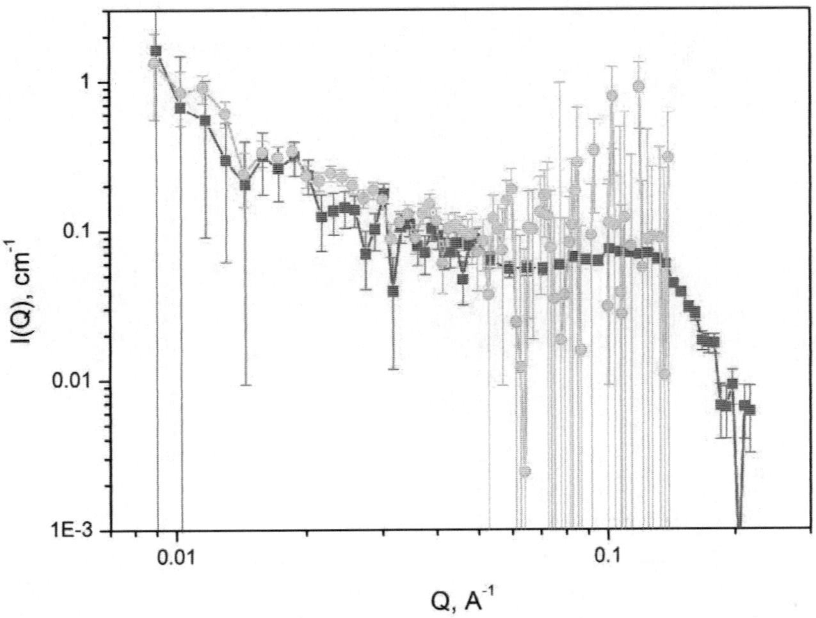

Figure 3. SANS curves for polyelectrolyte membranes Nafion 105 measured in 1999 with one detector (green circles) and two-detector system in 2000 (red squares).

During the experiment, the parameters and all spectrometer motors states are tracked and the protocol is logged. The results obtained have systematic errors excluded and there is always the possibility to watch the logs when incorrect work of the spectrometer takes place. Thereby the accuracy of obtained results increased significantly.

Qualitatively another level of experiment control is provided by the automated nodes of the spectrometer.

Finally, there are papers that present the results of the upgrade, which became possible because of the changes in spectrometer. The part of them is [30, 37-65]. For example, in [39, 40] the investigation of influence of substances interesting in terms of science and pharmacy on lipid membranes is described. In [41] and [42-46] – fractal arrangement in biological objects and in soils respectively. In all cases, it was the wide dynamic q-range that allowed obtaining the further mentioned results. Theoretical studies [66-70] show that for determined fractals investigation it is required not only wide dynamic q-range but also good resolution.

Authors of the work [41] studied intact chicken erythrocyte nuclei on several installations (Dubna, Juelich). It is shown that on the scale of 15 nm – 1.5 um the scattering curve could be interpreted in terms of mass fractal. Fractal dimension of the protein parts of cell nucleus is constant and approximately equals 2.5. In comparison, the DNA arrangement is two-phase with the value of 2 for the scale less than 300 nm and tends to 3 on the larger scale.

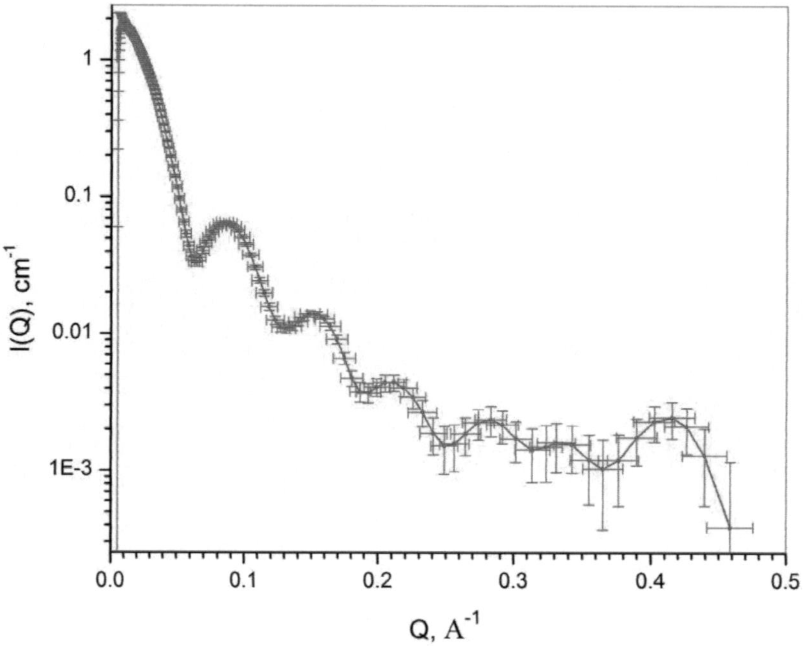

Figure 4. SANS curves for water solution of apoferritin measured in 2017 with two-detector system and smoothing procedure [25]. It is shown low level of background and good resolution.

Methodic value of this work is to check the coincidence of scattering curves obtained on upgraded spectrometer YuMO and installations KWS-2 and KWS-3 (Forschungzentrum, Juelich, Germany). Obtained lines in log-log scale coincide in area of overlapping ranges of transferred momentum.

Fractal arrangement of colloidal compartments of soil is considered in [42-46]. Scattering curves obtained on these samples have a dynamic q-range 90 (q-range is from 0.007 to 0.6Å^{-1}). More than 20 types of soils and horizons are considered. It is shown that fractal dimension depends on the type of soil and soil horizon, degree of water saturation and temperature. Conclusions about the fractal structure were made because of wide dynamic q-range.

It is worth to point out that future nanotechnology breakthrough will be probably connected with creating monodisperse objects. The question of small-angle scattering curves in case of monodisperse deterministic fractals is theoretically considered in [66-69, 71]

It is shown that at a given iteration m, both surface fractal models can be represented as a sum of mass fractals at iterations from zero to m. This confirms that in general, any surface fractal can be represented as a sum of mass fractals [70].

The work [47] is about studying the kinetics of bacteriorhodopsin crystallization in lipidic cubic phase. The beginning of crystallization processes when adding salt implies significant decrease of lattice parameters. The symmetry type and the lattice parameters were determined and the kinetics of parameters change was studied. In terms of method a good resolution of spectrometer is demonstrated in that work. The time of one measurement for studying kinetics was 3 min.

The questions of phase decay in bismuth superconductor are considered in [48].

The object of studies in polyelectrolyte hydrogels was a phenomenon of charge-induced microphase separation [49, 50]. The aggregation number was determined. It was shown that the hydrophobic part decreases when the charge increases. When adding salt the microphase separation vanishes. The model was suggested for such behavior of polyelectrolyte hydrogels. In addition, the wide dynamic q-range is used in that work because there is a structural arrangement of polyelectrolyte hydrogels at different scales.

One of the direction in polymers physical chemistry is searching and creation of controlled gels. In the frameworks of this direction the work about studying of rigid polyelectrolyte rods in polyacrylamide gel was done [51]. It was shown that as inside gel as in water solution polyelectrolyte rods are self-organizing in cylindrical aggregates containing 8-9 single polymer chains. The chains are oriented along the axis of such aggregate. When the aggregates occurred, the gel becomes more homogenous at large scales. Conclusions are based on systematic rods studies immersed into gel or solvent with a screening of gel (solvent) percentage relatively to the rods and studying the influence of salt addition on structural arrangement of the system in general.

The effect of insertion of antibiotic AmB in membranes of egg phosphatylcholine in concentration range from 0.01 to 5 molar percent per a structure and dynamic properties of lipid bilayers were studied by three methods – SANS (YuMO spectrometer), X-ray diffraction (DRON-4) and infrared spectroscopy (FTIR) [52]. The results showed that antibiotic is predominantly placed near head groups of membrane under the concentration lower than 1 molar percent and under higher concentrations AmB inserts the hydrophobic membrane part.

Started at the beginning of 90-th works on magnetic liquids [12] where it was shown how to distinguish nuclear and magnetic counterparts in scattering intensity on the installation without magnetic field and polarized neutrons have found their continuation on upgraded YuMO spectrometer [53-55]. Authors [12] showed that by contrast variation method in experiment of small-angle scattering of non-polarized neutrons one can distinguish nuclear and magnetic counterparts. This, as it was mentioned, was absolutely new result at that time.

In review [55] the questions of relation between magnetic and nuclear counterparts and influence of the solvent type on the behavior of SANS curve are considered. The size of magnetic particle and surfactant external shell are given for several types of deuterated solutions and surfactants. In particular it is pointed out that in water solutions (that is important for practice) the particles with surfactant shell associate into clusters. The conclusion was based on the power law of small-angle curve for water solutions. In the work the comparative table for four SAS instruments is shown: YuMO, IBR-2, Dubna; Yellow Submirine, KFKI, Budapest, Hungary; SANS, PSI, Willigen, Switzerland; SANS-1, FRG-1, GKSS, Geschacht, Germany. Under approximately the same q-range spectrometer YuMO makes a single measurement when the other spectrometers require three-four independent measurements (and consequently three-four times more the measurement time). At the same time it is noted that all non-Russian spectrometers equipped with magnetic devices (filed from 1.2 to 2.5 T). The project of installation for creating magnetic field (up to 2.6 T) was successfully implemented in Dubna on small-angle YuMO spectrometer and the first test experiments were already done.

In structural investigation of biological objects it is about to jump from single protein, membranes and even complexes investigation to investigation of these objects as a part of the whole organelles. In works [56-58] made on upgraded YuMO spectrometer the structures of membranes of intact mitochondria were investigated. Small-angle experiments showed that when the system of volume regulation is switched on the ultrastructural rearrangements in mitochondria from liver and heart of mouse take place. Peaks positions showed that if liver mitochondria are put under hypotonic conditions the cristae rearrange from disordered package to ordered two-membrane package with distance between membranes centers 190 Å. Under the same conditions in heart mitochondria the cristae rearrange from lamellar package with distance 220 Å to presumably hexagonal with lattice parameter 250 Å.

The spectrum of the beam on YuMO spectrometer includes also fast neutrons and gamma quants [6]. For decreasing the destructive radiation of the object the time of measurement should be decreased. This was achieved in the frameworks of two-detector system of tracking scattered neutrons. After the measurements a respiratory control of the mitochondria was carried out, which showed their resistance to radiation within 0.76 Sv for the gamma-ray flux and 49.9 Sv for fast and thermal neutrons. Thus, the possibility of using the method of small-angle neutron scattering to study the structure of functioning mitochondria was demonstrated.

Development and installation on the spectrometer the device with high hydrostatic pressure on the sample allowed investigating the phase transitions in micellar solutions of tetradecyl trimethyl ammonium bromide / heavy water (TTAB / D2O) in wide temperature range, pressure and surfactant concentration [34,35]. The kinetics of the phase transition micelles – solid phase was studied. The phase diagram of solvents was calculated and the temperature, pressure and surfactant concentration dependency of the transition velocity was shown. Two-phase coexistence of micelles and precipitates (solid phase) under certain conditions on concentration, pressure, and temperature was observed.

Recently the unique PVT installation was created. With its help, one can simultaneously conduct neutron structural investigations and measure the volume change[72]. Experiments made on upgraded reactor IBR-2M on the PVT installation for lipid membrane mixture for the first time showed that in the mixture of DPPC/POPC in excess of water the simultaneous jump of isothermal compressibility and minimum in the derivative of the periodicity [73]. Dynamic q-range is extremely important when studying lipid membranes. Indeed, in case of wide dynamic q-range it becomes possible to track the changes of periodicity of lipid membranes and the thickness of lipid bilayer at the same time. Taking into account the kinetics of solvent this consequence allows avoiding systematical errors [72].

In the series of works [30, 31, 59-61], aimed at investigation of the structure of polyallylcarbosilane dendrimers, the specificities of interactions such nanodisperse objects with solvent and their arrangement inside the solution, some quantitative parameters characteristics were obtained, such as sizes of several generations and average scattering density. Appliance of modern data treatment methods and the usage of good resolution of upgraded spectrometer YuMO allowed characterizing the form of dendrimers of several generations for 3 and 4-functional core. The fact of solvent penetrates inside dendrimer allows expecting the possibility of practical application of these new synthetic big molecules. The wide dynamic q-range (from 0.007 to 0.4$Å^{-1}$) and absolute intensity scale allowed solving the structure of dendrimers and understanding their interaction with solvent. Using modern mathematical data analysis algorithms to treat SANS data the spatial distribution of scattering density of investigated type of dendrimers was obtained [30]. SANS contrast variation showed that dendrimers molecules in solvents do not contain closed inner cavities where the solvent cannot penetrate into. Knowing the SANS intensity in absolute scale the partial dendrimer volume in solvent was obtained and the volume ratio (30-40%) of opened inner cavities inside effective dendrimer volume where the solvent can penetrate into was calculated. The anisometry of dendrimer macromolecules form that was inconsistent with AFM data [62] found the confirmation in results obtained by molecular dynamics [63].

Investigation of structure and solubilized behavior of three-layer nanoparticles obtained by using radiation polymerization from methyl methacrylate was done by two methods – NMR and SANS. SANS measurements were done on two installations – YuMO, IBR-2, Dubna and Yellow Submirine, KFKI, Budapest, Hungary. The absorption kinetics and two low molecular weight compounds (benzene and chloroform) equilibration were studied. SANS curves obtained on YuMO were in full consistency with the data obtained in neutron center in Budapest.

Lipid membranes is an adequate object for SANS structural studies. Using this method it is feasible to obtain as periodicity as thicknesses of single vesicles. Q-range plays a key role here because only in case it is widely enough these parameters could be obtained simultaneously. With the help of YuMO spectrometer the influence of DMSO $(CH_3)_2SO$ on the phospholipid (DMPC) membrane structure was investigated in wide DMSO concentration range $0.0 \leq X_{DMSO} \leq 1.0$ in excess of solvent at T=12.5 $^{\circ}C$ and T=55 $^{\circ}C$ [64]. The dependency of periodicity d of multi lamellar vesicles and thickness of single vesicles d_b in gel and liquid-crystalline phases versus X_{DMSO} in excess of solvent DMSO/water was determined. The complementary use of the values d and d_b made it possible to determine the intermembrane distance d_s. It is shown that the intermembrane distance decreases significantly with increasing DMSO concentration, and at a concentration of X_{DMSO}=0.4, the membranes in the neighborhood are practically in a steric contact with each other, which results in the fusion of single membranes into multilayer lamellar structures. The use of deuterated DMSO and the contrast variation method on water allowed determining for the first time the number of DMSO molecules strongly bounded to the membrane. The number is 6.9, and the total volume of such molecules is 820 Å^3, which is comparable with the volume of the polar head of the lipid molecule itself and explains the previously observed phase transition of the lipid membrane to the phase with mutual penetration of chains.

The investigation of influence Ca^{2+} on structure DMPC multilamellar vesicles in [74] was done.

The study of binary mixtures of branched polyethyleneimine and cationic surfactants aggregation in aqueous solutions was also carried out by several methods [65]. In that work, the dependency of shape and size of polymer-colloid aggregates, as well as their catalytic activity on the polymer concentration, the structure of surfactants and substrates was obtained. It has been found out that in some aqueous compositions it is possible to achieve significant acceleration (up to 10^4 times) of decomposition of esters of tetrakodonated phosphorus acids in comparison with their hydrolysis reaction in water at the same pH in both cases. The methodological feature of this experiment is a combination of conductometry, dynamic light scattering and SANS methods, which in this case are complementary and give self-consistent data about the sample structure.

During the SANS experiment the sample is placed in the air gap between the neutron guide and the detector tube. Thus, it is possible to illuminate the sample with a laser beam or a specific beam of light. In a sample environment system, a Schott KL 1500 electronic (SCHOTT Glas) source of visible light with a high and constant light intensity was used on the YuMO spectrometer. Light is directed to the sample with a flexible fiber. During the experiment, the sample is not exposed to excessive stress heat. With the help of this light source, the influence of light on the structure of bacteriorhodopsin (bR) protein on the YuMO spectrometer was studied.

Mentioned above several examples of investigations from material science to biology show new qualitative state of spectrometer after the upgrade.

Further instrument development is connected first of all with the method of thermal neutrons detection. Since March 2002 the project connected with creating the contemporary 2D position-sensitive detector (PSD) was launched. The detector is created, preliminary tests were done in LLB (France) and the detector was tested on YuMO[32]. PSD will make it possible to start working at usage of important advantage of pulsed reactor in comparison with the stationary one – good resolution on the momentum transfer. When PSD start working it will become possible to obtain additional information about the structure and consequently about the properties and processes occurring in samples under the action of temperature, pressure, light illumination and humidity.

4. Conclusion

Let us sum up the main advantages of upgraded YuMO spectrometer in comparison with the old instrument and the advantages of new nodes and devices of the spectrometer. Firstly, it is a high rate of experimental data collection and a wide dynamic q-range because of a two-detector system, secondly, it is a high level of service and remote control; thirdly, there is the possibility of quick commissioning of practically any new device on the spectrometer and integrating them into the existing control program.

The scheme of the experiment makes it possible to obtain the intensity curve in absolute units immediately after the experiment is completed and during the experiment. This gives an undeniable advantage and additional information when studying the molecular and supramolecular structures before the commonly used scheme (after the experiment additional measurements of water, carbon or other normalizers are made). New programs for the initial processing of experimental data and the fitting of obtained curves were updated, revised, and created.

The expanded opportunities led to the increase in the number and quality of experiments, and, correspondingly, to the increase of scientific publications.

Thus, the main result of the upgrade was the appearance of a small-angle world-class neutron spectrometer.

We dedicate this work to the memory of Yu.M. Ostanevich, our colleague and creator of the unique small-angle neutron spectrometer at the IBR-2 rector, the 80th anniversary of his birth was recently celebrated[75].

Acknowledgments

The authors grateful A.S. Kirilov, T.B. Petukhova, S.M. Murashkevich, A.P. Sirotin, A.A. Bogdzel and the staff of the small-angle scattering group for practical implementation and assistance during the work. The authors thank G. Eckold, M. Balashou, A. Raevska, J. Pleštil, P. Balgavi for their support in receiving grants from BMBF (Germany), Plenipotentiary Representatives from the participating countries: Romania, Poland, the Czech Republic and Slovakia.

We are also grateful to the heads of the department and laboratory, the Program Advisory Committee for constant attention and assistance in the implementation of the spectrometer modernization program, and Vlasov A.V. for valuable comments.

The works on high hydrostatic pressure and the creation of the new unique position-sensitive detector were supported by the grant No. 03DU03G2 of the Federal Ministry for Education and Science of Germany.

References

[1] Ostanevich YM, Serdyuk I. 1982 Neutronographic studies of biological macromolecule structures *Uspekhi Fizicheskih Nauk*.**137(5)**:85-116.

[2] Roko M, Williams R, Alivisatos P. 2000 *Nanotechnology Research Directions: Vision for Nanotechnology in the Next Decade*: New York: Springer.

[3] Sergeev G. 2003 Nanokhimiya (Nanochemistry) *Mosk Gos Univ*:287.

[4] Gordeliy V, Kuklin A. 2002 Small angle neutron scattering on IBR-2 reactor *Preprint/JINR (Dubna)*.

[5] May RP. 1994 Geometrical optimization of neutron small-angle scattering instruments *Journal of applied crystallography*.**27(3)**:298-301.

[6] Kuklin A, Bobarikina G, Bogdzel' A, Gordelii V, Islamov A, Konovalov V, Rogov A, Florek M. 2002 *Preprint JINR*.**P13-2002-249**.

[7] Vagov V, Kunchenko A, Ostanevich Y, Salamatin I. 1983 *Soobsheniya JINR*.**P14-83-898**.

[8] Gorski N, Ostanevich Y. 1993 Small-angle neutron scattering (SANS) determination of the volume occupied by a single water molecule in the inverted micellar systems AOT+ X (H, D) 2 O+ C10D22 by internal contrast variation technique *Le Journal de Physique IV*.**3(C8)**:C8-149-C148-152.

[9] Pavlov MY, Rublevskaya I, Serdyuk I, Zaccai G, Leberman R, Ostanevich YM. 1991 Experimental verification of the triple isotopic substitution method in small-angle neutron scattering *Journal of applied crystallography*.**24(3)**:243-254.

[10] Bezzabotnov VY. VI Gordely, AA Kornyshev, AK Kostin, Yu. M. Ostanevich. On the fractal nature of carbonate and apatite rocks *JINR Preprint, El*:4-89.

[11] Borély S, Smirnov L, Kovneristii YK, Nartova T, Tarasova O, Ostanevich YM. 1991 Thermal stability of Ti41Zr41Si18 metallic glass studied by small-angle neutron scattering *physica status solidi (b)*.**164(2)**:343-350.

[12] Grabcev B, Balasoiu M, Bica D, Kuklin A. 1994 Magnetohydrodynamics *Riga, Latvia*.**30**:156-162.

[13] Birkholz W, Apel PY, Danziger M, Häuβler F, Hempel M, Kuklin A, Stetsenko S, Baumbach H. 1995 Investigation of latent and short etched heavy ion tracks in solids *Radiation measurements*.**25(1-4)**:57-62.

[14] Pleštil J, Ostanevich YM, Bezzabotnov VY, Hlavata D. 1986 Small-angle scattering by polyelectrolyte solutions. Hydration and conformation of poly (methacrylic acid) *Polymer*.**27(8)**:1241-1246.

[15] Kalus J, Kostromin S, Shibaev V, Kunchenko A, Ostanevich YM, Svetogorsky D. 1988 Small Angle Neutron Scattering Study of Comb-Like Liquid Crystal Polysiloxane Macromolecule Conformation in Smectic Phase *Molecular Crystals and Liquid Crystals*.**155(1)**:347-356.

[16] Gordeliy V, Ivkov V, Ostanevich YM, Yaguzhinskij L. 1991 Detection of structural defects in phosphatidylcholine membranes by small-angle neutron scattering. The cluster model of a lipid bilayer *Biochimica et Biophysica Acta (BBA)-Biomembranes*.**1061(1)**:39-48.

[17] Merta J, Garamus VM, Kuklin AI, Willumeit R, Stenius P. 2000 Determination of the structure of complexes formed by a cationic polymer and mixed anionic surfactants by small-angle neutron scattering *Langmuir*.**16(26)**:10061-10068.

[18] Kuklin AI, Islamov AK, Gordeliy VI. 2005 Two-Detector System for Small-Angle Neutron Scattering *Neutron News*.**16**:16-18.

[19] Kirilov A, Litvinenko E, Astakhova N, Murashkevich S, Petukhova T, Yudin V, Gordelii V, Islamov AK, Kuklin A. 2004 Evolution of the SONIX Software Package for the YuMO Spectrometer at the ИБР-2 Reactor *Instruments and Experimental Techniques*.**47(3)**:334-345.

[20] Soloviev A, Solovieva T, Stadnik A, Islamov A, Kuklin A. 2003 SAS. Program for first data treatment for small-angle scattering spectrum *Soobsheniya JINR*.**P10-2003-86**.
[21] Kuklin AI, Rogov AD, Gorshkova YE, Utrobin PK, Kovalev YS, Rogachev AV, Ivankov OI, Kutuzov SA, Soloviev DV, Gordeliy VI. 2011 Analysis of Neutron Spectra and Fluxes Obtained with Cold and Thermal Moderators at IBR-2 Reactor: Experimental and Computer Modelling Studies *Physics of Particles and Nuclei Letters*.**8(2)**:200-217.
[22] Kuklin A, Islamov AK, Gordeliy V. 2005 Scientific reviews: Two-detector system for small-angle neutron scattering instrument *Neutron News*.**16(3)**:16-18.
[23] Kuklin A, Islamov AK, Kovalev YS, Utrobin P, Gordelii V. 2006 *Surface*.**6**:73-84.
[24] Kuklin AI, Soloviov DV, Rogachev AV, Utrobin PK, Yu SK, Balasoiu M, Ivankov OI, Sirotin AP, Murugova TN, Petukhova TB et al. 2011 New opportunities provided by modernized small-angle neutron scattering two-detector system instrument (YuMO) *Journal of Physics: Conference Series*.**291(1)**:012013.
[25] Morkovnikov IA, Ivankov OI, Kirilov AS. 2012 The Program for Measurement Registration at the YuMO Instrument *Communication of the JINR*.**P10-2012-134**.
[26] Kuklin A, Sirotin A, Kirilov A, Islamov A, Petukhova T, Astakhova N, Utrobin P, Kovalev YS, Gordelii V. 2004 Avtomatization and envioriment for sample on modernized instalation YuMO *Preprint JINR P13-2004-77 Dubna*.**11**.
[27] Soloviev A, Litvinenko E, Ososkov G, Islamov AK, Kuklin A. 2003 Application of wavelet analysis to data treatment for small-angle neutron scattering *Nuclear Instruments and Methods in Physics Research Section A: Accelerators, Spectrometers, Detectors and Associated Equipment*.**502(2)**:500-502.
[28] Bezzabotnov VY, Ostanevich YM. 1989 Solution of some experimental problems met in the time-of-flight small angle neutron scattering *Physica B: Condensed Matter*.**156**:595-597.
[29] Ostanevich Y. 1988 Time-of-flight small-angle scattering spectrometers on pulsed neutron sources *Makromolekulare Chemie Macromolecular Symposia*.**15(1)**:91-103.
[30] Ozerin A, Svergun D, Volkov V, Kuklin A, Gordelyi V, Islamov AK, Ozerina L, Zavorotnyuk D. 2005 The spatial structure of dendritic macromolecules *Journal of applied crystallography*.**38(6)**:996-1003.
[31] Kuklin A, Ignat'eva G, Ozerina L, Kh A. 2002 Is lamov, RI Mukhamedzyanov, NA Shumilkina, VD Myakushev, E. Yu. Sharipov, VI Gordeliy, AM Muzafarov, and AN Ozerin *Polym Sci, Ser A*.**44(12)**:1.
[32] Kuklin A, Kutuzov S, Gabriel A, Eckold G, Utrobin P, Smirnov A, Ivankov A, Islamov AK, Kovalev YS, Feoktistov A. 2007 Measurements with New Type of Position Sensitive Detector in Backscattering and Small-angle Setup *Annual report FLNP*:129-132.
[33] Anan'ev B, Kunchenko A, Lazin V, Ostanevich Y, Pikel'ner E. Manywires ring detector for low neutrons based on He-3 *Soobsheniya JINR*:3-11502.
[34] Kuklin A, Islamov AK, Kovalev YS, Utrobin P, Gordelii V. Poverkhnost', No. 6, 74 (2006) *Google Scholar*.

[35] Nyam-Osor M, Soloviov DV, Kovalev YS, Zhigunov A, Rogachev AV, Ivankov OI, Erhan RV, Kuklin AI. 2012 Silver behenate and silver stearate powders for calibration of SAS instruments *Journal of Physics: Conference Series*.**351(1)**:012024.
[36] Gorski N, Ivanov A, Kuklin A, Smirnov L. 1995 Small angle neutron scattering setup for high pressure measurements at IBR-2 *International Journal of High Pressure Research*.**14(1-3)**:215-220.
[37] Islamov A, Haramagatti C, Gibhardt H, Kuklin A, Eckold G. 2006 Pressure-induced phase transitions in micellar solutions *Physica B: Condensed Matter*.**385**:791-794.
[38] Haramagatti C, Islamov A, Gibhardt H, Gorski N, Kuklin A, Eckold G. 2006 Pressure induced phase transitions of TTAB-micellar solutions studied by SANS and Raman spectroscopy *Physical Chemistry Chemical Physics*.**8(8)**:994-1000.
[39] Uhrıková D, Kučerka N, Islamov A, Kuklin A, Gordeliy V, Balgavý P. 2003 Small-angle neutron scattering study of the lipid bilayer thickness in unilamellar dioleoylphosphatidylcholine vesicles prepared by the cholate dilution method: n-decane effect *Biochimica et Biophysica Acta (BBA)-Biomembranes*.**1611(1)**:31-34.
[40] Gallová J, Uhríková D, Islamov A, Kuklin A, Balgavy P. 2004 Effect of cholesterol on the bilayer thickness in unilamellar extruded DLPC and DOPC liposomes: SANS contrast variation study *General physiology and biophysics*.**23**:113-128.
[41] Lebedev D, Filatov M, Kuklin A, Islamov AK, Kentzinger E, Pantina R, Toperverg B, Isaev-Ivanov V. 2005 Fractal nature of chromatin organization in interphase chicken erythrocyte nuclei: DNA structure exhibits biphasic fractal properties *FEBS letters*.**579(6)**:1465-1468.
[42] Fedotov G, Pakhomov E, Pozdnyakov A, Kuklin A, Islamov A, Putlayev V. 2007 Structure and properties of organic-mineral soil gel *Pochvovedenie*(**9**):1071-1077.
[43] Fedotov G, Tret'yakov YD, Ivanov A, Kuklin A, Pakhomov E, Islamov A, Pochatkova T, editors. Fractal coloid structures in other zone soils. Doklady akademii nauk; 2005.
[44] Fedotov G, Tret'yakov YD, Ivanov A, Kuklin A, Islamov A, Putlayev V, Garshev A, Pakhomov E, editors. Fractal structures coloid formations in soils. Doklady Akademii nauk; 2005.
[45] Fedotov G, Tret'yakov YD, Pakhomov E, Kuklin A, Islamov AK, Pochatkov T, editors. The Effect of Moisture on Fractal Properties of Soil Colloids. Dokl Akad Nauk; 2006.
[46] Fedotov G, Tret'yakov YD, Pakhomov E, Kuklin A, Islamov AK, editors. Temperature effect on the evolution of soil gels. Doklady Chemistry; 2006: Springer.
[47] Efremov R, Shiryaeva G, Bueldt G, Islamov A, Kuklin A, Yaguzhinsky L, Fragneto-Cusani G, Gordeliy V. 2005 SANS investigations of the lipidic cubic phase behaviour in course of bacteriorhodopsin crystallization *Journal of Crystal Growth*.**275(1)**:e1453-e1459.
[48] Knotko A, Garshev A, Makarova M, Putlayev V, Trtiakov Y, Kuklin A. 2004 Phase dacay in Pr-included solid solvents based on semiconductor Bi [2] Sr [2] CaCu [2] O [8] *Materialovedenie*(**2**):2-8.
[49] Philippova OE, Andreeva AS, Khokhlov AR, Islamov AK, Kuklin AI, Gordeliy VI. 2003 Charge-induced microphase separation in polyelectrolyte hydrogels with associating

hydrophobic side chains: Small-angle neutron scattering study *Langmuir*.**19(18)**:7240-7248.

[50] Andreeva AS, Philippova OE, Khokhlov AR, Islamov AK, Kuklin AI. 2005 Effect of the mobility of charged units on the microphase separation in amphiphilic polyelectrolyte hydrogels *Langmuir*.**21(4)**:1216-1222.

[51] Zaroslov YD, Gordeliy V, Kuklin A, Islamov A, Philippova O, Khokhlov A, Wegner G. 2002 Self-assembly of polyelectrolyte rods in polymer gel and in solution: small-angle neutron scattering study *Macromolecules*.**35(11)**:4466-4471.

[52] Hereć M, Islamov A, Kuklin A, Gagoś M, Gruszecki WI. 2007 Effect of antibiotic amphotericin B on structural and dynamic properties of lipid membranes formed with egg yolk phosphatidylcholine *Chemistry and physics of lipids*.**147(2)**:78-86.

[53] Grabcev B, Balasoiu M, Tirziu A, Kuklin A, Bica D. 1999 Application of contrast variation method in SANS experiments with ferrofluids *Journal of magnetism and magnetic materials*.**201(1)**:140-143.

[54] Balasoiu M, Avdeev M, Kuklin A, Aksenov V, Hasegan D, Garamus V, Schreyer A, Bica D, Vekas L, Almasan V. 2006 Nuclear and magnetic structures of non-polar ferrofluids by small-angle neutron scattering *Romanian Reports in Physics*.**58(3)**:305.

[55] Balasoiu M, Avdeev M, Kuklin A, Aksenov V, Bica D, Vékás L, Hasegan D, Torok G, Rosta L, Garamus V. 2004 Structural studies of ferrofluids by small-angle neutron scattering *Magnetohydrodynamics*.**40(4)**:359-368.

[56] Murugova T, Gordeliy V, Kuklin A, Kovalev YS, Yurkov V, Nurenberg A, Islamov AK, Yaguzhinskii L. 2006 Detection of new double-membrane structures in native mitochondria by the method of small-angle neutron scattering *Biophysics*.**51(6)**:882-886.

[57] Murugova T, Gordeliy V, Islamov AK, Kuklin A, Solodovnikova I, Yaguzhinsky L. 2006 Detection of three-dimensional structures of inner mitochondrial membrane under low-amplitude swelling by small angle neutron scattering *Biochim Biophys Acta*.**14**:524-525.

[58] Murugova T, Gordeliy V, Kuklin A, Solodovnikova I, Yaguzhinsky L. 2007 Study of three-dimensionally ordered structures of intact mitochondria by small-angle neutron scattering *Crystallography Reports*.**52(3)**:521-524.

[59] Kuklin A, Ozerin AN, Islamov AK, Muzafarov AM, Gordeliy VI, Rebrov EA, Ignat'eva GM, Tatarinova EA, Mukhamedzyanov RI, Ozerina LA. 2003 Complementarity of small-angle neutron and X-ray scattering methods for the quantitative structural and dynamical specification of dendritic macromolecules *Journal of applied crystallography*.**36(3)**:1-683.

[60] Ozerin A, Muzafarov A, Kuklin A, Islamov AK, Gordelyi V, Ignat'eva G, Myakushev V, Ozerina L, Tatarinova E, editors. Determination of the shape of dendrimer macromolecules in solutions from small-angle neutron scattering data. Doklady Chemistry; 2004: Springer.

[61] Ozerin AN, Muzafarov AM, Gordeliy VI, Kuklin AI, Ignat'eva GM, Krykin MA, Ozerina LA, Shumilkina NA, Islamov AK, Sharipov EY, editors. Structure and dynamics of dendritic macromolecules. Macromolecular Symposia; 2003: Wiley Online Library.

IOP Conf. Series: Journal of Physics: Conf. Series **848** (2017) 012010 doi:10.1088/1742-6596/848/1/012010

[62] Tatarinova E, Rebrov E, Myakushev V, Meshkov I, Demchenko N, Bystrova A, Lebedeva O, Muzafarov A. 2004 Synthesis and study of the properties of the homologous series of polyallylcarbosilane dendrimers and their nonfunctional analogs *Russian chemical bulletin*.**53(11)**:2591-2600.

[63] Maiti PK, Goddard WA. 2006 Solvent quality changes the structure of G8 PAMAM dendrimer, a disagreement with some experimental interpretations *Journal of Physical Chemistry B*.**110(51)**:25628.

[64] Gorshkova Y, Gordeliy V. 2007 Investigation of DMSO/LIPID membranes interaction by small-angle neutron scattering *Kristalographya*.**52(3)**:560-563.

[65] Kosacheva E, Kudravcev D, Bakeeva R, Kuklin A, Islamov A, Kudravceva L, Sopin V, Konovalov A. 2006 *Colloid J*.**68**:713.

[66] Cherny AY, Anitas E, Kuklin A, Balasoiu M, Osipov V. 2010 Small-angle scattering from the deterministic fractal systems 1 *Journal of Surface Investigation: X-ray, Synchrotron and Neutron Techniques*.**4(6)**:903-907.

[67] Cherny AY, Anitas E, Osipov V, Kuklin A. 2011 Deterministic fractals: extracting additional information from small-angle scattering data *Physical Review E*.**84(3)**:036203.

[68] Cherny AY, Anitas EM, Osipov VA, Kuklin AI. 2014 Small-angle scattering from multiphase fractals *Journal of Applied Crystallography*.**47(1)**:198-206.

[69] Cherny AY, Anitas E, Osipov V, Kuklin A. 2015 Scattering from surface fractals in terms of composing mass fractals *arXiv preprint arXiv:150707376*.

[70] Cherny AY, Anitas EM, Osipov VA, Kuklin AI. 2017 Small-angle scattering from the Cantor surface fractal on the plane and the Koch snowflake *Physical Chemistry Chemical Physics*.

[71] Cherny AY, Anitas E, Kuklin A, Balasoiu M, Osipov V. 2010 Scattering from generalized Cantor fractals *Journal of Applied Crystallography*.**43(4)**:790-797.

[72] Kuklin AI, Soloviev DV, Rogachev AV, Utrobin PK, Kovalev YS, Balasoiu M, Ivankov OI, Sirotin AP, Murugova TN, Petukhova TB et al. 2012 New opportunities provided by modernized small-angle neutron scattering two-detector system instrument (YuMO) *Journal of Physics: Conference Series*.**291(1)**:012013.

[73] Bulavin L, Savenko V, Lebovka N, Kuklin A, Soloviov D, Ivankov O. 2013 Small-angle neutron scattering of multiwalled carbon nanotubes in aqueous suspensions in presence of laponite platelets or cetyltrimethylammonium bromide *Nuclear Physics and Energy*.**14(4)**:372-376.

[74] Gorshkova YE, Kuklin A, Gordeliy V. 2017 Structure and phase transitions of DMPC multilamellar vesicles in the presence of Ca2+ ions *Journal of Surface Investigation: X-ray, Synchrotron and Neutron Techniques*.**11(1)**:27-37.

[75] Kučerka N, Balasoiu M, Kuklin AI. 2016 III International Conference on Small Angle Neutron Scattering Dedicated to the 80th Anniversary of Yu. M. Ostanevich *Neutron News*.**27(4)**:14-16.

Small-Angle Scattering on Magnetoferritin Nanoparticles

L Balejčíková[1], VI Petrenko[2,3], MV Avdeev[2], VM Garamus[4], L Almásy[5,6] and P Kopčanský[1]

[1]Institute of Experimental Physics Slovak Academy of Sciences, Watsonova 47, 04001 Košice
[2]Joint Institute for Nuclear Research, Joliot-Curie 6, 141980 Dubna, Moscow region, Russia
[3]Taras Shevchenko National University of Kyiv, Volodymyrska Street 64, Kyiv, 01033 Ukraine
[4]Helmholtz-Zentrum Geesthacht: Centre for Materials and Coastal Research, Max-Planck-Street 1, 21502 Geesthacht, Germany
[5]State Key Laboratory Cultivation Base for Nonmetal Composites and Functional Materials, Southwest University of Science and Technology, Mianyang 621010, China
[6]Wigner Research Centre for Physics, Hungarian Academy of Sciences, H-1525 Budapest, POB 49, Hungary

balejcikova@saske.sk

Abstract. Magnetoferritin is a synthetically prepared magnetic bio-complex, consisting of apoferritin shell and iron-based nanoparticles. Superparamagnetic behaviour, nanoscale size (about 12 nm) and biological origin allow to use magnetoferritin in various applications. In this report, we present a general overview about basic physicochemical properties of magnetoferritin, as determined by small-angle X-ray and neutron scattering experiments and some interesting references on their potential bio-applications.

1. Introduction

Iron-based magnetic nanoparticles are interesting in wide range of applications. In liquid dispersions, they tend to agglomerate due to attractive magnetic interactions. To secure their colloidal stability and homogeneity in aqueous or oily solutions, magnetic nanoparticles should be coated by various organic polymers. Such prepared magnetic fluids are used especially in biomedicine, nanotechnology and industry. For minimization of their toxicity and increasing of biocompatibility in possible biomedical applications, magnetic nanoparticles can be bound also through DNA (e.g. plasmids), phospholipids, or proteins [1]. In 1991 it was shown that the empty cavity of the iron-storage protein ferritin, entitled apoferritin, could be used as a reaction medium for *in vitro* chemical synthesis of iron-based nanoparticles [2]. The first synthesis of such material, called magnetoferritin, was published in 1992 [3]. Since then, many physicochemical properties have been examined and its synthesis technology is continuously improving. In the recent years, we carried out systematic investigations on magnetoferritin prepared at various synthesis conditions and at various loading factors (i.e. the average number of iron atoms per one apoferritin). The overview of basic physicochemical properties investigated using available techniques, significant results obtained by small-angle X-ray (SAXS) and neutron (SANS) scattering, and potential bio-applications of magnetoferritin are described in this paper. Numerous experimental methods were used in our studies in order to better understand physicochemical properties of magnetoferritin. Synthesis technology modification allow to obtain magnetoferritin samples with different loading factors, colloidal stability, size distribution and magnetisation, demanding for thorough characterization of magnetoferritin using advanced physico-chemical methods. For the first time we investigated magneto-optical and structural properties of magnetoferritin colloidal solutions using structure-sensitive methods such as Cotton-Mouton and Faraday rotation measurements and SAXS/SANS. The obtained results had significant contribution for better understanding the magnetoferritin behavior, and first of all, its structure, under well defined physico-chemical conditions. Characterization of the structural changes of magnetoferritin at various conditions is indispensable for understanding its behavior and interactions with biological media in the perspective biomedical applications.

2. Basic physicochemical properties of magnetoferritin

Magnetoferritin aqueous colloidal solution prepared by controlled chemical synthesis with a procedure described previously [4] was characterized by zeta (ζ) potential measurements, dynamic light scattering (DLS), transmission electron microscopy (TEM) and SQUID magnetometry. The ζ-potentials of the apoferritin and magnetoferritin at comparable concentrations were –25.5 and –21.9 mV, respectively. These measurements confirmed negative surface charge of the protein and good colloidal stability. The hydrodynamic size of magnetoferritin was investigated by dynamic light scattering technique and compared with apoferritin. The average hydrodynamic diameter for magnetoferritin increases ($\langle D_{HYDR} \rangle$ = 19.5 nm) in comparison with pure apoferritin hollow sphere ($\langle D_{HYDR} \rangle$ = 14.1 nm). This increase of hydrated/solvated protein macromolecules can be related to partial destruction and agglomeration of a small fraction of the proteins upon loading iron. The majority of the iron oxide phase is still inside the inner cavity of apoferritin while the presence of nanocrystallites with an average diameter of 5 nm have been revealed by TEM [5],[6] .

Magnetic properties of magnetoferritin were studied using a SQUID magnetometer in magnetic fields up to 4000 kA/m. Magnetic-field dependencies have shown superparamagnetic behaviour of system without hysteresis at room temperature. Hysteresis was observed below a blocking temperature (T_b = 26 K) at 2 K with a coercive field of 20 kA/m [4] -[6].

Sensitive and non-invasive magneto-optical methods allowed to distinguish between magnetoferritin and ferritin. Magneto-optical Cotton-Mouton effect is generated thanks to the optical anisotropy of magnetic nanoparticles and orientation of their magnetic moments in an applied magnetic field. The electromagnetic radiation which pass through the sample causes rotation of the polarization plane of light and change the angle, which is proportional to the birefringence. Direct measurements of magnetically induced optical linear birefringence normalised to iron concentration as a function of the applied magnetic field intensity square H^2 at the wavelength 632.8 nm and 295 K indicated differences in magneto-optical behaviour of aqueous dispersion of magnetoferritin in comparison with pure ferritin solution. Cotton-Mouton constant for magnetoferritin, calculated at low region of magnetic field below 200 Oe was about 4 orders of magnitude higher than for ferritin that was related to the presence of iron containing core inside both biomacromolecules. This magneto-optical study allows identification of magnetic core of the unknown sample of ferritin and its biogenic or synthetic derivatives [7], [8].

The Faraday magneto-optical effect, similar as the Cotton-Mouton effect, is based on the rotation of the polarization plane of light depending on the intensity of the applied magnetic field. Faraday rotation spectra for magnetoferritin in comparison with commercial ferritin in the applied magnetic field with the intensity of H = 2970 Oe in the spectral range of the used light with the wavelength from 300 to 680 nm have shown significant differences, which are related to the various phase composition of iron cores. The dependence of Faraday rotation on magnetic field has shown superparamagnetic behaviour of magnetoferritin [9], what is in good agreement with our previous investigations [4] -[5]. The calculated Verdet constant could, after standardization, serve as a suitable parameter for quantitative determination of the iron oxidation state of various nanoparticles with a specific size. Cotton-Mouton and Faraday magneto-optical effects could be useful in biomedical applications for the detection of iron oxides in pathological tissues (e.g. in diagnosis of neurodegenerative diseases) [9].

3. Small-angle scattering on magnetoferritin nanoparticles

Small-angle scattering of X-rays (SAXS) and neutrons (SANS) was applied for the structural study of magnetoferritin nanoparticles in colloidal solution. The first experiments, performed with full loading of iron, have shown large difference between scattering curves of magnetoferritin in comparison with pure apoferritin solution [5],[6], as a result of the presence of the new component. Therefore, further investigations were focused on the behaviour with relatively low iron content (i.e. loading factor, the average iron atoms per one apoferritin) expecting only a weak change in the protein structure. Nevertheless, scattering curves of magnetoferritin with loading factor 160 in comparison with pure apoferritin were markedly different (Fig. 1). *Ab initio* analysis points to partial destruction of the magnetoferritin shell (Fig. 1). It should be noted that the structure obtained in the modelling represents the average shape and does not exclude the existence of complete shells.

For better understanding the effect of iron loading on the magnetoferritin structure, SAXS measurements were performed for different loading factors. Here, the scattering curves and the calculated pair distance distribution functions have shown an increase of aggregation with the loading factor growth. Comparisons of radius of gyration and hydrodynamic diameter obtained by dynamic light scattering, have shown that the magnetoferritin particles size increases with the loading factor and a strong aggregation occurs at the loading factor over 600 [10].

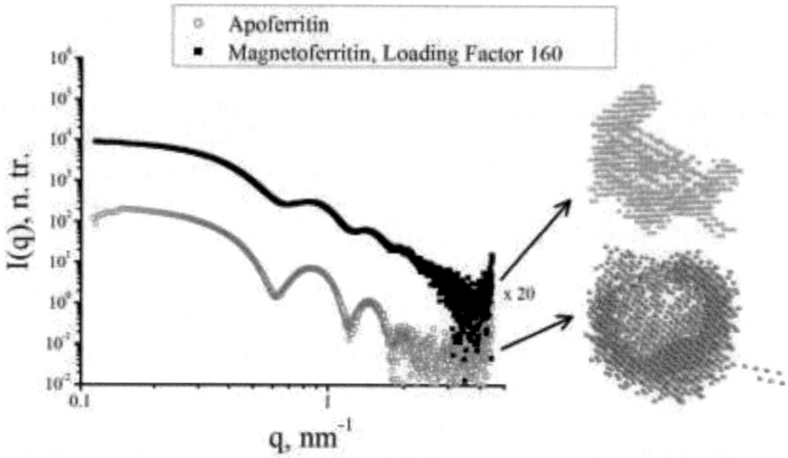

Figure 1. SAXS scattering curves for apoferritin and magnetoferritin at the loading factor 160 with the corresponding *ab initio* models. Figure taken from [10] with permission.

Next investigation was SANS contrast variation study of magnetoferritin with different loading factors. The difference between the neutron scattering length density for light (ρ_{H2O} = -0.559 × 10^{10} cm^{-2}) and heavy water (ρ_{D2O} = 6.34 × 10^{10} cm^{-2}) can be used for the variation of contrast by varying the light and heavy water ratio in the solution. Forward scattering intensities measured at the different heavy water content allowed to obtain match points at 0.434, 0.484 and 0.495 volume fractions of D_2O, for loading factors of 160, 510 and 770, respectively. Assuming homogeneous and monodisperse approximation of the magnetoferritin core containing magnetite, the calculated scattering length densities were obtained as 2.46 × 10^{10} cm^{-2}, 2.79 × 10^{10} cm^{-2} and 2.86 × 10^{10} cm^{-2} , for loading factors 160, 510 and 770, respectively, in comparison with SLD of apoferritin 2.34 × 10^{10} cm^{-2}. Volume fractions of magnetic part in magnetoferritin were 0.026, 0.099 and 0.114 for loading factor 160, 510 and 770, assuming magnetite or maghemite form of the iron oxide phase. SANS contrast variation indicated an abnormally high ratio of the amount of the magnetic component. The reason could be associated with partial shell destruction, observed also from SAXS data. On the other hand, the non-zero intensity in the match point indicated the size and structural polydispersity growth with increasing of loading factor, what could be explained also by the presence of unknown iron phases in magnetoferritin [10],[11]. Protein shell destroying could be caused by the specific effect of the iron oxide nanoparticles.

Many studies confirmed that various nanoparticles with specific physicochemical properties affect proteins in different way. Understanding of these interactions is the main subject of intensive research in recent years. Explanation of this phenomenon requires further detailed studies, aimed to optimize the synthesis technology and use of suitable separation techniques for future bio-applications of magnetic nanoparticles complexes with proteins.

In the latest SAXS study the effect of magnetoferritin on the structure of lysozyme amyloid aggregates was investigated. It was observed that magnetoferritin is able destroy the lysozyme amyloid fibrils. The lysozyme amyloid fibrils size reduction, confirmed by the pair-distance distribution function analysis (Fig. 2) was supported by the fluorescence spectrophotometry measurements as the decreasing of lysozyme amyloid fibrils amount after incubation with magnetoferritin. Magnetoferritin and lysozyme amyloid fibrils due to their biological origin are suitable pathological model systems for better understanding neurodegenerative diseases mechanism, and the presence of magnetic nanoparticles in magnetoferritin seems to play an important role in the amyloid destruction [12]. Moreover, magnetoferritin can be useful for potential applications as therapeutic agent against neurodegenerative diseases development. The precise mechanism of interaction of magnetic nanoparticles with lysozyme amyloid fibrils was not established yet. Such interactions could depend on nanoparticles phase composition, concentration, surface, charge distribution, physicochemical condition of surrounding medium (e.g. pH, temperature, ionic strength). In our case we assume that lysozyme amyloid fibrils could be destroyed by the partially uncovered magnetic nanoparticles with redox potential affecting chemical bounds in lysozyme amyloid fibrils. At present, it is not known whether Fe_3O_4 magnetite core is fully clamped inside apoferritin protein shell. However, we have demonstrated, that a partial destruction of apoferritin shell occurred, which can lead to partial exposure of magnetite core to the solvent. Thus, it follows from the experimental data, that the influence of magnetite core on anti-amyloid activity is a possibility that should be considered. We propose targeted transport and thus minimization of toxic effect of magnetoferritin nanoparticles on surrounding medium. Up to now, the effect of magnetoferritin on biological membrane/cells is still unknown. It is not excluded chemical surface modification of magnetoferritin nanoparticles for increasing its biocompatibility for various another applications. We are conscious that further characterization and possible therapeutic effect as well as cytotoxicity and possible side-effects on cell and animal models will be necessary to prepare magnetoferritin nanoparticles suitable for various bio-applications.

IOP Conf. Series: Journal of Physics: Conf. Series **848** (2017) 012011 doi:10.1088/1742-6596/848/1/012011

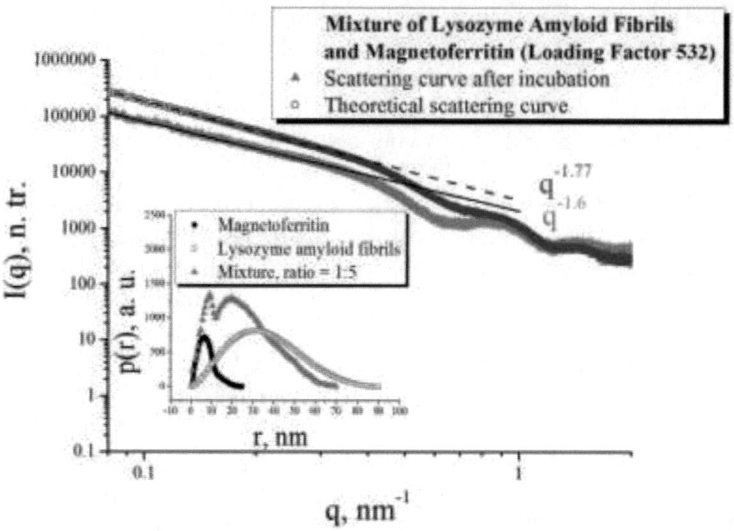

Figure 2. SAXS data and pair-distance distribution functions for magnetoferritin, lysozyme amyloid fibrils and their mixture in ratio 1:5. Figure taken from [12] with permission.

4. Potential bio-applications of magnetoferritin

Extensively studied physicochemical properties of magnetoferritin allow to use this unique magnetofarmaceutical nanomaterial in various application, as nanotechnology, industry or nanoelectronics (biosensors). The biocompatibility and high colloidal stability enhances the potential use in biomedicine [13] -[15]. In addition, magnetoferritin structure can be modified and functionalised by different chemical substances [16], drugs, surfactants, signal molecules or antibodies [17], [18]. Magnetoferritin can also serve as contrast agent in magnetic resonance imaging [19] or cell labeling [20], as a standard for diagnosis of various diseases [7] - [9], in cell separation [18], in hyperthermia therapy or in targeted transport of drugs [13] -[15]. Magnetoferritin characterised by peroxidase-like activity allows visualization of pathological tissues of the organism [21]. Recently it was shown, that magnetoferritin is able to decompose hydrogen peroxide in the presence of the substrate, N,N-diethyl-p-phenylenediamine sulfate, as evidenced by colour change after the reaction. This effect was stronger with loading factor increasing and could be useful in nanocatalytic chemistry [22]. Varying the synthesis technology, in combination with detailed fundamental research can advance and enlarge the range of potential applications of magnetoferritin.

5. References

[1] Laurent S, Forge D, Port M, Roch A, Robic C, Elst LV and Muller RN 2008 *Chem. Rev.* **108** 2064
[2] Meldrum FC, Heywood BR and Mann S 1991 *Nature* **349** 684
[3] Meldrum FC, Heywood BR and Mann S 1992 *Science* **257** 522
[4] Mitróová Z, Melniková L, Kováč J, Timko M and Kopčanský P 2012 *Acta Phys. Pol. A* **121** 1318
[5] Melníková L, Mitróová Z, Timko M, Kováč J, Koralewski M, Pochylski M, Avdeev MV, Petrenko VI, Garamus VM, Almasy L and Kopčanský P 2013 *Magnetohydrodynamics* **49** 293
[6] Melníková L, Mitróová Z, Timko M, Kováč J, Avdeev MV, Petrenko VI, Garamus VM, Almásy L and Kopčanský P 2014 *Mendeleev Commun.* **24** 80
[7] Koralewski M, Pochylski M, Mitróová Z, Timko M, Kopčanský P and Melníková L 2011 J. *Magn. Magn. Mater.* **323** 2413.
[8] Koralewski M, Pochylski M, Mitróová Z, Melníková L, Kováč J, Timko M and Kopčanský P 2012 *Acta Phys. Pol. A* **121** 1237.

[9] Koralewski M, Klos JW, Baranowski M, Mitróová Z, Kopčanský P, Melníková L, Okuda M and Schwarzacher W 2012 *Nanotechnology* **23** 355704.
[10] Melníková L, Petrenko VI, Avdeev MV, Garamus VM, Almásy L, Ivankov OI, Bulavin LA, Mitróová Z and Kopčanský P 2014 *Colloids Surf., B* **123** 82
[11] Melnikova L, Petrenko VI, Avdeev MV, Ivankov OI, Bulavin LA, Garamus VM, Almásy L, Mitroova Z and Kopcansky P 2015 *J. Magn. Magn. Mater.* **377** 77
[12] Kopcansky P, Siposova K, Melnikova L, Bednarikova Z, Timko M, Mitroova Z, Antosova A, Garamus VM, Petrenko VI, Avdeev MV and Gazova Z 2015 *J. Magn. Magn. Mater.* **377** 267
[13] Jutz G, van Rijn P, Santos Miranda B and Böker A 2015 *Chem. Rev.* **115** 1653
[14] Zeth K, Hoiczyk E and Okuda M 2015 *Trends Biochem. Sci.* **41** 190
[15] Theil EC, Behera RK and Tosha T 2013 *Coord. Chem. Rev.* **257** 579
[16] Urban I, Ratcliffe NM, Duffield JR, Elderb GR and Pattona D 2010 *Chem. Commun.* **46** 4583
[17] Dehal PK, Livingston CF, Dunn CG, Buick R, Luxton R and Pritchard DJ 2010 *Biotechnol. J.* **5** 596
[18] Zborowski M, Fuh CB, Green R, Baldwin NJ, Reddy S, Douglas T, Mann S and Chalmers JJ 1996 *Cytometry* **24** 251
[19] Jordan VC, Caplan MR, Bennett KM 2010 *Magn. Reson. Med.* **64** 1260
[20] Uchida M, Terashima M, Cunningham CH, Suzuki Y, Willits DA, Willis AF, Yang PC, Tsao PS, McConnell MV, Young MJ and Douglas T 2008 *Magn. Reson. Med.* **60** 1073
[21] Fan K, Cao C, Pan Y, Lu D, Yang D, Feng J, Song L, Liang M, Yan X 2012 *Nat. Nanotechnol.* **7** 459
[22] Melníková L, Pospiskova K, Mitroova Z, Kopcansky P and Safarik I 2014 *Microchim. Acta* **181** 295

Acknowledgments

This work was supported by the project of the Slovak Scientific Grant Agency VEGA (project No. 0045) and by the Ministry of Education Agency for European Structural funds (projects No. 26220120033 and No. 26220220186). L.A. thanks the Hungarian Fellowship Board for supporting a short research stay in the Institute of Physics, Kosice.

Molecular dynamics studies on the interaction and encapsulation processes of the nucleotide and peptide chains inside of a carbon nanotube matrix with inclusion of gold nanoparticles

Kholmirzo Kholmurodov[1,2,3,*], Eric Dushanov[1], Mirzoaziz Khusenov[3], Khaiyom Rahmonov[3,4], Tatyana Zelenyak[2], Alexander Doroshkevich[1] and Subrata Majumder[5]

[1]Joint Institute for Nuclear Research,141980, Dubna, Moscow Region, Russian Federation

[2]Dubna State University, 141980, Dubna, Moscow Region, Russian Federation

[3]S.U. Umarov Physical-Technical Institute, Academy of Sciences, Republic of Tajikistan

[4]Moscow State University "MEPhI", 115409, Moscow, Russian Federation

[5]National Institute of Technology, N.I.T., Patna: 800 005, India

E-mail: mirzo@jinr.ru

Abstract. Studying of molecular systems as single nucleotides, nucleotide and peptide chains, RNA and DNA interacting with metallic nanoparticles within a carbon nanotube matrix represents a great interest in modern research. In this respect it is worth mentioning the development of the electronics diagnostic apparatus, the biochemical and biotechnological application tools (nanorobotic design, facilities of drug delivery in a living cell), so on. In the present work using molecular dynamics (MD) simulation method the interaction process of small nucleotide chains (NCs) and elongated peptide chains with different sets of metallic nanoparticles (NPs) on a matrix from carbon nanotube (CNT) were simulated to study their mechanisms of encapsulation and folding processes. We have performed a series of the MD calculations with different NC,peptides-NP-CNT models that were aimed on the investigation of the peculiarities of NC,peptide-NP interactions, the formation of bonds and structures in the system, as well as the dynamical behavior in an environment confined by the CNT matrix.

1. Introduction

The fundamental understanding of the chemical interactions and dynamics of the nucleotides, RNA and DNA molecules at the dry-wet interface has become a key scientific issue in nanoscience due to a great number of potential applications (in electronics, biochip design, diagnostic tools, future computer architectures with massive memories, so on). For example, immobilization of DNA onto various surfaces, the conformations and the dynamics of DNA molecule adsorbed onto a metal and/or

[1] To whom any correspondence should be addressed.

semiconductors, or more complicated effects, the influence of energetic ions on DNA/surface interfaces are the hot topics of today research. The interaction of DNA or its fragments as single nucletodies with the carbon nanotubes (CNT), metallic and/or semiconductor matrices are the immense targets in modern research and applied sciences [1-3]. The immobilization of a biomolecule onto a desired surface is a crucial step for its application as it rapidly loses its biological activity in an external environment. In literature, there are several studies exists for developing patterned surface for successful functionalization of biological molecules, preferably through self-assembly techniques [4,5]. Most of the studies are performed on gold surfaces with thiol modified DNA molecules [6]. In this regard, the effect of low energy exchange (VdW interactions) as well as the molecular level orientation of the NC, peptides, RNA or DNA molecules adsorbed onto an CNT matrix or oxide surfaces have a great interest to understand the adsorption phenomena on the surface more clearly.

The CNTs as drug delivery vehicles have shown a potential interest due to a targeting of specific cancer cells with a lower dosage rather than conventional drugs have. With regard to the different aspects of the DNA-CNT application in the DNA nanotechnology there have been a great discussions from the point of view of molecular recognition processes, as a candidate material in cell drug delivery, as nucleic acid selection method (DNA aptamer) in SELEX (Systematic Evolution of Ligands by Exponential enrichment), so on. The conformational transition of aptamers (single chain DNA or RNA molecules that possess specific spatial structure) around CNT may cause some modification of the charge distribution on the CNT surface. It is worth noting that CNT surface has extremely sensitive to even a small change of the electrical charge of its environment. The replacement of even a single nucleotide for the DNA or RNA structure can modify, on the other hand, the charge environment around CNT. As a result, the CNT charge conductivity will be changed as well. So far, the DNA or RNA interactions with CNT could result to an essential modification for the charge distribution and consequent charge transfer through by the CNT surface. In the physical measurement the DNA-CNT charge distribution can be estimated trivially. This simple scheme from the point of view of application and diagnostic purposes has considered being one of the promising technologies in the DNA-CNT interaction processes with target proteins (say, of blood cells in human body). The DNA or RNA interaction mechanism with metallic nanoparticles and carbon nanotube is an important target of a great interest in modern nano- and bio-technologies, electronics industry and medicine. For example, one has to mention the design and development of the electronic mobile diagnostic facilities for the express blood analysis, the chemical or drug delivery inside living cell, and so on, where the NC-NP-CNT system are in use. The molecular structures like as nucleotide chain (NC) - gold nanoparticles (NPs) - carbon nanotube (CNT) is an important stage in the understanding of interaction mechanism of a whole DNA or RNA molecule with NP and CNT [7-18].

Metallic nanoparticle of gold, silver, etc. are of great research interest in modern biomedical applications [19-37]. In recent experiments [29], for example, such nanoparticles have been used, which consisted of porous gold discs with a diameter of 400 nanometers. By using the phenomenon of plasmon resonance (oscillation special electronic nanoparticles on the surface resulting from the irradiation with light at a particular angle) on can achieve on the surface of the nanoparticles an efficient radiation absorption and release of large amounts of heat. Further, on the disks of gold nanoparticles were placed three types of bacteria: *E.coli* (*Escherichia coli*) and heat-resistant *Bacillus subtilis* and *Exiguobacterium*, after which the nanoparticles are irradiated with infrared light. As the results show, the nanoparticles were heated up to 200 °C - and it is a temperature that exceeds the heat resistant bacteria. At the same time, *E.coli* colony completely die within five seconds, and *Bacillus subtilis* and *Exiguobacterium* 25 seconds. In another study designed nanoparticles generators - porous silicon disks, which penetrate the cancerous tumors and releasing the drug molecule. This method is extended half life of mice with breast cancer. Thus, the study of metal nanoparticles is increased interest primarily in terms of their interaction with biological materials. Applications include diagnostic purposes electronic tools for analyzing the composition of biochemical reagent, drug delivery inside the living cell, etc., where the carbon nanotubes are active implementation as a carrier for the biomaterials [19-25]. Therefore, the study of the interaction of biomolecules (nucleic acids,

IOP Conf. Series: Journal of Physics: Conf. Series **848** (2017) 012012 doi:10.1088/1742-6596/848/1/012012

nucleotide chains of DNA, proteins) with metal bass is very important to determine the extent of their links, which are in demand in the design and development of drug transport or analysis of biochemical reactions (DNA - NP) (Figures 1 and 2).

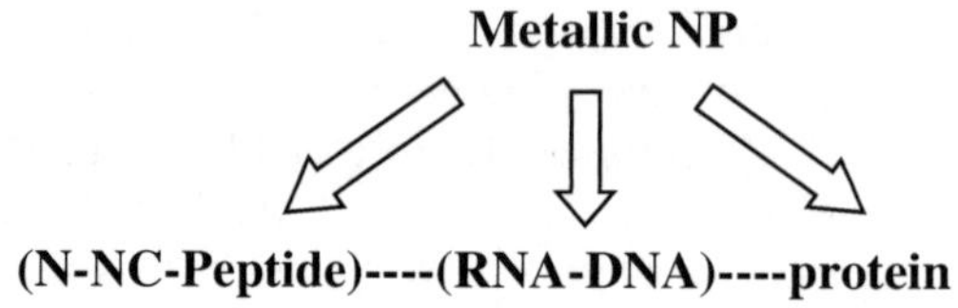

Fig. 1. A schematic view of the molecular dynamics models regarding on the interactions of metallic nanoparticles with single nucleotides, nucleotide or peptide chains, RNA, DNA and proteins.

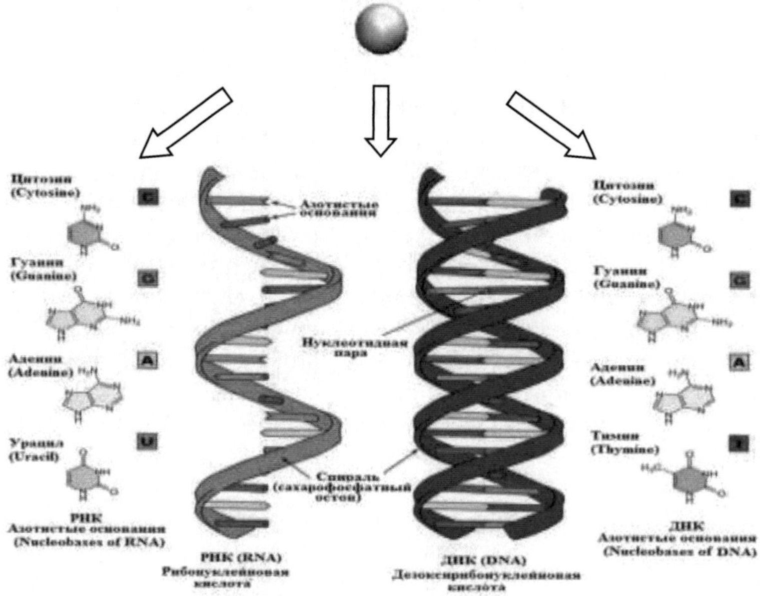

Fig. 2. A schematic view of the molecular dynamics models on the interactions of metallic nanoparticles with single nucleotides, nucleotide and peptide chains, RNA and DNA molecules.

Increased computing power with molecular calculations every year can effectively investigate the above-mentioned objects at the atomic level and receive important information about the structure, dynamics and functional properties. But the important aspect of the application and that is of great practical interest in the above mentioned systems and facilities, primarily with a view to their use in biomedical applications [38-50]. For example, of great interest to the study of structures and aquatic toxicity cluster solutions C60 is connected with the possibility of their use for controlled drug delivery, diagnosis and treatment of cancer. Fullerene endohedral complexes with rare earth metals and a number of other metals (and endo-metall-fullerenes, fullerenols-soluble forms), Gd @ C82 (OH) X40, can be used in biomedical applications as an effective and non-toxic agents (Fig. 3).

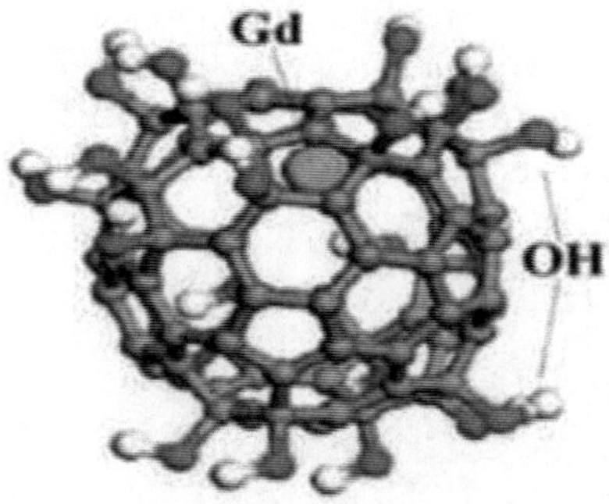

Fig. 3. A fullerenol model with atoms of rare earth elements [38-49].

As tests shown, by animal studies, they can also provide a therapeutic effect thereby inhibiting the development of tumor - in mouse brain glioma. It should be noted that the CNTs, fullerenes, endofullerene, fullerenols with various biological molecules, also intensively studied by neutron scattering at the Laboratory of Neutron Physics IM Frank, JINR (Joint Institute for Nuclear Research), Dubna, Moscow Region. and the National Research Centre "Kurchatov Institute" PNPI "Petersburg Nuclear Physics Institute. BP KONSTANTINOVA" in Gatchina, St. Petersburg [38-50].

Recent experimental and simulation studies involve the DNA interaction with highly localized proton beams or metallic NPs (such as Ag, Au, etc.), aimed on targeted cancer therapy through the injection of metal micro- or nanoparticles into the tumor tissue with consequent local microwave or laser heating. Along with DNA-NP also the DNA-CNT system represents a great interest in today biomedicine applications due to diagnostic and treatment of oncology diseases. Cancer, in which cells grow and divide abnormally, is one of the primary diseases with regards to how it responds to CNT drug delivery. Representing a revolutionarily potential for the biochemistry and medicine the use of CNTs in drug delivery has based on the enhancing of sufficient solubility and allowing of efficient tumor targeting. These aspects prevent CNTs from being cytotoxic and altering the function of immune cells. For today, cancer therapy involves surgery, radiation therapy, and chemotherapy. The experimental and simulation studies involve the interaction of DNA with highly localized high power beams and various nanoparticles (Ag, Au, etc.). These studies are aimed on targeted cancer therapy through the injection of metal micro- or nanoparticles into the tumor tissue with consequent local microwave or laser heating. Due to their good heat conductivities of NPs (Ag, Au, and so on) the experiments reveal that the only tumor cells to destroy, remaining normal cells undamaged. Nevertheless, such kind treatment methods are usually painful and kill normal cells in addition to producing adverse side effects [7-18].

So far, the understanding of the nucleotides and peptides – nanoparticles – carbon nanotubes dynamics and binding, as an important stage of the DNA interaction mechanism with metallic nano-particles and metallic surfaces embedded by the CNT environment, represents a potential target of modern research. In this work, the molecular dynamics (MD) simulations were performed on a small nucleotide chains (purines and pyrimidines) and peptides to investigate the interaction and binding processes with gold nanoparticles that happen within a carbon nanotube matrix. The behavior of a small NC, peptides models interacting with the NP-CNT system have to possess a lot of similarities with a full DNA or RNA molecule interacting in the NP and CNT environment.

2. Models and simulation method

A primary DNA or RNA structure consists of a linear sequence of nucleotides that are linked together by phosphodiester bonds. Nucleotides consist of 3 components: 1). Nitrogenous base - A (Adenine), G (Guanine), C(Cytosine), T (Thymine, present in DNA only) and U (Uracil, present in RNA only); 2). 5-carbon sugar which is called deoxyribose (found in DNA) and ribose (found in RNA); 3). One or more phosphate groups.

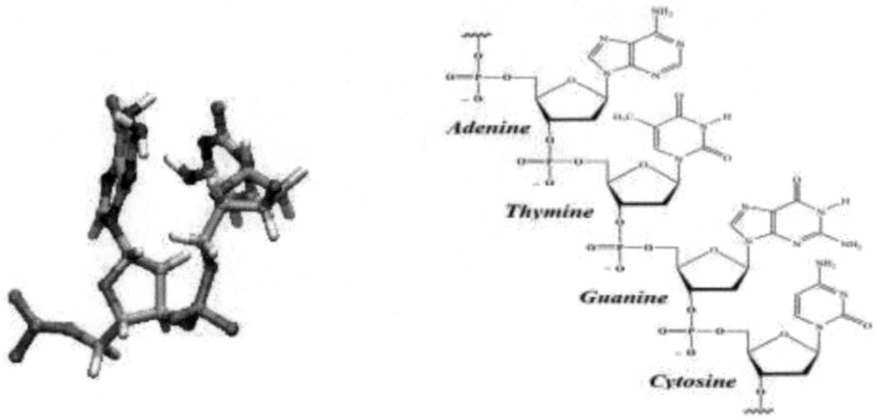

Fig. 4. (Left) A nucleotide chain (NC), consisting of a pyrimidine (C, cytosine) and a purine (G, guanine). (Right) A linear chain sequence of nucleotides that make up the primary structure of DNA or RNA.

We have simulated a three component molecular system consisting of a two nucleotide chain (NC), gold nanoparticles (NPs) and a carbon nanotube (CNT) under different temperature conditions (T=100, 200 and 300 K). Two nucleotides (one pyrimidine and one purine) as in the primary structure of DNA were relaxed in the vicinity of gold particles. In Fig. 5 (*top*) a schematic view of the simulation system consisting of a two nucleotide chain (NC), gold nanoparticles (NPs) and a carbon nanotube (CNT) is shown. In Fig. 5 (*bottom*) a two nucleotides chain, consisting of one pyrimidine (C, cytosine) and one purine (G, guanine) as in the primary structure of DNA is separately presented.

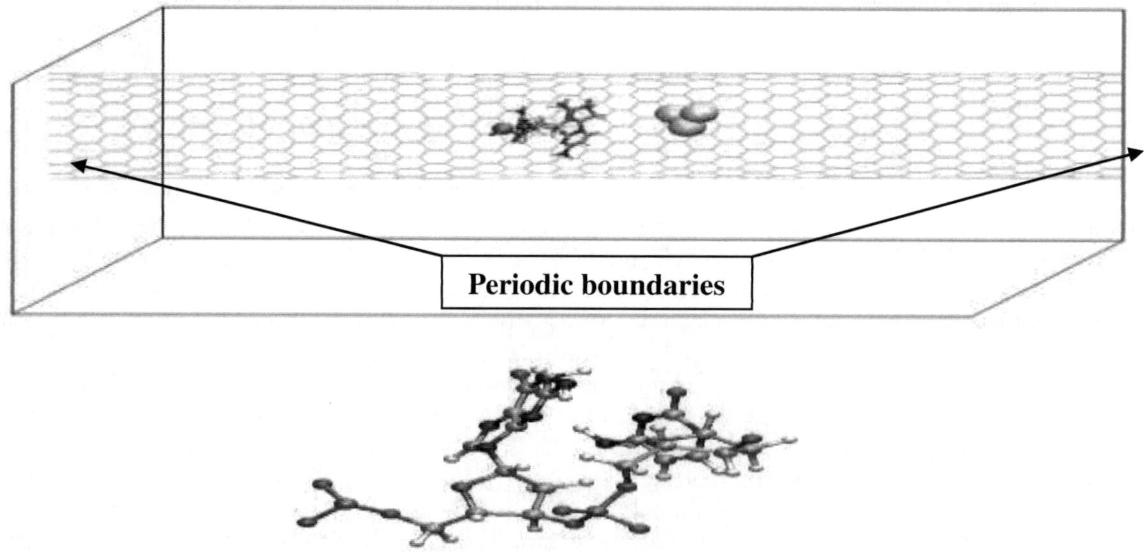

Fig. 5. (*top*) A schematic view of the simulation system consisting of a two nucleotide chain (NC), gold nanoparticles (NPs) and a carbon nanotube (CNT). (*bottom*) A two nucleotides chain, consisting of one pyrimidine (C, cytosine) and one purine (G, guanine) as in the primary structure of DNA is shown. A primary DNA or RNA structure consists of a linear sequence of nucleotides that are linked together by phosphodiester bonds. Nucleotides consist of 3 components: (1)Nitrogenous base - A (Adenine), G (Guanine), C(Cytosine), T (Thymine, present in DNA only) and U (Uracil, present in RNA only); (2) 5-carbon sugar which is called deoxyribose (found in DNA) and ribose (found in RNA); (3) One or more phosphate groups.

IOP Conf. Series: Journal of Physics: Conf. Series **848** (2017) 012012 doi:10.1088/1742-6596/848/1/012012

The nucleotide chain was located from gold atoms at distances [5-10] Å, i.e. within a range of Van der Waals (VdW) forces (Figs. 6). During the MD simulations, first, we have generated three positional slightly different gold particles enable to interact with the NC as NC-1Au, a one gold particle and a nucleotide chain NC, NC-2Au, a two gold particle and a nucleotide chain, and NC-3Au, a three (or more) gold particle and a nucleotide chain. Thus, we have made several small gold clusters and simulated the NC-NPs system as NC-2,3,…Au. Next, we extended the MD simulations for the elongated peptide chains interacting with metallic nanoparticles inside of a carbon nanotube matrix. So far, we have investigated for the single nucleotides, nucleotide and peptide chains interacting with NPs-CNT system multiple model configurations under the same simulation and temperature conditions (Figs. 6 and 7):

(1) **Model 1**, NC-1Au,
(2) **Model 2**, NC-2Au,
(3) **Model 3**, NC-3Au.

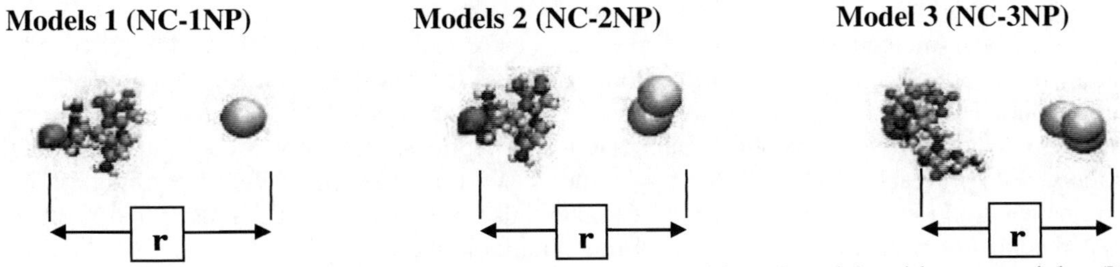

Fig. 6. The models of the nucleotide chain (NC) interacting with 1, 2 and 3 gold nanoparticles (NPs). The positions of one of two phosphorus (P) atoms of the NC and gold NPs are shown respectively by the dark brown and yellow spheres.

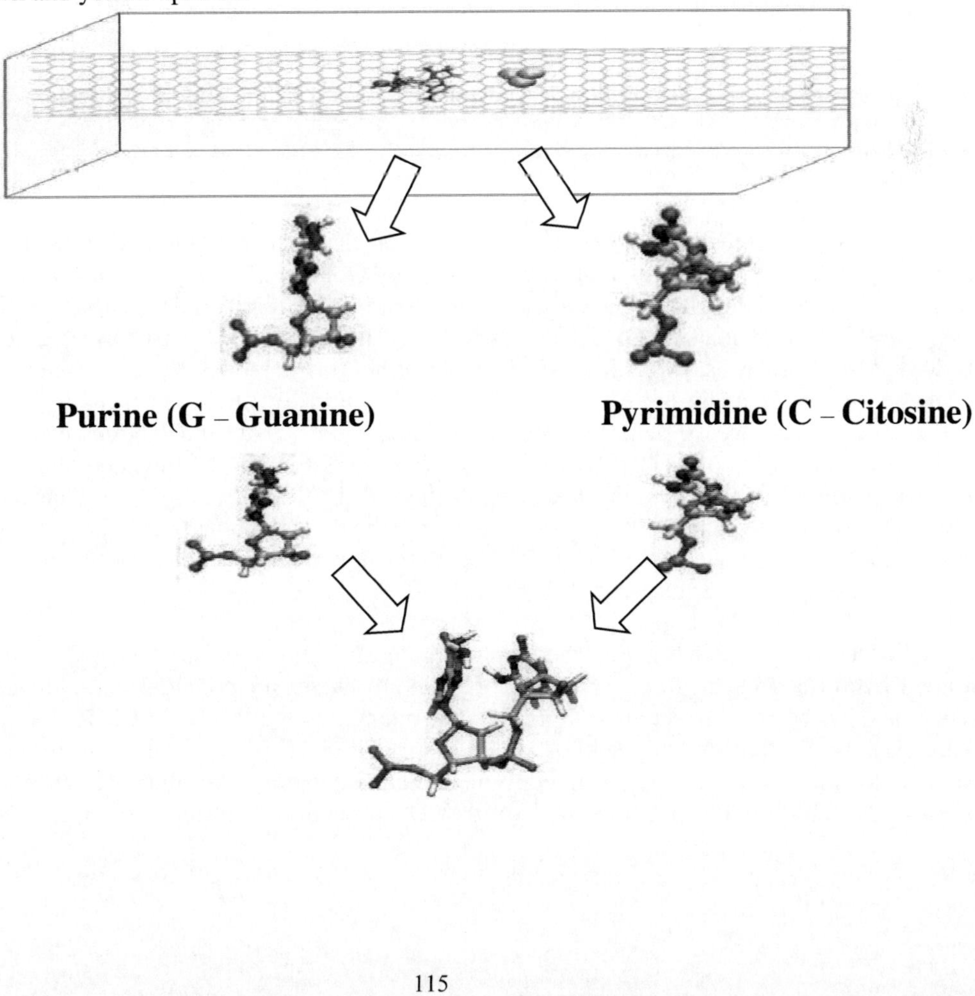

Fig. 7. The hierarchy of MD models for the single nucleotides, nucleotide chains and long peptides, interacting with metallic nanoparticles inside of a carbon nanotube matrix.

A classical molecular dynamics study was performed using the DL_POLY_2.20 [51-52] general-purpose code. The NVT ensemble with a Berendsen thermostat and a Verlet leapfrog scheme were employed. The integration time step of the dynamical equations of motion was 1 fs. The entire system (the nucleotide (N), nucleotide chain (NC) and peptides, gold atoms and carbon nanotube) were allowed to interact with each other via the VdW potential only. For describing of VdW interactions we used Lennard-Jones (LJ or 12-6) potential, which is commonly in use for simulation of liquids and condensed phases. The LJ and 12-6 potential look like:

$$V(r) = 4\varepsilon\left[\left(\frac{\sigma}{r}\right)^{12} - \left(\frac{\sigma}{r}\right)^{6}\right], \qquad U = \frac{A}{r^{12}} - \frac{B}{r^{6}}$$

where ε, σ are LJ interaction parameters and the cross-section interaction parameters were defined using the Lorentz-Berthelot mixing rule: $\varepsilon_{ij} = (\varepsilon_{ii}\varepsilon_{jj})^{\frac{1}{2}}$ and $\sigma_{ij} = \frac{1}{2}(\sigma_{ii} + \sigma_{jj})$.

For the LJ interaction parameters we have used the data from the literature as reported, for example: ε(Au-Au)=0.039 kcal/mol, σ(Au-Au)=2.9 Å by Q. Pu *et al.* [53]; ε(C-Au)=0.29256 kcal/mol, σ(C-Au)=2.99 Å by Song Hai Yang and Zha Xin Wei [54]; ε(C-C)=0.1055 kcal/mol, σ(C-Au)=3.851 Å, ε(O-O)=0.156 kcal/mol, σ(O-O)=3.166 Å, by J.H. Walther *et al.* [55]; ε(N-N)=0.072 kcal/mol, σ(C-Au)=3.31 Å by P. Kowalczyk [56]; ε(P-P)=0.40 kcal/mol, σ(P-P)=3.33 Å by P. Ballone, R.O. Jones [57], so on. Some potentials and parameters are also shown in Appendix (Tables 1-2), where C, O, N and P stands for carbon, oxygen, nitrogen and sulphur atoms of the nucleotides and DNA and C_C denotes the carbon atom of CNT. The force field parameters for the NC (nucleotide chain) and CNT (carbon nanotube) molecules were chosen from the DL_FIELD database [51]. The intramolecular interactions for the NC chain were described using the LJ, combined with angular and dihedral bonding potentials. The LJ, cross-interaction and bond parameters for the NC, peptide-NP-CNT system were the same as in our papers [18, 58-66].

For the describing CNT we have employed the quantum chemistry Tersoff potential, so far a co-called hybrid approach (the combination of quantum chemistry potential and classical trajectory calculations) has been realized to investigate the NC interaction with CNT. In CNTs we have chemical bonding is hybridization sp^2 (as graphite), which is stronger than sp^3 bond (of diamond). The nature of chemical bonding in CNTs is described by quantum chemistry, through the process of orbital hybridization. The Tersoff potential in hybrid MD simulations correctly describes the nature of covalent bonding in carbon nanotube; it allows the breaking and formation of chemical bonds, that it is associated with hybridization process. Tersoff potential is pair wise potential, but coefficient in attractive term depends on local environment, thus, Tersoff potential possesses a many body nature (see, [18, 51-52] and references therein).

3. Results and discussion

The MD simulations results for the single nucleotides (N), nucleotide chains (NC), peptides – metallic nanoparticles (NP) – carbon nanotube (CNT) models are shown below in three parts. During the MD statistics analysis we have focus on a comparative behaviour of two main physical characteristics:

(A) The N, NC – NP distance distributions,

(B) The N, NC – NP bond energy distributions.

The calculation results below illustrate the dependence of the total potential energy, the N,NC-NP interaction distances, the N,NC angular and torsion (dihedral) bond energies along with the configuration snapshots at the temperatures T=100 (thin lines), 200 (dotted lines) and T=300 K (thick lines).

At first, in Fig. 8 the total potential energy *vs* simulation time are presented for the NC-NP-CN system. The total potential energy are presented for all models (1-3: NC-1Au, NC-2Au and NC-3Au), respectively. Fig. 8 demonstrates the change on the interaction potential energies with adding another gold atom to the system at different system temperatures.

IOP Conf. Series: Journal of Physics: Conf. Series **848** (2017) 012012 doi:10.1088/1742-6596/848/1/012012

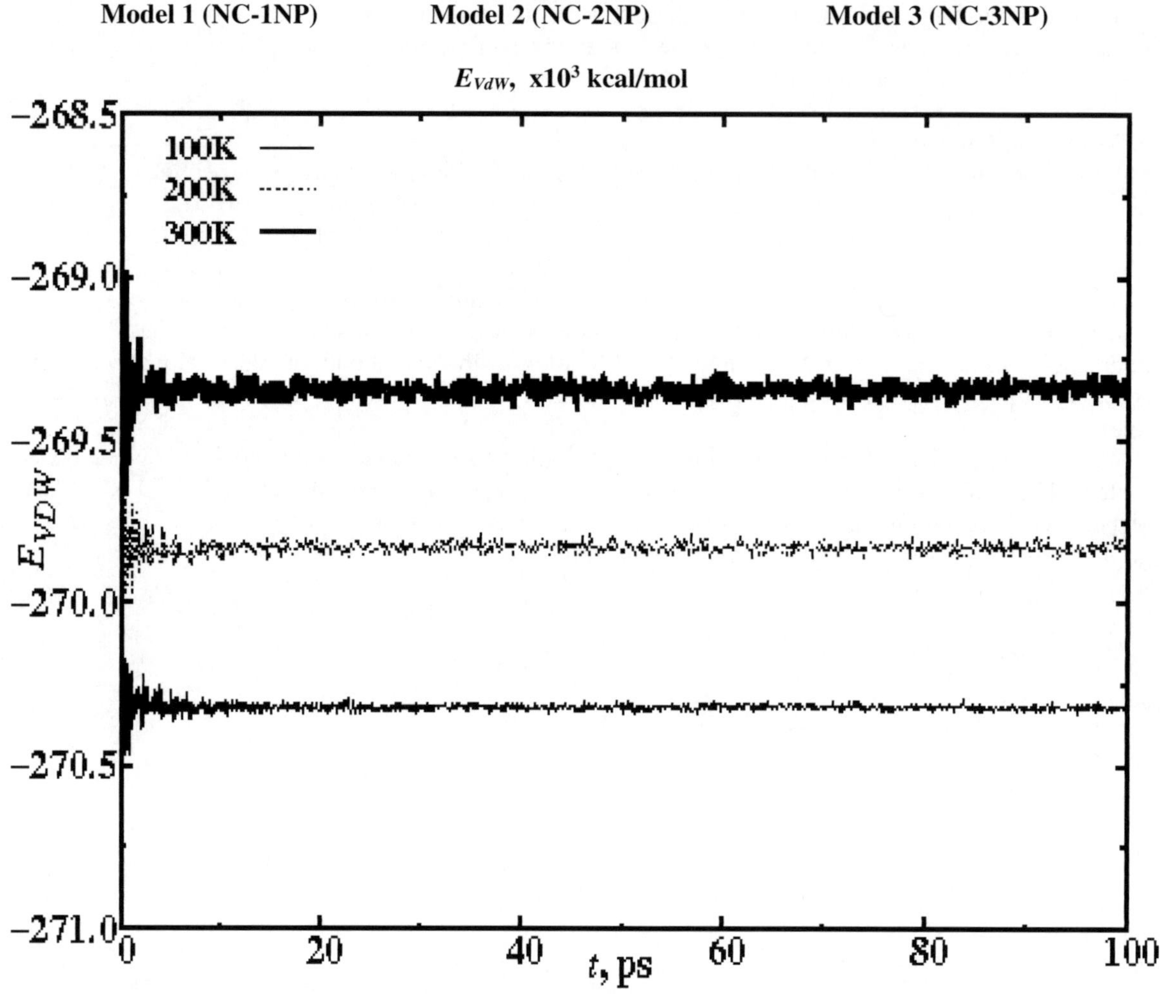

Fig. 8. The NC-NP-CNT total potential energy for the models 1-3: NC-1Au, NC-2Au and NC-3Au at the temperatures T=100, 200 and 300 K.

3.1. Single nucleotides (N) - metallic nanoparticles (NP) - carbon nanotube (CNT) system

In Figs. 9-11 the MD simulation results are shown for the single nucleotide (N) – NP interaction distances (*top*), the nucleotide' angular (*middle*) and torsion (dihedral) (*bottom*) bond energies *vs* time for all models (1: Nucleotide–1NP–CNT, 2: Nucleotide–2NP–CNT and 3: Nucleotide–3NP–CNT), Black lines indicate Purine (G – Guanine), light lines Pyrimidine (C – Citosine). The N–NP interaction and binding processes that happen within a CNT matrix estimated from the MD data thereby built the distance distributions, the angular and dihedral (torsional) bond energy graphs versus simulation time at different temperatures. The MD simulation results show that with the temperature increase (from T=100 up to 300 K) many peculiarities of the Nucleotide–NP dynamics remain, though the Nucleotides exhibit larger oscillation that effect the Nucleotide–NP close bond formation for all Nucleotide–NP–CNT models.

IOP Conf. Series: Journal of Physics: Conf. Series **848** (2017) 012012 doi:10.1088/1742-6596/848/1/012012

The N – NP distance distributions

Model 1 (Nucleotide–1NP–CNT)

In Fig. 9 (*top*) the Nucleotide–NP interaction distances *vs* time are presented for Purine (G – Guanine) (black line) and Pyrimidine (C – Citosine) (light line) (T=100 K). It should be stressed out that the Nucleotide–NP distance, as shown in Fig. 9 (*top*), represents a larger distance between the nucleotide' phosphorus (P) and gold (Au) atoms. (All other Nucleotide–NP interatomic distances lie within *d* [Nucleotide(P)–NP(Au)] values and are smaller than *d* [Nucleotide(P)–NP(Au)]). From Figs. 9 (*top*) for the model 1 we observe that Pyrimidine (C – Citosine) forms a strong bond with the gold NP, while for the Purine (G – Guanine) the disruption of the bond is clear. That is, even at low temperature (T=100 K) for the nucleic acid (Purine) the intramolecular vibrations dominate over the weak VdW non-bonding forces between the Nucleotide–NP.

Model 2 (Nucleotide–2NP–CNT)

Next, we simulated three different combinations of two gold atoms and a nucleotide. In Fig. 10 (*top*) the Nucleotide–2NP interaction distances are shown for Purine (G – Guanine) (black line) and Pyrimidine (C – Citosine) (light line), respectively. As like as previous section, for the model 2 we can observe the Pyrimidine (C – Citosine) –NP strong bond formation, while the bond disruption for the Purine (G – Guanine) to exist.

Model 3 (Nucleotide–3NP–CNT)

The three gold particle system (Nucleotide–3Au–CNT) for the next analyzing step could be associated with a small atomic cluster that possesses its own oscillation modes as a collective motion between the gold atoms. Even for such a small cluster one can assume the existing of more intense VdW interactions with the Nucleotides within of a CNT matrix. In Fig. 11 (*top*) the Nucleotide–NP interaction distances are shown for Purine (G – Guanine) (black line) and Pyrimidine (C – Citosine) (light line), respectively. The distance distribution diagrams of Fig. 11 (*top*) have shown that the Nucleotide–3Au close contact has kept for both, Purine (G – Guanine) (black line) and Pyrimidine (C – Citosine).

The N – NP bond energy distributions

In Figs. 9-11 the Nucleotide' angular (*middle*) and torsion (dihedral) (*bottom*) bond energies *vs* time are presented for the Nucleotide–1Au–CNT (model 1), Nucleotide–2Au–CNT (model 2) and Nucleotide–3Au–CNT (model 2), respectively (T=100 K). As it can be observed, the curves of the angular and torsion bond energies for the Pyrimidine (C – Citosine) (light line) lies below than that of Purine (G – Guanine) (black line) ones. That is the comparison of the curves in Figs. 10-12 indicates that Purine (G – Guanine) possesses higher intramolecular bond oscillation modes than that of the Pyrimidine (C – Citosine). That is a characteristics feature related the Nucleotide–NP bond formation or disruption. As the Nucleotide' internal vibrations dominate over weak VdW nonbonding forces with the gold NP, it makes a strong bond formation with the gold atom difficult. Comparing Figs. 11 (Nucleotide–3NP–CNT) with Figs. 9-10 (Nucleotide–1NP–CNT and Nucleotide–2NP–CNT), one can conclude that the behavior of the Nucleotide' angular and torsion bond energies at early stages of the dynamics define the systems final states. Thus, the initial interaction stages between nucleotides and gold nanoparticles could be a triggering point for the type of the bond formation within a CNT. This is another characteristics feature of the Nucleotide–NP dynamics in an environment, confined by a CNT matrix. From Figs. 9-11 on the Nucleotide' angular and torsion (dihedral) bond energy distribution one can see that for the Pyrimidine (C – Citosine) making its strong bond with the 1Au, 2Au and 3Au NP seem are always preferable, while for the Purine (G – Guanine), having more higher internal vibration modes, the only larger gold cluster can keep it on a close distant as long-lived contact. With the temperature increase (from T=100 up to 300 K) many peculiarities of the Nucleotide–NP dynamics, as described above, remain, though the Nucleotide' angular and torsion (dihedral) bond energies exhibit larger oscillation that distort the Nucleotide–NP binding for all Nucleotide–NP–CNT models.

Model 1 (Nucleotide–1NP–CNT)

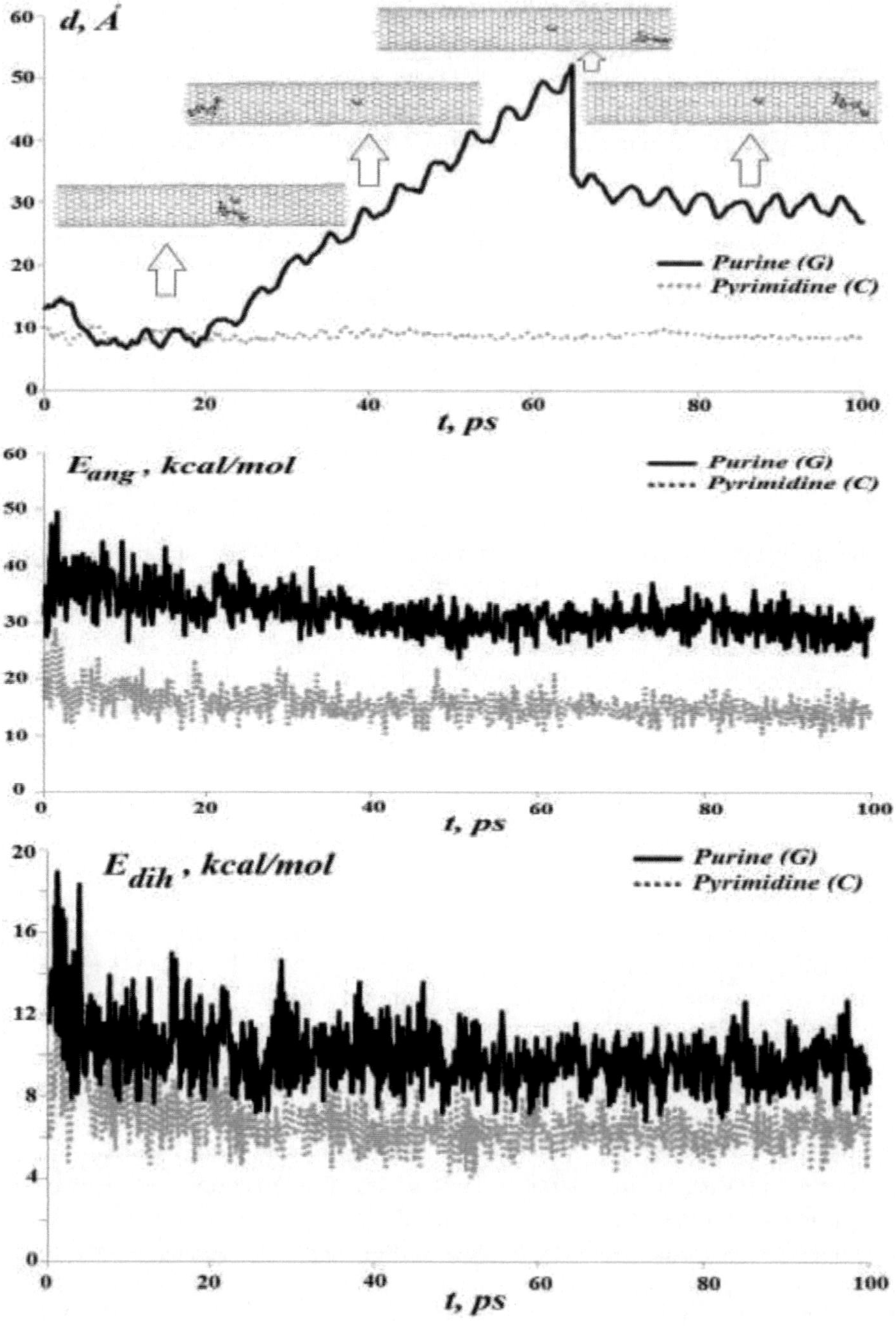

Fig. 9. The single nucleotide (N) – NP interaction distances (*top*), the nucleotide' angular (*middle*) and torsion (dihedral) (*bottom*) bond energies *vs* time for the **Model 1 (Nucleotide–1NP–CNT)**. Black line: Purine (G – Guanine), Light lines: Pyrimidine (C – Citosine) (T=100K).

IOP Conf. Series: Journal of Physics: Conf. Series **848** (2017) 012012 doi:10.1088/1742-6596/848/1/012012

Model 2 (Nucleotide–2NP–CNT)

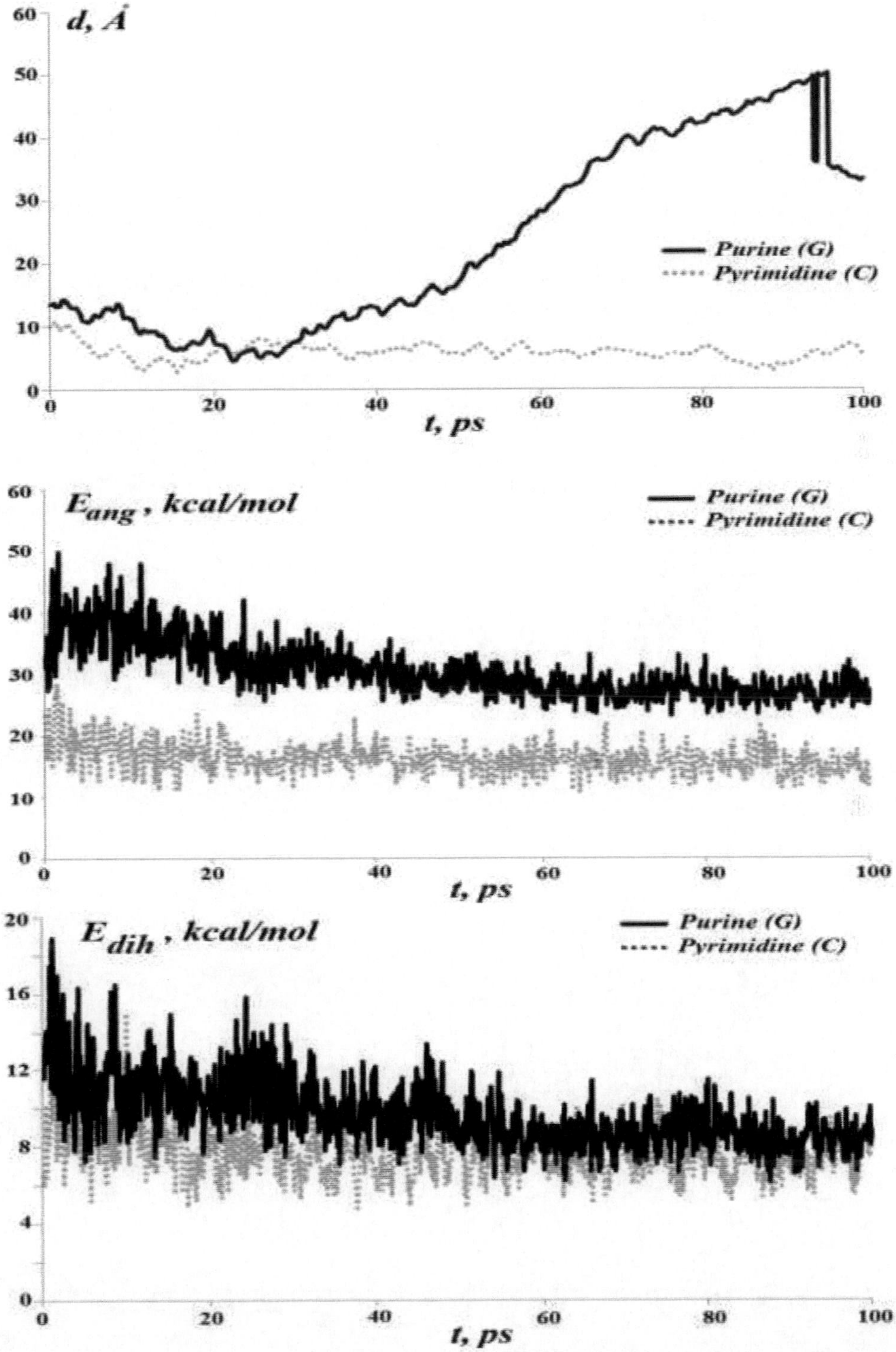

Fig. 10. The single nucleotide (N) – NP interaction distances (*top*), the nucleotide' angular (*middle*) and torsion (dihedral) (*bottom*) bond energies *vs* time for the **Model 2 (Nucleotide–2NP–CNT)**. Black line: Purine (G – Guanine), Light lines: Pyrimidine (C – Citosine) (T=100K).

Model 3 (Nucleotide–3NP–CNT)

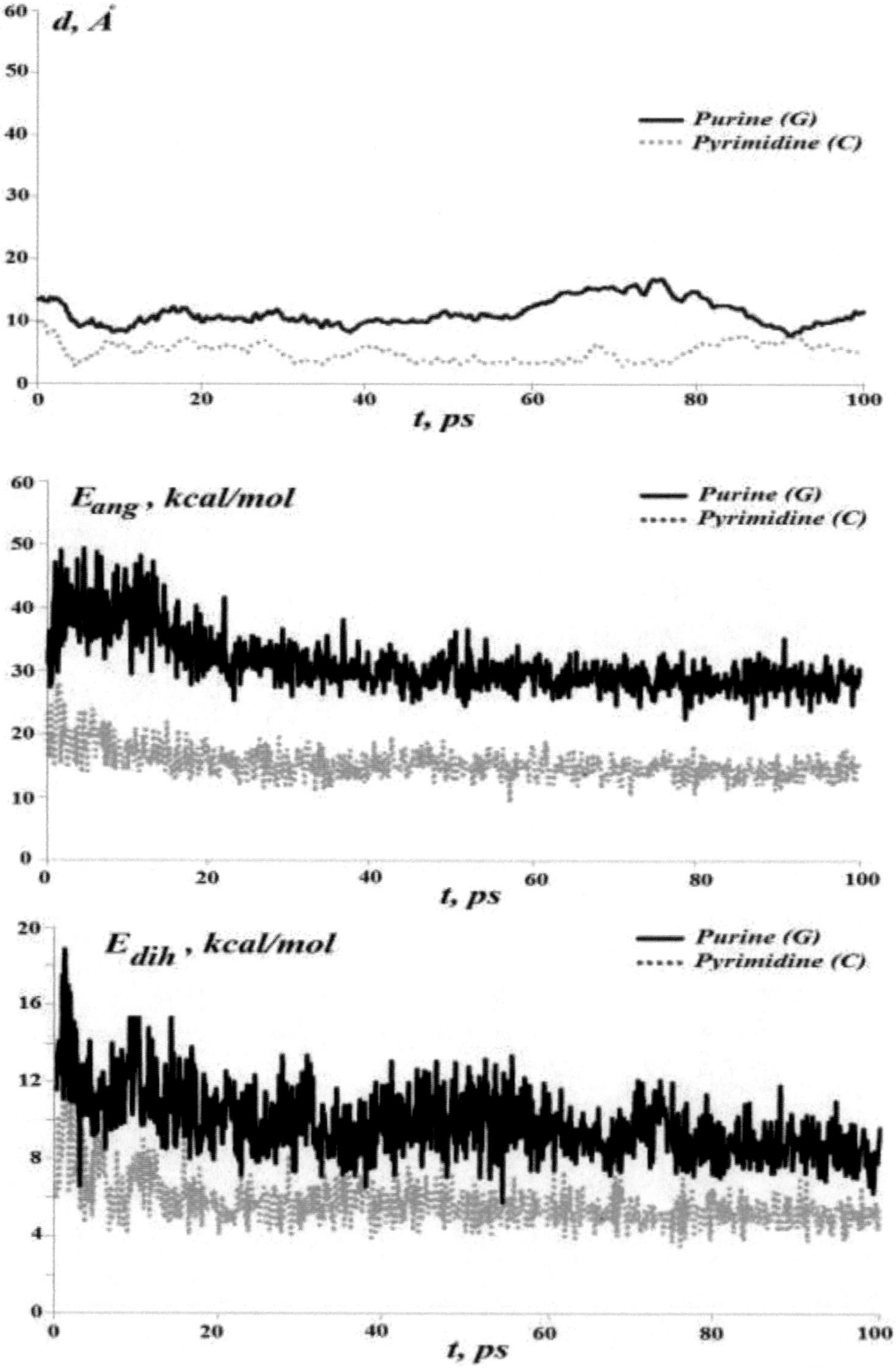

Fig. 11. The single nucleotide (N) – NP interaction distances (*top*), the nucleotide' angular (*middle*) and torsion (dihedral) (*bottom*) bond energies *vs* time for the **Model 3 (Nucleotide–3NP–CNT)**. Black line: Purine (G – Guanine), Light lines: Pyrimidine (C – Citosine) (T=100K).

3.2. Nucleotide chains (NC) - metallic nanoparticles (NP) - carbon nanotube (CNT) system

The NC – NP distance and bond energy distributions

Model 1 (NC-1NP-CNT)

In Figs. 12 for the model 1 the NC-NP interaction distances, the NC angular and torsion (dihedral) bond energies vs time are presented for the temperatures T=100, 200 and 300 K. It should be stressed out that the NC-NP distance, as shown in Fig. 12 (*top*), represents a larger distance between the NC phosphorus (P) and gold (Au) atoms. (All other NC-NP interatomic distances lie within r[NC(P)-NP(Au)] values and are smaller than r[NC(P)-NP(Au)]). From Fig. 12 (*top*) for the model 1 we observe at high temperature T=300 K the disruption of the NC-NP bonds. That is, with the temperature growth the NC intramolecular vibrations begun to dominate over the weak NC-NP non-bonding VdW forces.

In Figs. 12 (*middle* and *bottom*) the NC angular and torsion (dihedral) bond energies vs time are presented for the temperatures T=100, 200 and 300 K. At low temperatures (T=100 and 200 K) the curves of the angular and torsion energies lie significantly below than that at T=300 K ones. One can also observe that, for NC-1Au the angular and torsion energy curves correlate with the distance distributions shown above. That is for the model 1 we observe a very close contact between the NC-1Au only at low temperatures; formation of a preferably strong NC-1NP bond is possible at T<200 K, which disrupts with the temperature increasing. The behavior of the angular and dihedral bond energies, Figs. 12 (*middle* and *bottom*), with the distance distribution characteristics as presented in Fig. 12 (*top*), typically show a weak nature of VdW interactions owe for one metallic particle system.

Models 2 (NC-2NP-CNT)

As described above, the NC interaction with gold NP inside a CNT matrix can produce preferably the weak NC-1NP bonds. Next, we simulated the two gold atom and a nucleotide chain as NC-2Au. In Fig. 13 (*top*) the NC-2NP-CNT interaction distances are shown for model 2, NC-2Au at the temperatures T=100, 200 and 300 K. From Fig. 13 (*top*) one can see that for NC-2NP-CNT system at low temperatures the formation of a strong bonds between the NC-2Au are mostly possible.

Figs. 13 (*middle* and *bottom*) demonstrate the NC angular and torsion (dihedral) bond energies *vs* time for the NC-2NP-CNT system. Now for the two gold atomic systems the weak VdW forces make an additional contribution to the NC angular and torsion bond energy profiles, thereby influencing the final NC-2NP binding states. Comparing Figs. 13 (NC-2NP model) and Figs. 12 (NC-1NP model), one can conclude that the behavior of the NC angular and torsion bond energies at early stages of the dynamics define the systems final states. Thus, the initial interaction stages between nucleotides and gold nanoparticles could be a triggering point for the type of the bond formation within a CNT. This is a characteristics feature of the NC-NP dynamics in an environment, confined by a CNT matrix.

Model 3 (NC-3NP-CNT)

The three gold particle system, for the next analyzing step, could be associated with a small atomic cluster that possesses the own oscillation modes as a collective motion. Even for such a small cluster one assumes the existing of more intense VdW interactions with the NC and CNT matrix. In Fig. 14 (*top*) the interaction distances, the NC angular (*middle*) and dihedral (*bottom*) bond energies are presented for the NC-3Au system at the temperatures T=100, 200 and 300 K, respectively. The distance distribution diagrams in Fig. 14 (*top*) show that the NC-3Au close contact has kept even at high temperatures. The comparison of the angular and torsion (dihedral) bond energy curves with the NC-1NP-CNT and NC-2NP-CNT models are straightforward.

IOP Conf. Series: Journal of Physics: Conf. Series **848** (2017) 012012 doi:10.1088/1742-6596/848/1/012012

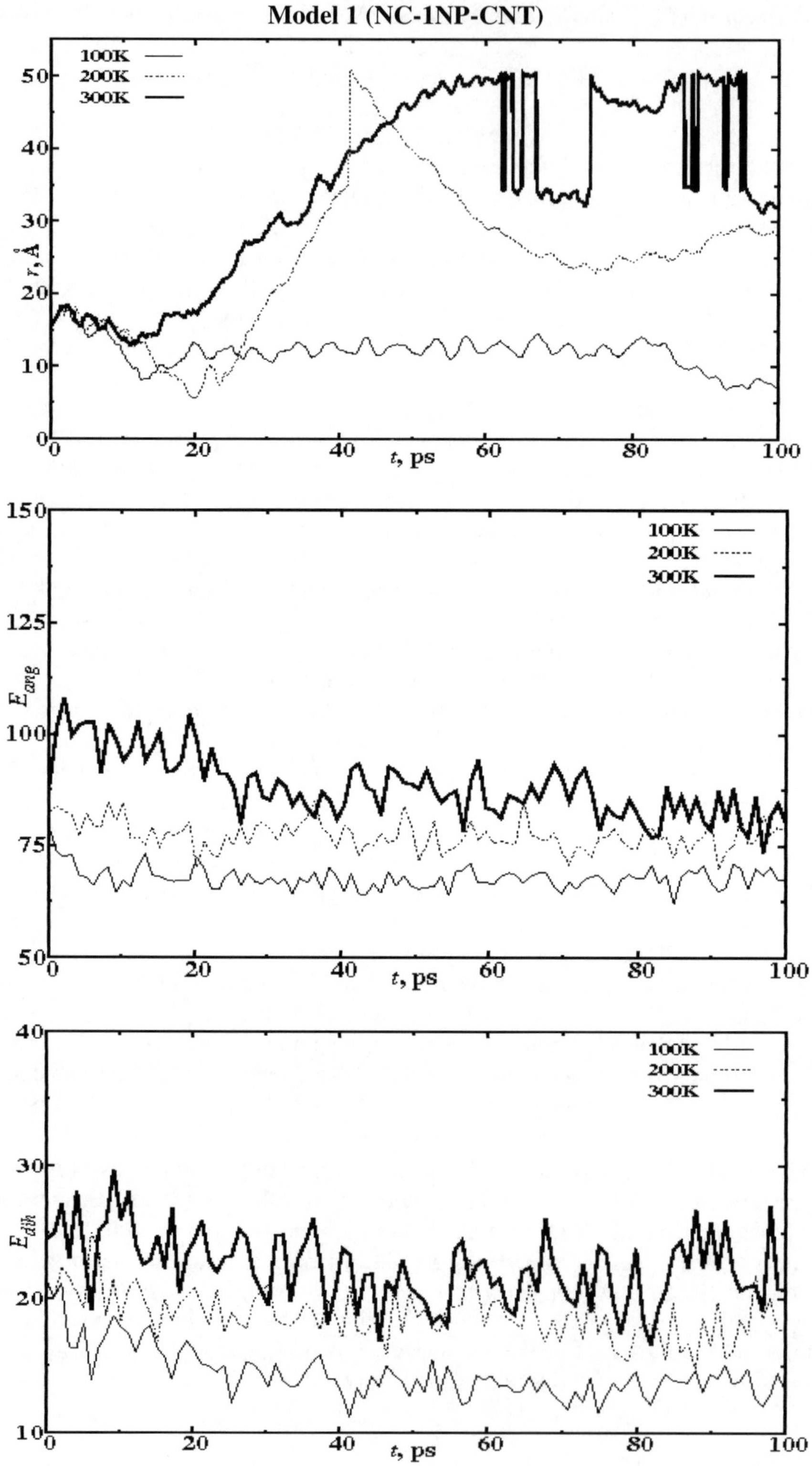

Fig. 12. The NC-1NP-CNT interaction distances (*top*), the NC angular (*middle*) and dihedral (*bottom*) bond energies vs time at the temperatures T=100, 200 and 300 K.

Model 2 (NC-2NP-CNT)

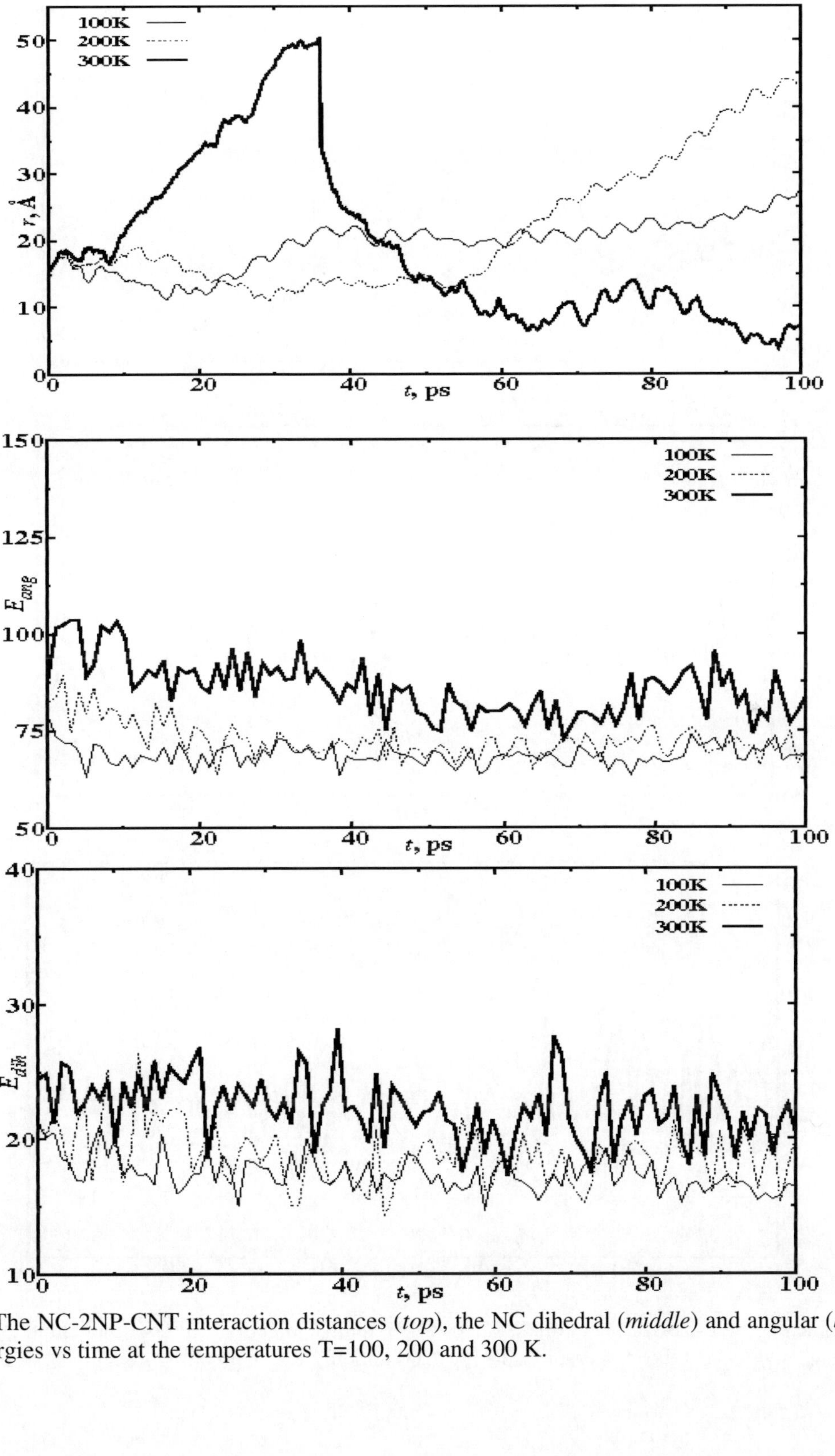

Fig. 13. The NC-2NP-CNT interaction distances (*top*), the NC dihedral (*middle*) and angular (*bottom*) bond energies vs time at the temperatures T=100, 200 and 300 K.

Model 3 (NC-3NP-CNT)

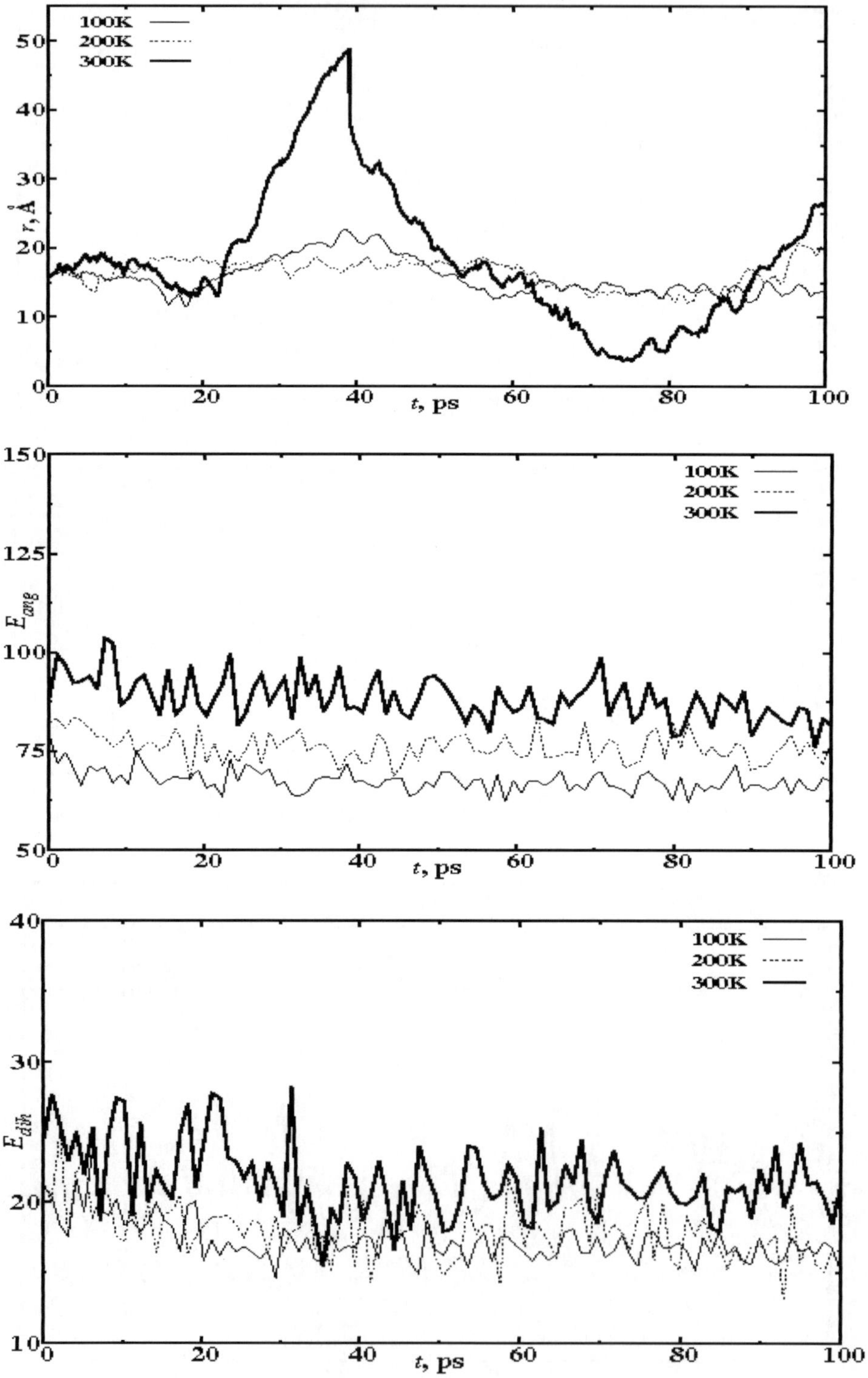

Fig. 14. The NC-3NP-CNT interaction distances (*top*), the NC dihedral (*middle*) and angular (*bottom*) bond energies vs time at the temperatures T=100, 200 and 300 K.

The configuration snapshots in Figs. 15 demonstrate the NC-1Au (*left*) and NC-3Au (*right*) positional changes at the initial (t=0), intermediate (t=50 ps) and final (t=100 ps) simulation states. Obviously, for the three gold particle system the NC-3NP strong bond formation inside a CNT matrix is a comparably most probable event.

Model 1 (NC-1NP-CNT) **Model 3 (NC-3NP-CNT)**

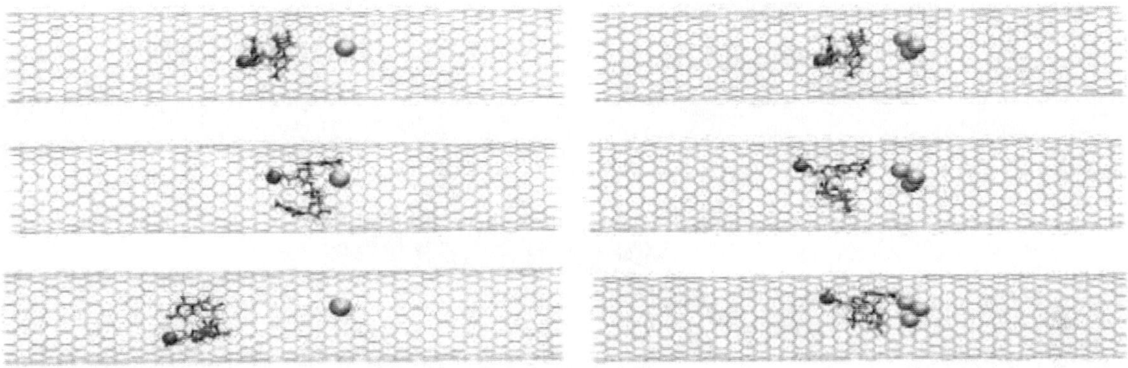

Fig. 15. Snapshots of the interaction process for the models 1 (NC-1Au; *left*) and 3 (NC-3Au; *right*) inside a CNT matrix at the t=0 (top), t=50 ps (middle) and t=100 ps (bottom) states (T=100 K).

3.3. Peptide chains (PC) - metallic nanoparticles (NP) - carbon nanotube (CNT) system

Below the results of MD simulations are presented for the extended molecular structures (single-filament peptide chains) and the mechanisms of their encapsulation and folding within the CNT. The peptide molecule (a chain with many bound nucleotides) in CNT environment modeled with different sets and positional variations gold NP. Initially, the system was investigated peptide-CNT without the presence of any metal particles (Fig. 16).

The study of the molecular mechanisms of the interaction and folding of peptide chains, and a variety of biomolecular compounds (fragments of chemical molecules, drugs) with metal nanoparticles (NPs) in a matrix of carbon nanotubes (CNT) and the spherical bodies (fullerenes) in today's nano-bio-technology and medical applications is a task of great importance [1-40]. Undoubtedly, the process of folding (encapsulation, folding) biomolecules inside CNTs can rank as the most exciting scientific spectacle (see Figs. 16-18 below). Especially when in the process of folding of biomolecules (peptides, proteins), taking place in a CNT matrix, make their own adjustments with metallic NP (Figs. 17-18), where gold nanoparticles contribute to the full encapsulation of the peptide chain within the CNT. This phenomenon is similar to the process of folding (dense packing) viruses within the foreign cells (the capsid) [1-40].

The results of Fig. 16 show the process of spontaneous packing peptide molecule from a linear to a globular-like structures as a result of intramolecular vibrations, studied in the previous sections for the individual nucleotides and small nucleotide chains.

However, the inclusion of a small cluster of gold NP for consideration in the peptide-CNT system as seen from Figs. 17-18, also did not significantly alter the final result of their interaction (compare Figs. 17-18 with Fig. 16 above).

So far, the results of MD simulations and graphical visualization demonstrate various scenarios of the effect of metal nanoparticles in the final conformation (twisting or folding) of the peptides within the CNT matrix. In some cases, the effect of a weak VdW forces acting between the peptide chain and metallic NP leads to the full package (encapsulation) of the peptide chain within the

IOP Conf. Series: Journal of Physics: Conf. Series **848** (2017) 012012 doi:10.1088/1742-6596/848/1/012012

CNT. In other cases, the metal nanoparticles (in this case gold NP) may hinder the process of encapsulation of the peptide inside the CNT matrix.

(a)

(b)

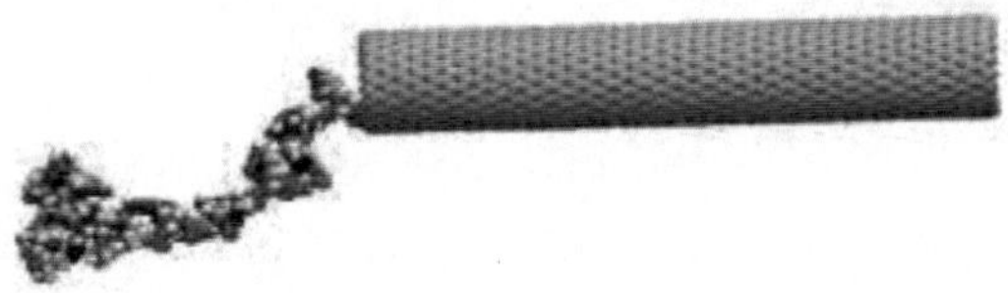

(c)

(d)

Fig. 16 (a-d, from top to bottom). The sequential configurations (snapshots), illustrating the folding process (encapsulation) peptide molecule (a chain of several nucleotides) inside carbon nanotubes.

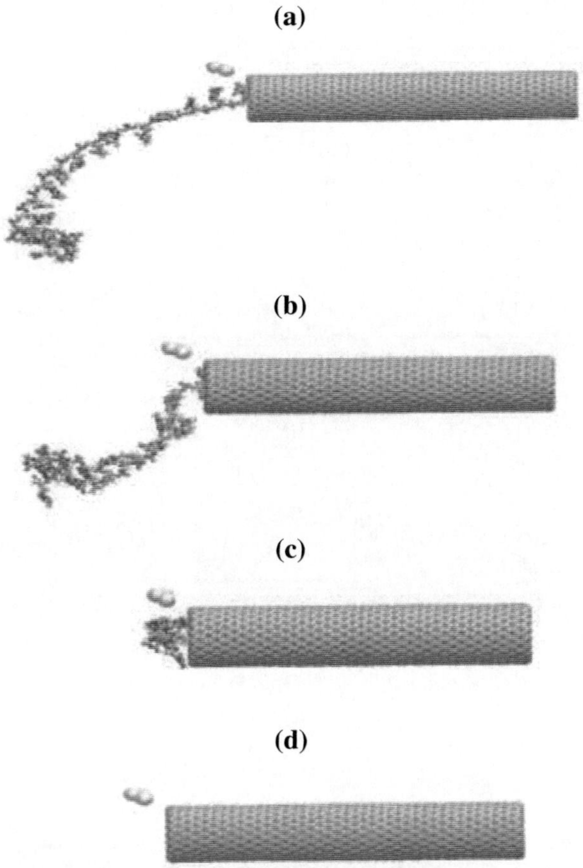

Fig. 17 (a-d, from top to bottom). The sequential configurations (snapshots), illustrating the folding process (encapsulation) peptide molecule (a chain of several nucleotides) inside the carbon nanotube in interactions with nanochastitsvmi gold ((a-d), from top to bottom). Here gold nanoparticles contribute to the full encapsulation of the peptide chain within the CNT.

IOP Conf. Series: Journal of Physics: Conf. Series **848** (2017) 012012 doi:10.1088/1742-6596/848/1/012012

a)

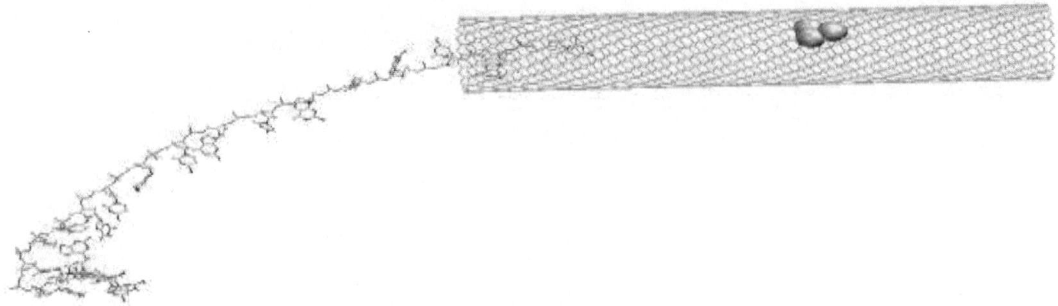

(b)

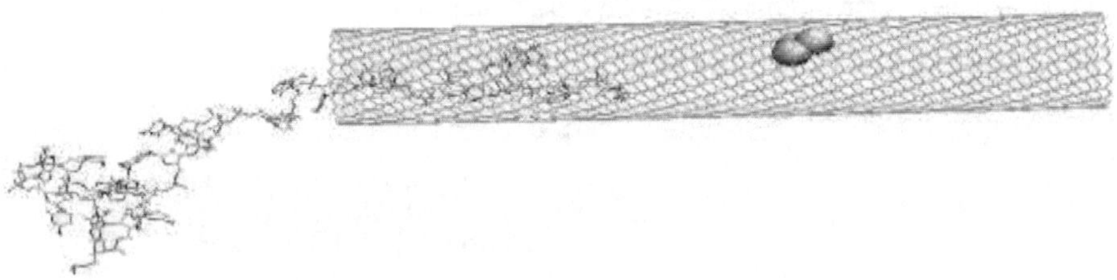

(c)

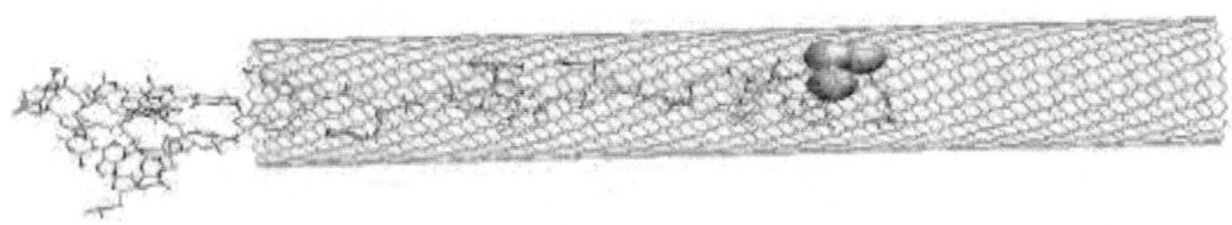

(d)

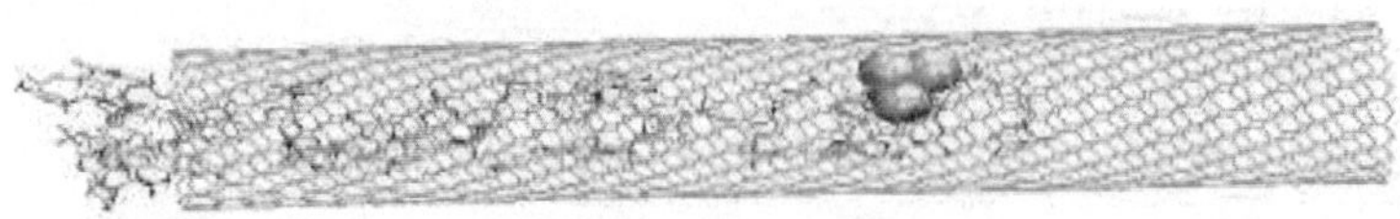

Fig. 18 (a-d, from top to bottom). The sequential configurations (snapshots), illustrating the folding process (encapsulation) peptide molecule (a chain of several nucleotides) inside the carbon nanotube in interactions with nanochastitsvmi gold ((a-d), from top to bottom). Here gold nanoparticles contribute to the full encapsulation of the peptide chain within the CNT.

4. Conclusion

The multiple molecular dynamics (MD) simulations have been performed for the single nucleotides (N), nucleotide (NC) and peptide chains (PC), interacting with gold nanoparticles (NP) inside of a carbon nanotube (CNT) with periodic boundaries. We have been aimed to investigate the peculiarities of Van der Waals (VdW) interactions, a nature of the N,NC,peptide-NP bonding in a confined environment by the CNT matrix. The small nucleotide chain is an important stage in the understanding of interaction mechanism of a whole DNA or RNA molecule with the NPs and CNT. It is well known that a primary DNA or RNA structure consists of a linear sequence of nucleotides that are linked together by phosphodiester bonds. With regard to application aspects of the DNA-NP-CNT system one has to mention the development of the DNA-CNT mobile electronic devices for the purposes of diagnostic applications, for the chemical or drug delivery inside the living cells and related nanorobotic design. The NC intermolecular motions were estimated from MD data thereby building the distance distributions, the angular and dihedral (torsional) bond energy graphs versus simulation time at different temperatures from T=100 K up to 300 K. The MD simulation results have shown that depending on the relative NC-NP position a different scenario of bonding between the N,NC-NP within a CNT matrix is possible. We have observed the possibilities of formation of weak, strong and intermediate bonds between the N,NP-NC, which are overestimated by a presence of CNT matrix as a confining environment. The NC chain can form with a particular gold atom a close contact, while with another under the same positional and temperature conditions the weak resultant bonding formation might be possible. At high temperatures, as the NC intramolecular oscillations have grown, the picture of the NC-NPs bonding inside a CNT matrix has essentially modifies. We observe more fluctuations in the NP-NC bonding processes, not only for a single gold atomic case (model 1, N,NC-1NP-CNT) but also for the two (model 2, N,NC-2NP-CNT) and three (model 3, N,NC-3NP-CNT) gold particle ones. Thus, we observe a concurrent effect between the NC intramolecular vibrations and a weak VdW interaction between the N,NC and gold NP, thereby defining their final binding in CNT matrix. In conclusion, the peculiarities of the N,NC-NP-CNT structural and dynamical behaviour, investigated in the present study, could be important stage for the understanding of binding mechanism of a full DNA molecule with NP and CNT.

The behavior of the peptide chain or the DNA inside the nanotube is an important link in understanding the folding mechanism (encapsulation) of biological macromolecules in a living cell. In all cases, the competition show a concurrent effect of the intra-molecular vibrations with weak VdW forces. The effect of the cross-correlation of the two main types of atomic and molecular interactions determine the formation or destruction of relationships gold nanoparticles with the peptide or nucleotide chains. It is well known, for example, a common feature - the encapsulation of virus process in the restricted area (capsids, nano-objects and micro-sized) in guest cells, leading eventually to various diseases (as the parent cell will not recognize a foreign gene and allow its penetration, incubation and further reproduction). In this respect, a variety of mechanisms to encapsulate a single-chain-peptide chain, freely fit within the CNT indicates the implementation and design of the new micro-nano-devices for transportation of biomaterials and drug delivery inside the living cell and the blood of those or other diagnostic purposes of analysis.

IOP Conf. Series: Journal of Physics: Conf. Series **848** (2017) 012012 doi:10.1088/1742-6596/848/1/012012

Appendix

Table 1. The 12-6 potential parameters of the NC chain.

Pair	A,Å^{12} kcal/mol	B,Å^6 kcal/mol	Pair	A,Å^{12} kcal/mol	B,Å^6 kcal/mol
C-C	1171340	667.5	P-P	8350780	3369.4
C-P	3145970	1481.6	P-O	1476420	1016.6
C-O	535958	452.2	P-N	1985180	1117.9
C-N	730530	500.7	N-N	450301	373.4
O-O	232116	198.1	N-O	325886	334.9

Table 2. The intermolecular potential parameters of the NC chain.

Harmonic bonds: $U(r) = k(r - r_0)^2/2$

Bonds	k	r_0
O – H	300	1.724

Valence angles: $U(\theta) = k(\theta - \theta_0)^2/2$

Bonds	k	θ_0	Bonds	k	θ_0
O-P-O	100.33	93.300	C-N^*-C	109.00	106.700
P-O-C	106.70	104.510	C-N-H	109.00	106.700
C-O-C	106.70	104.510	O-C-H	112.50	109.471
N-C-N	133.33	120.000	H-C-H	112.50	109.471
N-C-C	133.33	120.000	O-C-C	112.50	109.471
N-C-H	133.33	120.000	H-C-C	112.50	109.471
C-N-C	133.33	120.000	C-C-C	112.50	109.471

Dihedral angles: $U(\varphi) = A[1+\cos(m\varphi - \delta)]$

Bonds	A	δ	m
P-O-C-H	0.3333	0	3
P-O-C-C	0.3333	0	3
C-O-C-H	0.3333	0	3
C-O-C-N	0.3333	0	3
C-O-C-C	0.3333	0	3
C-O-P-O	0.5000	0	3
C-N-C-C	2.5000	180	2
N-C-C-O	2.5000	180	2
N-C-C-N	2.5000	180	2
C-C-C-O	2.5000	180	2
C-C-C-N	2.5000	180	2
C-N-C-N	2.5000	180	2
H-C-N-C	11.250	180	2
N-C-N-C	11.250	180	2
N-C-C-H	5.6250	180	2
N-C-C-C	5.6250	180	2
H-C-C-H	5.6250	180	2
H-C-C-C	5.6250	180	2

IOP Conf. Series: Journal of Physics: Conf. Series **848** (2017) 012012 doi:10.1088/1742-6596/848/1/012012

References

[1] Majumder S *et al* 2013 *Appl. Phys. Lett.* **103** 063103
[2] Boudaïffa B, Cloutier P, Hunting D, Huels M A 2000 *Science*, **287** 1658
[3] Ray S G, Daube S S and Naaman R 2005 *PNAS* **102** 15
[4] Rosenberg R A *et al* 2014 *Phys. Chem. Chem. Phys.* **16** 15319
[5] Boland T and Ratner B 1994 *Langmuir* **10** 3845
[6] Crain J N *et al* 2001 *J. App. Phys.* **90** 3291
[7] Dunford R *et al* 1997 *FEBS Letters* **418** 87
[8] Monti S, Carravetta V, Zhang W and Yang J 2007 *J. Phys. Chem. C* **111** 7765
[9] SantaLucia J (Jr) 1998 *Biochemistry* **95** 1460
[10] Tulpan D, Andronescu M and Leger S 2010 *BMC Bioinformatics* **11** 105.
[11] Breslauert K J, Franks R, Blockers H and Markyt L A 1986 *Biochemistry* **83** 3746
[12] Freyre-Fonseca V *et al* 2011 *Toxicology Letters* **202** 111
[13] Trouiller B *et al* 2009 *Cancer Res.* **69** 22
[14] Saquib Q *et al* 2012 *Toxicology in Vitro* **26** 351
[15] Dunford R *et al* 1997 *FEBS Letters* **418** 87
[16] Hilder T A and Hill J M 2008 *Curr. App. Phys.* **8** 258
[17] Kholmurodov Kh 2011 *Molecular Dynamics of Nanobiostructures* (Nova Science Publishers Ltd ISBN 978-1-61324-320-6)
[18] Khusenov M, Dushanov E and Kholmurodov K 2014 *J. Mod. Phys.* **5** 137
[19] Birchall J *et al* 2006 *Int. J. Pharm.* **312** 15
[20] Chugunov A O 2008 *Cosmetics and medicine* **2** 72
[21] Lawson L B, Freytag L C and Clements J D 2007 *Clin. Pharmacol. Ther.* **82** 641
[22] Vogt A *et al* 2006 *J. Invest. Dermatol.* **126** 1316
[23] Lademann J *et al* 2007 *Eur. J. Pharm. Biopharm.* **66** 159
[24] Bramson J *et al* 2003 *Gene Ther.* **10** 251
[25] Cans A S *et al* 2003 *Proc. Natl. Acad. Sci. U.S.A.* **100** 400
[26] Djalali R *et al* 2004 *J. Am. Chem. Soc.* **126** 7935
[27] Niemeyer C M 2001 *J. Biotechnol.* **82** 47
[28] Moghimi S M and Szebeni J 2003 *Prog. Lipid. Res.* **42** 463
[29] Greggy M 2016 *Opt. Mat. Exp.* **6** 1217
[30] Yang H and Kao W J 2006 *J. Biomater. Sci. Polymer* **17** 3
[31] Gupta U, Agashe H B, Asthana A, Jain N K 2006 *Nanomedicine* **2** 66
[32] Shi Kam N W, Jessop T C, Wender P A, Dai H 2004 *J. Am. Chem. Soc.* **126** 6850
[33] Choi J H *et al* 2007 *Nano Lett.* **7** 861
[34] Trzaskowski B, Jalbout A F and Adamowicz L 2006 *Chem. Phys. Lett.* **430** 97
[35] Cui D *et al* 2004 *Mech. Chem. Biosyst.* 1 113
[36] Hughes G A 2005 *Nanomedicine* **1** 22
[37] Baroli B *et al* 2007 J. Invest. Dermatol. **127** 1701
[38] Nikolaev I V *et al* 2012 *Nanotubes, and Carbon Nanostructures* **20** 345
[39] Lebedev V T *et al* 2014 *Solid State Physics* **56** 176
[40] Lebedev V T *et al* 1993 *J. Magn. Magn. Mater.* **122** 83
[41] Sears V F 1992 *Neutron News* **3** 26
[42] Avdeev M V, Aksenov V L, Tropin T V 2010 *Journal of Chemical Physics (Rus)* **84** 1405
[43] Kyzyma O A *et al* 2010 *Chem. Phys. Lett.* **493** 103
[44] Kyrey T O *et al* 2012 *Fullerenes, Nanotubes, and Carbon Nanostructures* **20** 341
[45] Kyzyma O A *et al* 2013 *Chem. Phys. Lett.* **556** 178
[46] Kulvelis V A *et al* 2010 *Solid State Physics* **52** 974
[47] Svergun D J 1992 *J. Appl. Cryst.* **25** 495
[48] Svergun D J and Feigin L A 1986 *X-Ray and neutron small angle scatterings* (Nauka, Moscow)
[49] Shliomis M I 1974 *Uspekhi Fizicheskikh Nauk* **112** 427

IOP Conf. Series: Journal of Physics: Conf. Series **848** (2017) 012012 doi:10.1088/1742-6596/848/1/012012

[50] Eremin R A *et al* 2012 *Int. J. Chem.* **4** 14
[51] Forester T R and Smith W 1996 *J. Mol. Graph.* **14** 136
[52] Yong C W 2010 *DL_FIELD - a force field and model development tool for DL_POLY* (CSED Science and Technology Facilities Council, STFC Daresbury Laboratory)
[53] Qing P *et al* 2007 *Nanotechnology* **18** 424007
[54] Song Hai Yang and Zha Xin Wei 2015 *Molecular dynamics study of mechanical properties of carbon nanotube-embedded gold composities* (manuscript)
[55] Walther J H, Jaffe R, Halicioglu T and P.Koumoutsakos P 2000 *Proc. Summer School* (NASA Ames/Stanford Univ)
[56] Kowalczyk P 2012 *Electronic Supplementary Material (ESI) for Physical Chemistry and Chemical Physics*
[57] Ballone P and Jones R O 2014 *J. Chem. Phys.* **121** 8147
[58] Khusenov M A, Dushanov E B, Kholmurodov Kh T 2015 *Materials Transactions* **56** 1390
[59] Khusenov M A, Dushanov E B, Kholmurodov Kh T 2015 *British Journal of Applied Science&Technology* **8** 313
[60] Eremin R A, Kholmurodov Kh T, Petrenko V I, Avdeev M V 2013 *Russian Journal of Physical Chemistry A.* **87** 745
[61] Eremin R A, Kholmurodov Kh T, Petrenko V I, Rosta L, Avdeev M V 2013 *Journal of Applied Crystallography* **46** 372
[62] Eremin R A, Kholmurodov Kh T, Petrenko V I, Avdeev M V 2013 *Journal of Surface Investigation. X-ray, Synchrotron and Neutron Techniques* **7** 1128
[63] Eremin R A, Kholmurodov Kh T, Petrenko V I, Rosta L, Avdeev M V 2014 *Physics of the Solid State* **56** 81
[64] Eremin R A *et al* 2013 *Chapter 3: In Models in bioscience and materials research: molecular dynamics and related techniques* (Nova Science Publishers Ltd New York ISBN: 978-1-62808-052-0)
[65] Kholmurodov Kh T 2009 *Molecular Simulation in Material and Biological Research* (Nova Science Publishers Ltd New York ISBN 978-1-60741-553-4)
[66] Kholmurodov Kh T 2015 *Computational Materials and Biological Sciences* (Nova Science Publishers Ltd New York ISBN 978-1-63482-541-2)

Nanocomposite composed of charged wormlike micelles and magnetic particles

V S Molchanov[1], V A Pletneva[1], A I Kuklin[2], O E Philippova[1]

[1]Lomonosov Moscow State University, 119991 Moscow, Russia
[2]Joint Institute for Nuclear Research, 141980 Dubna, Russia

E-mail: molchan@polly.phys.msu.ru

Abstract. Wormlike micelles solutions of cationic surfactant were examined as surfactant concentration in the presence KCl salt. Dilute and semidilute regimes were determined. Concentration power law dependencies of the zero-shear viscosity and plateau modulus indicate on charged wormlike micelles network. SANS data prove long flexible wormlike micelles formation. Nanocomposite composed of the chains and magnetite particles was prepared. It was shown that particles addition leads to increase rheological parameters of the system which is in good agreement with suggestion that micellar chains can link to particles surface covered by surfactant to decrease number of end-caps.

1. Introduction

Surfactant molecules in aqueous solutions can aggregate into enormously long flexible cylindrical micelles so-called wormlike micelles or micellar chains [1]. Three dimensional network of these entangled polymerlike chains impact viscoelastic properties to solutions. Micellar network of entangled wormlike micelles are attract attention of many authors due to linear rheological properties, including zero-shear viscosity plateau at steady-shear measurements and elastic plateau at oscillation measurements. These properties are highly sensitive to many factors, for example, concentration of salt, presence of hydrocarbons or temperature due to shape and length changes as result of weak noncovalent binding the molecules in the micelles.

On the other hand surfactants are extensively used and investigated as stabilizers of colloidal particles that require usually low concentrations close to ccm. But the use of relatively high surfactant concentrations provided wormlike micelles can induce a loss of the stability of the dispersion. It was found [2] that negatively charged polystyrene particles suspended in zwitterionic/anionic wormlike micelles solution at significantly low colloid and micelle concentrations at the presence and at the absence salt, but the use of high surfactant concentrations can induce a loss of the stability of the dispersion. The depletion mechanism is believed to be responsible for this instability. Asakura and Oosawa [3] shown that the depletion force in the presence of polymer chains can be much greater than in solutions of spherical objects: the configurational entropy of the chains is reduced next to the particle surfaces. Therefore, the presence of wormlike micelles in solution at high surfactant concentrations should augment the depletion potential. Also the instability was observed at both negative and positive charge of the colloidal latex particles in mixture with both negative and positive surfactant at high concentrations [4].

Although described studied there are several works the colloidal particles improving rheological properties of viscoelastic wormlike micelles of surfactant at high concentrations [5-7]. In the work [5] the common system of wormlike micelles of cationic surfactant and negatively charged silica particles was investigated. The dependencies of viscosity, elastic modulus, relaxation time went through maximum that was explained in terms of electrostatic interactions similarly salt effect and adsorption of sufficient part of surfactant molecules on the particles surface at high particles content. The improvement of viscoelastic properties reduces with increasing surfactant concentration so that only fall of rheological parameters was indicated. In the other works [6] similarly charged silica particles and cationic wormlike micelles system was studied and electrostatic effect of particles was excluded. It was demonstrates that the power law exponents of the viscosity as function of surfactant concentration do not change upon addition of nanoparticles at three salt concentrations. These power law exponents are highly sensitive to the effects of electrostatic screening therefore the presence of nanoparticles does not significantly alter the electrostatic interactions between micelles. In the system homogeneity region was limited by high surfactant concentration and expanded with salt concentration.

Thus, described effects in the suspensions of colloidal particles and wormlike micelles contrary to each other and phenomena remains incompletely understood. Nevertheless, one can conclude that colloidal particles actively interact with micelles and they may play role of modifier and improve viscoelastic properties of wormlike micelles solutions. At the presence paper novel nanocomposite material composed of particles of magnetite and oppositely charged wormlike micelles of surfactant are prepared and studied. The submicron magnetite colloidal particles embedded into entangled wormlike micelles of cationic surfactant solution. The erucyl bis-(hydroxyethyl)methylammonium chloride (EHAC) was selected as cationic surfactant with long unsaturated hydrocarbon tail providing microns wormlike micelles [8] and formation of layer on negatively charged magnetite particles. Potassium chloride salt was added to induce growth of cylindrical micelles at relatively small surfactant concentrations as result of screening electrostatic repulsions on the surface of micelles.

2. Materials and methods

Aqueous solution of a cationic surfactant erucyl bis-(hydroxyethyl)methylammonium chloride (EHAC) with a long mono-unsaturated C22 tail initially containing 25 wt.% 2-propanol was provided by Akzo Nobel (Netherlands). For purification from 2-propanol crude EHAC was diluted by deionized water (1:10) and freeze-dried. The absence of 2-propanol in the EHAC thus obtained was proved by ^{1}H NMR. Magnetite particles from Aldrich was used without further purification. Average size of the particles is 250 nm estimated from TEM.

Steady and dynamic shear rheological experiments were carried out using a stress-controlled rheometer Physica MCR 301 Anton Paar (Austria) with cone-plate geometry (diameter 50 mm, cone angle 2^0) and a solvent trap.

SANS measurements were carried out with two detector system on the time-of-flight YuMO high-flux pulsed reactor IBR-2M at the Frank laboratory of neutron physics, Joint Institute for Nuclear Research, Dubna, Russia. The experiments were performed at scattering vectors Q ranging from 0.006 to 0.14 $Å^{-1}$. All data were treated according to standard procedures of small-angle isotopic scattering. The samples were prepared using heavy water D_2O to get higher contrast.

3. Results and Discussion

Rheological properties of EHAC solutions were studied both in steady-state and in dynamic regimes at different concentrations at the presence 1 wt.% KCl. Micellar chains formation can be assigned to the lengthening of cylindrical micelles due to the screening of electrostatic repulsion between similarly charged groups of surfactant molecules on the surface of micelles by the ions present in a solution [9]. Figure 1 shows typical results of steady-state measurements. One can see a Newtonian plateau at low shear rate called zero-shear viscosity η_0 followed by a pronounced shear thinning, which can be due to the alignment of micelles along the direction of the flow.

IOP Conf. Series: Journal of Physics: Conf. Series **848** (2017) 012013 doi:10.1088/1742-6596/848/1/012013

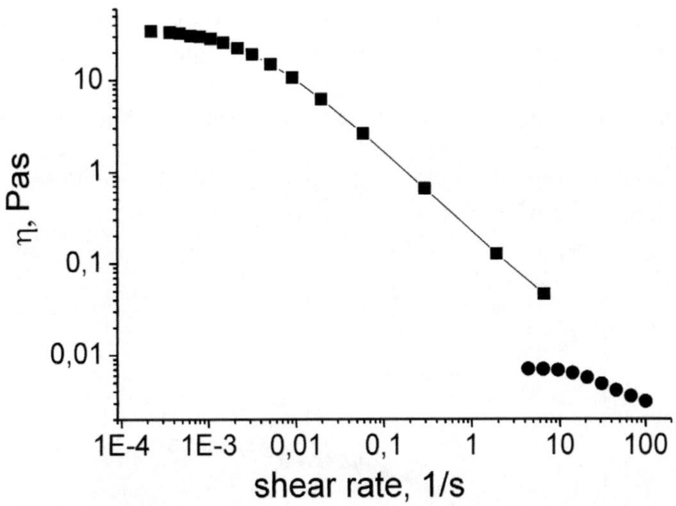

Figure 1. Flow curves of 0.1 wt% (circles) and of 0.6 wt.% (squares) aqueous solutions of EHAC at the presence 1 wt.% salt KCl at 25 °C.

One can see that zero-shear viscosity rises dramatically with the surfactant concentration. The concentration curve of zero-shear viscosity shown on figure 2.

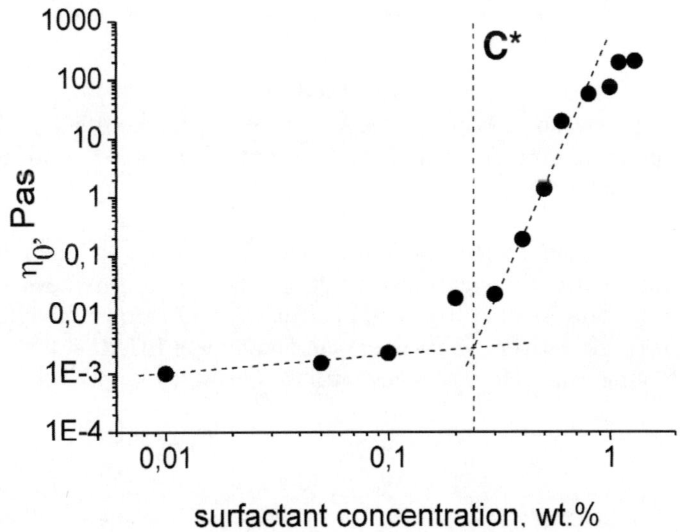

Figure 2. The dependence of zero-shear viscosity of aqueous solutions of EHAC on surfactant concentration at the presence 1 wt.% salt KCl at 25 °C.

In the region of low concentrations (0.01–0.25 wt %), solution viscosity is close to that of water. This region corresponds to the dilute regime at which surfactant micelles are not yet overlapped. Beginning with concentration C* = 0.25 wt %, solution viscosity sharply increases. A rise in the viscosity can be explained by the transition of solution to a semidilute regime at which cylindrical micelles begin to be overlapped.

IOP Conf. Series: Journal of Physics: Conf. Series **848** (2017) 012013 doi:10.1088/1742-6596/848/1/012013

The semidilute region is characterized by the following dependence of viscosity on the surfactant concentration: ~ $C^{9.5\pm0.5}$. The exponent value is much higher than 5.25 theoretically predicted for a system of uncharged chains. According literature such strong dependence can be obtained in case of charged micellar chains [10]. Thus, the obtained dependence of solution viscosity on the surfactant concentration in the semidilute region indicates the presence of charged micelles. It can be assumed that, at these conditions the amount of salt is moderately large and they are insufficiently screen the electrostatic repulsion on surface of the micellar chains.
The dependences of storage G' and loss G" moduli of the complex elastic modulus on the frequency of external action (shear) on the sample were obtained on figure 3.

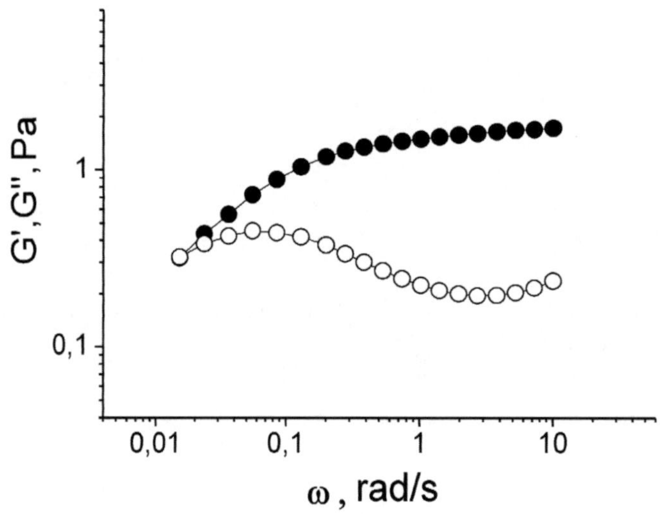

Figure 3. Result of frequency sweep rheological measurement – storage (filled symbols) and (open symbols) loss modulus dependences on frequency of shear stress for 1.1 wt % aqueous solutions of EHAC at the presence 1 wt.% salt KCl at 25 °C.

It is also seen from Fig. 3 that, in a wide range of frequencies, the storage modulus is larger than loss modulus. In this region of the surfactant concentrations, the solution represents a physical gel composed of entangled long wormlike micelles. The plateau value of storage modulus called plateau modulus or elastic modulus of the network and varies with an increase EHAC concentration by the G_0 ~ $C^{2.3}$ law (Fig. 4), which quite satisfactorily agrees with the theoretically predicted dependence ($G_{0\ theor}$ ~ $C^{2.25}$ [1]).

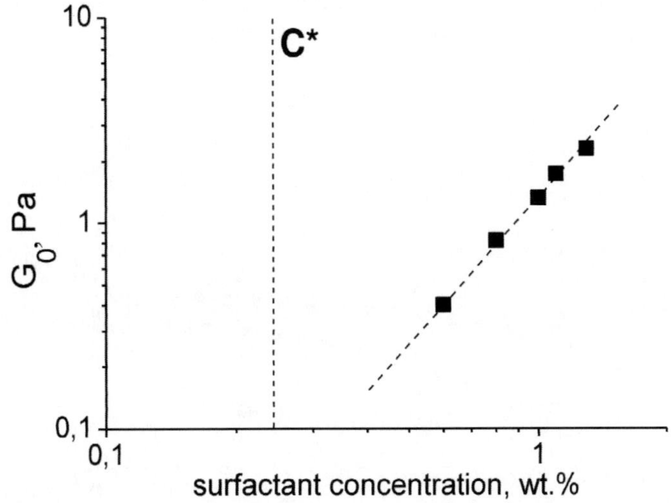

Figure 4. Concentration dependence of the plateau modulus of elasticity for aqueous EHAC solutions at the presence of 1 wt % salt KCl at 25°C.

An increase in the elastic modulus is related with the fact that, as the surfactant concentration increases, the number of micelle topological entanglements rises due to micelle lengthening.

SANS measurement was carried out for 0.6 wt.% solution that corresponds to semidilute regime. Figure 5 shows the SANS scattering curves obtained for these concentrations. They show the dependence between the intensity of scattering and scattering vector.

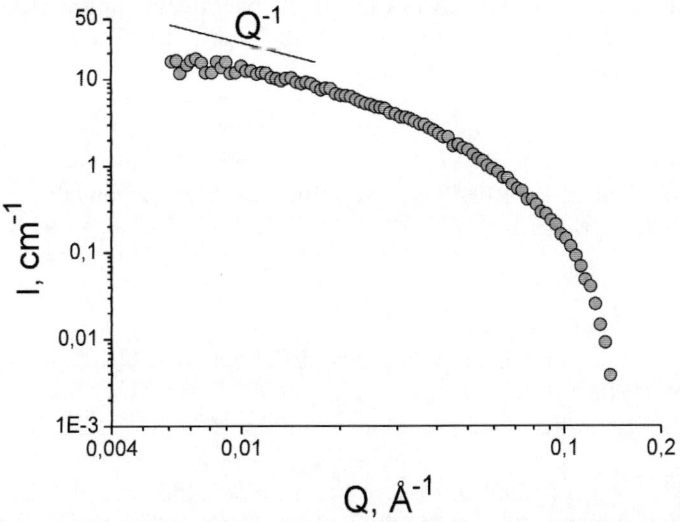

Figure 5. SANS profile for 0.6 wt% aqueous solutions of EHAC at the presence of 1 wt % salt KCl at 25°C. Solid line shows the Q^{-1} law of increase of scattering intensity in low-Q region.

On scattering plots in low-Q region the shape of the aggregates can be revealed [1]. In the case of the long enough wormlike micellar chains, the intensity of I decreases as Q^{-1} at low Q values. In case of short rodlike micelles having numbers spherical end-caps dependence I(Q) is weaker [11].

IOP Conf. Series: Journal of Physics: Conf. Series **848** (2017) 012013 doi:10.1088/1742-6596/848/1/012013

Now the magnetite particles was embedded to network of wormlike micelles at the concentration of EHAC 0.6 wt.% at the presence 1 wt.% KCl, e.i. viscoelastic solution of charged micellar chains. On figure 6 effect of particles on storage and loss moduli is represented.

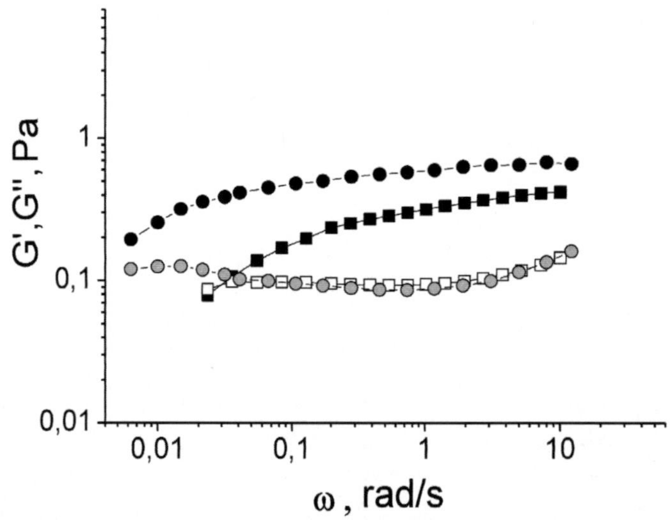

Figure 6. Frequency dependences of storage modulus G' (filled symbols) and loss modulus G" (open symbols) for 0.6 wt% EHAC solutions in the presence of 1 wt% KCl without particles (squares) and with 1 wt.% magnetite particles (circles) at 25^{0}C.

In the presence of particles the system continues to display behavior typical for entangled wormlike micelles solutions. Moreover, the particles induce an increase in the the storage G' and loss G" moduli and the frequency at the G'-G" crossover. All these effects may be due to the appearance of additional cross-links as a result of adsorption of energetically unfavorable micellar end-caps on the surfactant monolayer formed on the surface of the particles.

4. Acknowledgements

The reported study was funded by RFBR and Moscow city Government according to the research project №15-33-70046 «mol_a_mos».

References

[1] Dreiss C A 2007 *Soft Matter* **3** 956.
[2] Petekidis G, Galloway L A, Egelhaaf S U, Cates M E, Poon W C K 2002 *Langmuir* **18** 4248.
[3] Asakura S, Oosawa F 1958 *J. Polym. Sci.* **33** 183.
[4] Jódar-Reyes A B, Martín-Rodríguez A, Ortega-Vinuesa J L 2006 *J. Colloid Interface Sci.* **298** 248.
[5] Bandyopadhyay R, Sood A K 2005 *J. Colloid Interface Sci.* **283** 585.
[6] Helgeson M E, Hodgdon T K, Kaler E W, Wagner N J, Vethamuthu M, Ananthapadmanabhan K P 2010 *Langmuir* **26** 8049.
[7] Helgeson M E, Wagner N J 2011 *J. Chem. Phys.* **135** 084901-1.
[8] Shashkina J A, Philippova O E, Zaroslov Yu D, Khokhlov A R, Pryakhina T A, Blagodatskikh I V 2005 *Langmuir* **21** 1524.

IOP Conf. Series: Journal of Physics: Conf. Series **848** (2017) 012013 doi:10.1088/1742-6596/848/1/012013

[9] Khatory A, Kern F, Lequeux F, Appell J, Porte J G, Morie N, Ott A, Urbachs W 1993 *Langmuir* **9** 933.
[10] Kern F, Lequeux F, Zana R, Candau S J 1994 Langmuir **10** 1714.
[11] Sommer C, Pedersen J S, Egelhaaf S U, Cannavacciuolo L, Kohlbrecher J, Schurtenberger P 2002 *Langmuir* **18** 2495.

Structure and phase transitions of the multilamellar DMPC membranes in presence of the DMSO and DESO

Yu E Gorshkova[1], **O I Ivankov**[1,2,3]

[1] Joint Institute for Nuclear Research, Dubna, Russia
[2] Moscow Institute of Physics and Technology, Dolgoprudny, Russia
[3] Institute for Safety Problems of Nuclear Power Plants, Chornobyl, Ukraine

E-mail: gorshk@nf.jinr.ru

Abstract. The structure and phase transitions of the prepared and formed spontaneously multilamellar vesicles (MLVs) of 1,2-dimyristoyl-*sn*-glycero-3-phosphocholine (DMPC) in dimethyl sulfoxide (DMSO) and diethyl sulfoxide (DESO) were investigated using small angle neutron scattering (SANS). The both polar aprotic solvents increase the temperature of the main phase transition (T_m). The pre-transition does not observed at sulfoxides mole fraction $X = 0.2$. The transition of the MLVs DMPC in the presence DMSO from gel to liquid-crystalline phase occurs at lower temperature. The method of the MLVs preparation has directly effects on the temperature of the main phase transition and its structure. The value of T_m is higher with ~ 4.6 °C in case of the spontaneous forming MLVs from extruded ULVs. The thickness of the solvent layer for prepared MLVs is less by 4.0 Å in gel phase and by 5.6 Å in liquid-crystalline phase than the thickness of the solvent layer for spontaneously formed MLVs.

1. Introduction

The lipid bilayer forms the structural framework of cell membranes, which play an important role in many life processes. The lipid bilayer of the phospholipid membranes is constructed from amphiphilic molecules like phosphatidylcholine (PC) having polar-head groups and two non-polar hydrocarbon tails. The presence of the ions, small molecules and solvents affect the structure and phase behaviour of the lipid bilayer as well as do the temperature, pressure, and pH. For example, anesthetics, short alcohols, and carbohydrates reduce the phase transition temperature .On the contrary, the extension of non-polar lipids chains, external pressure, calcium ions, long alcohols, and hydrocarbons increase the phase transition temperature [1].

It is well known that neutral PC membranes under the excess of water condition exhibit two thermotropic lamellar phase transitions [2]: a gel to gel ($L_{\beta'} \to P_{\beta'}$) pre-transition and a gel to liquid-crystalline ($P_{\beta'} \to L_{\alpha}$) main transition can be observed with temperature increase. The main phase transition is always accompanied by melting of the hydrocarbon tails. From this point of view, it is not a surprise that the thickness of the lipid bilayer becomes smaller. The detailed analysis of the structure parameters for pure 1,2-dimyristoyl-*sn*-glycero-3-phosphocholine (DMPC) membranes in liquid-crystalline phase, using X-ray small-angle scattering, was done, for example, in [3]. Besides, a partial lipid volume (V_l) of DMPC bilayers is increased by 3 % in the main transition region as it was noticed from densimetric measurements (DM) [4]. At the same time, the enthalpy is modified as it was shown by differential scanning calorimetry (DSC) [5, 6]. The change in enthalpy and volume of the lipid molecules pointed out on the first-order phase transition.

In addition to the above mentioned methods, the density dilatometry [7, 8], electron spin resonance (ESR) [9, 10], permeability and fluorescence polarization [11], nuclear magnetic resonance (NMR) [12], small angle X-ray scattering (SAXS) [13] have been extensively used for identification of the lipid phases. However, diffraction methods have the advantage since they allow, in particular, to obtain information on the molecular packing [14].

In the current work, we present a small angle neutron scattering (SANS) investigation of the phase transition of the fully hydrated MLVs DMPC (2 wt %) dissolved in DMSO or DESO. The interest for these polar aprotic solvents is caused by several reasons. First, the distance between two planar lipid layers for uncharged PC membranes is determined by the superposition of the forces: long-range van der Waals attractive force, long-range undulation repulsion, and short-range repulsion («hydration») force [15]. In addition, the hydrophobic interactions can make a significant contribution to balance of the intermembrane interaction in the sulfoxides presence as it was proposed in [16]. Secondly, DMSO and DESO cause a fusion of the unilamellar vesicles (ULVs) DMPC as it was reported in previous work [17]. From this point of view, the above mentioned sulfoxides can be used as good agents for directional cell fusion *in vitro* with applications in biomedicine and biotechnology.

2. Materials and methods

2.1. Materials

14:0 PC (DMPC) – 1,2-dimyristoyl-*sn*-glycero-3-phosphocholine ($C_{36}H_{72}NO_8P$) were purchased from the Avanti (Birmingham, England). All lipids were used without further purification. DMSO ($(CH_3)_2SO$) and DESO ($(C_2H_5)_2SO$) over 99 % purity, were purchased from the JSC "Reachim" (Moscow, Russia). Heavy water (99.8 % D_2O) was from JSC "Isotop" (St. Petersburg, Russia).

2.2. Sample preparation

2.2.1. MLVs preparation. The lipid was dissolved in DMSO/D_2O or DESO/D_2O solution at predetermined mole fractions X_{DMSO} or X_{DESO} (X is mole fraction of sulfoxide in sulfoxide/water mixture) and mixed by a shaker. A homogeneous multilamellar vesicles (MLVs) were obtained by the freezing-thawing membranes in the range from – 80 ºC to + 55 ºC. The procedure was repeated several times. The final lipid concentration (C_l) for all prepared samples was 2 wt %.

2.2.2. Spontaneously formed MLVs. Large unilamellar vesicles (ULVs) were prepared by passing the MLVs of DMPC through two polycarbonate filters with 100 nm pores in diameter (Hamilton, Reno, Nevada, United States) of an extruder (Avanti, USA). The extrusion of the suspension was performed at the temperature above the temperature of the main phase transition (T_m). After 25 cycles we obtained unilamellar vesicles with average diameter determined by the size of the filter pores. An odd number of passages was chosen to avoid the MLVs, which might not penetrate through the filters, in the samples.

The vesicles in pure D_2O were stable for a sufficiently long time. The spontaneous MLVs formation at $X = 0.2$ was observed according to [17].

2.3. SANS technique

The phase transitions of the fully hydrated phospholipid membranes in the DMSO and DESO presence was investigated using small angle neutron scattering. The experiments were performed at YuMO time-of-flight spectrometer at the IBR-2 pulsed reactor (Dubna, Moscow region, Russia). The data were collected in two-detector configuration [18]. It gives a possibility to measure samples, twice faster, in the q-range of 0.007 - 0.4 Å^{-1}.

The raw data treatment was done using the SAS program [19]. The final small-angle neutron scattering curves are presented in the absolute scale with the background subtraction [20].

The studied samples were placed in 1 mm thick quartz cells (Hellma, Germany). During data collection, the samples were put in the temperature controlled holder (±0.2 °C) connected to the liquid thermostat (Lauda, Germany) in the range of the temperature from 10 °C to 60 °C. The waiting time between two temperature points was 15 min. The standard data acquisition time per sample was 30 min. The measurements were performed 12 hours after samples preparation.

3. Results and discussion

The first structural and phase transition investigations of the model phospholipid membrane (≥ 20 wt %) in the DESO presence were done by DSC [16] and X-Ray [21]. It was shown that the repeat distance (D) of the MLVs decreases and the temperature of the main phase transition (T_m) grows with increasing of the sulfoxide concentration. The similar effect was observed for MLVs DMPC at smaller lipid concentration (C_l).

In this paper we present the results for sulfoxides mole fraction, $X = 0.2$, which is nonrandom value. First, the dramatic changes in the structure of the PC membranes in excess solution take place at smaller DMSO mole fraction as it was shown in our previous work [22]. At the same time, at this concentration the fusion of the ULVs has been observed [17].

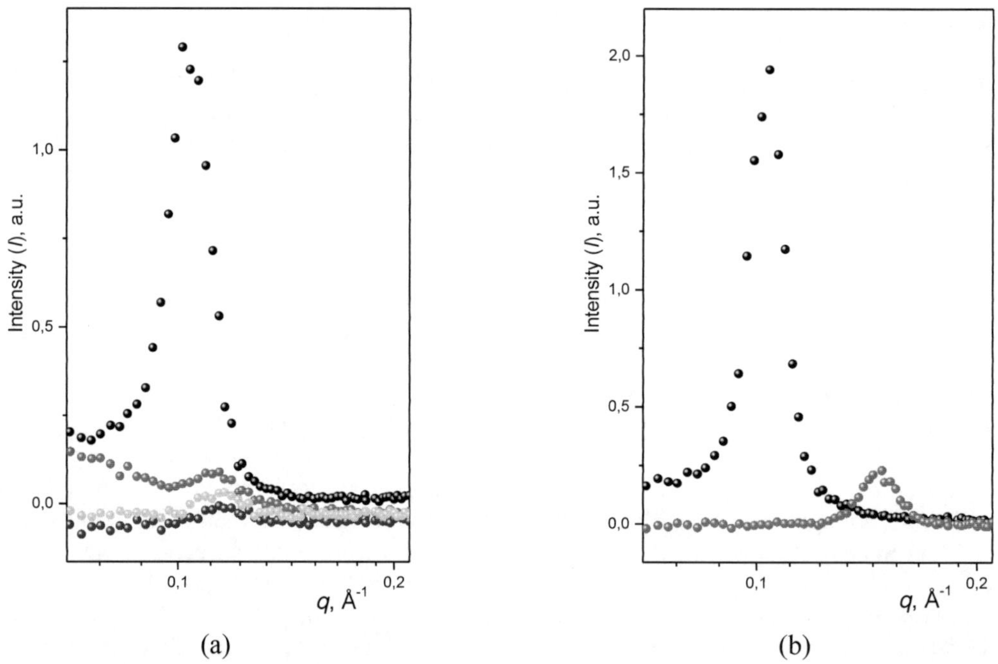

Figure 1. SANS curves for MLVs DMPC (2 wt %) in DESO/D_2O solution at different DESO mole fraction: $X_{DESO} = 0.0$ (black), 0.2 (red), 0.4 (blue) and 0.8 (green) in gel phase at $T = 14$ °C (a) and in liquid phase at $T = 64$ °C (b).

The investigations of the influence of the DESO mole fraction on the MLVs DMPC in excess solution condition in gel and liquid-crystal phases are presented in Figure 1. The decrease in the SANS curve intensity is caused by the reduction of the deuterium in the DESO/D_2O solution with X_{DESO} increase. Nevertheless, the obtaining of the D-values from the SANS spectra is possible. The repeat distance is calculated from the position of the diffraction peak q_0 at the intensity maximum using a relationship $D = 2\pi/q_0$. The D-value is a sum of two terms: a thickness of the lipid bilayer (D_b) and a thickness of the solvent layer – distance between two planar lipid layers (D_s). The parameter D is collected in Table 1 for MLVs DMPC (2 wt %) at the DESO mole fraction from 0.0 to 0.8. It is clear seen that the considerable decreasing of the repeat distance for MLVs DMPC from 60.6 ± 0.1 Å in pure D_2O to 56.9 ± 0.2 Å at X_{DESO} = 0.2 are observed in gel phase. At the same region of DESO mole fraction the repeat distance decreases more significantly from 60.9 ± 0.1 Å to 42.6 ± 0.1 Å in the liquid-crystalline phase. The future increasing of the DESO mole fraction in DESO/D_2O solution in the region 0.3 ≥ X_{DESO} ≥ 0.8 does not effect on the *D*-value in gel phase.

Table 1. The repeat distance (*D*) of the MLVs DMPC (2 wt %) in dependence on the DESO mole fraction (X_{DESO}) in gel phase at *T* = 14 °C.

X_{DESO}	0.0	0.2	0.3	0.4	0.5	0.7	0.8
***D* (Å)**	60.6±0.1	56.9±0.2	54.1±0.4	54.0±0.4	54.5±0.2	54.7±0.3	54.9±0.3

3.1. MLVs DMPC

The SANS spectra of the prepared MLVs DMPC (2 wt %) (a) and MLVs DMPC (20 wt %) (b) in DESO/D_2O solution at X_{DESO} = 0.2 are shown in Figure 2. The measurements were carried out in the temperature range 10 °C ≤ *T* ≤ 40 °C. The reduction in the lipid fraction of the samples resulted in a decrease in intensity of the scattering curves for MLVs DMPC (2 wt %). Nevertheless, the behaviour of the phase transitions is similar and two phase transitions were observed for both cases during heating at the same temperature.

Additionally, the changes on the membrane's structure during phase transitions can be observed using SANS technique. The dependences of the repeat distance on the temperature are presented in Figure 3 for MLVs DMPC (2 wt %) (a) and MLVs DMPC (20 wt %) (b) in DESO/D_2O at X_{DESO} = 0.2. Analysis of the data indicates that the *D*-value does not change in the gel phase and is equal to 55.5 ± 0.4 Å for C_l = 2 wt % and 53.8 ± 0.2 Å for C_l = 20 wt %. Further increase of the temperature led to decrease of the repeat distance for both lipid concentrations.

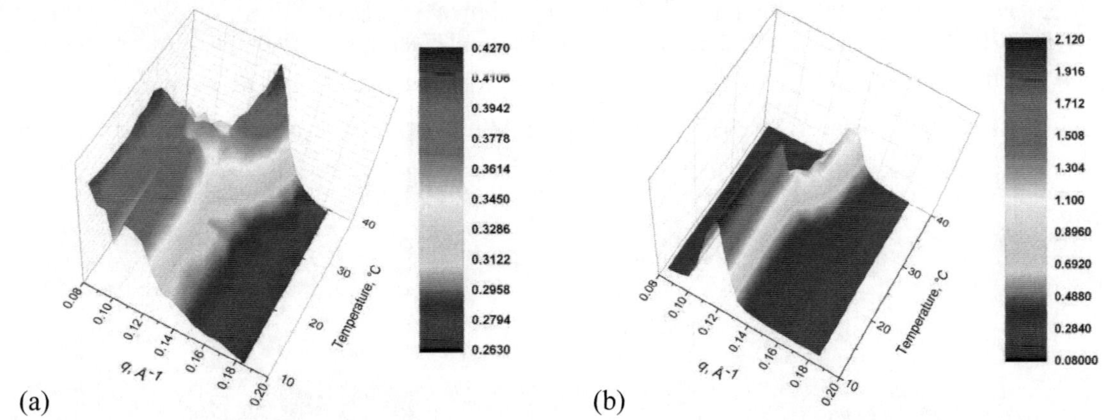

Figure 2. SANS spectra of the prepared MLVs DMPC (2 wt %) (a) and MLVs DMPC (20 wt %) (b) in DESO/D_2O at X_{DESO} = 0.2 with scales of the intensities.

The T_m values were obtained in two ways. One of them is to use the Boltzmann function for fitting experimental data as shown in Figure 3. Besides, the first derivatives, d*D*/d*T* (Figure 3, insets) can be successfully used for determination of the temperature of the main phase transition. Without going into the details, the following results were obtained. The transition from gel to liquid-crystal phase is observed at T_m = 28.9 °C for C_l = 2 wt %. Increasing the lipid concentration does not change the temperature of the main phase transition within the experimental errors. Thus, the transition is taking place at T_m = 28.4 ºC for sample with C_l = 20 wt %. This is in a good agreement with DSC data for MLVs DMPC (20 wt %) [16].

Thus, we can conclude that the temperature of the main phase transition of the MLVs DMPC in the presence of DESO does not depend on the lipid concentration for $C_l \leq 20$ wt. The situation is similar as for PC membranes in excess water [23].

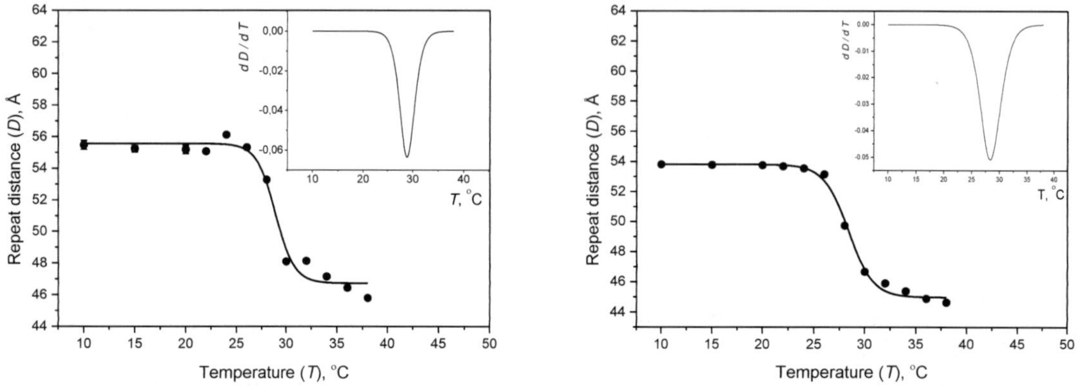

(a) (b)

Figure 3. Repeat distance vs. temperature for prepared MLVs DMPC (2 wt %) (a) and MLVs DMPC (20 wt %) (b) in DESO/D_2O at X_{DESO} = 0.2. Dots are the experimental data and lines are the fits by the Boltzmann function.

3.2. *Spontaneously formed MLVs*

The MLVs can be formed spontaneously from ULVs in the sulfoxides presence [17]. Here, we present the results for MLVs DMPC (2 wt %) produced by extruded ULVs in sulfoxide/water solutions at X = 0.2. The SANS spectra with intensity scales are shown in Figure 4 for X_{DMSO} = 0.2 (a) and X_{DESO} = 0.2 (b). It is clear that both sulfoxides induce the increasing of the temperature of the phase transitions. The transitions $L_{\beta'} \rightarrow P_{\beta'}$ and $P_{\beta'} \rightarrow L_\alpha$ were observed for MLVs prepared in the DESO/D_2O solution. However, for the system DMPC/DMSO/D_2O the pre-transition from gel to ripple phase disappeared and MLVs have the abrupt transition $L_{\beta'} \rightarrow L_\alpha$ only, as can be clearly seen in Figure 4 (a).

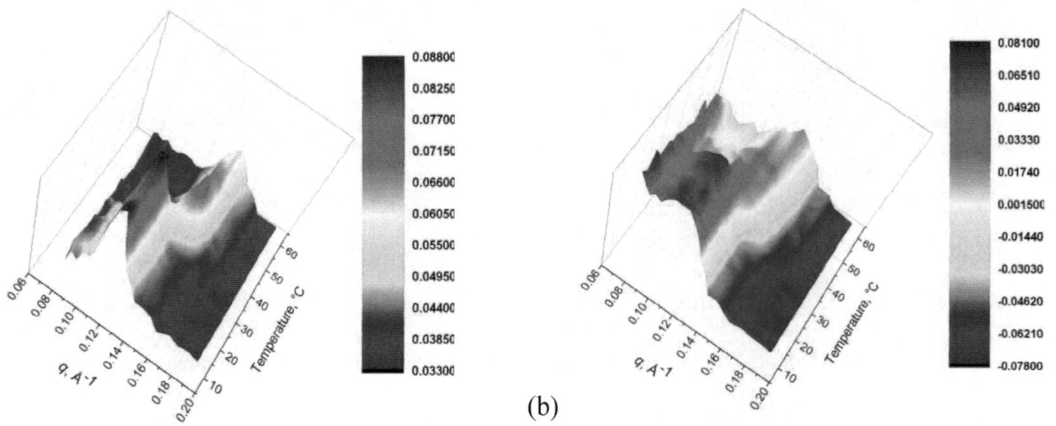

(a) (b)

Figure 4. SANS spectra of the spontaneous MLVs DMPC (2 wt %) in DMSO/D_2O at X_{DMSO} = 0.2 (a) and in DESO/D_2O at X_{DESO} = 0.2 (b) with scales of the intensities.

IOP Conf. Series: Journal of Physics: Conf. Series **848** (2017) 012014 doi:10.1088/1742-6596/848/1/012014

The transition temperatures were calculated as it was described in the previous section. As a result we obtained the following values: T_m = 35.2 °C for MLVs DMPC in DMSO/D_2O and T_m = 33.6 °C for MLVs DMPC in DESO/D_2O. It should be noted that the temperatures of the main phase transitions are higher with ~ 4.6 °C in case of the spontaneous forming MLVs from extruded ULVs.

Additionally, we estimated the region of the temperatures in which transition $L_{\beta'} \rightarrow L_\alpha$ was observed. With this purpose we analyzed the dependence of the first derivative, d*D*/d*T* *vs.* *T* (Figure 5, insets) with the background subtraction, d*D*/d*T* = 0. It turns out that the transition of the MLVs DMPC in the presence of DMSO from gel to liquid-crystal phase occurs in a narrow temperature range. It begins at T = 28 °C and ends at T = 42 °C, while for MLVs dissolved in DESO/D_2O, this transition is observed in the temperature range 20.6 °C ≤ T ≤ 47.3 °C.

The analysis of the dependence of the repeat distance *vs.* temperature for spontaneous MLVs DMPC in the sulfoxide/water solvent was done in the wide range of the T from 10 to 60 °C. The experimental data (dots) with fits by the Boltzmann function (lines) are presented in Figure 5 for MLVs DMPC (2 wt %) in DMSO/D_2O at X_{DMSO} = 0.2 and in DESO/D_2O at X_{DESO} = 0.2. It is clear that the *D*-values are constant in the gel phase. The repeat distance is equal to 57.5 ± 0.4 Å for DMPC/DMSO/D_2O system. The average magnitude of *D* is 59.5 ± 0.3 Å for spontaneous MLVs DMPC dissolved in DESO/D_2O. The difference in the repeat distances in the liquid-crystalline phase is much greater. For example, the D = 45.0 ± 0.2 Å and D = 50.3 ± 0.1 Å at T = 50 °C for MLVs DMPC in DMSO/D_2O and DESO/D_2O, respectively.

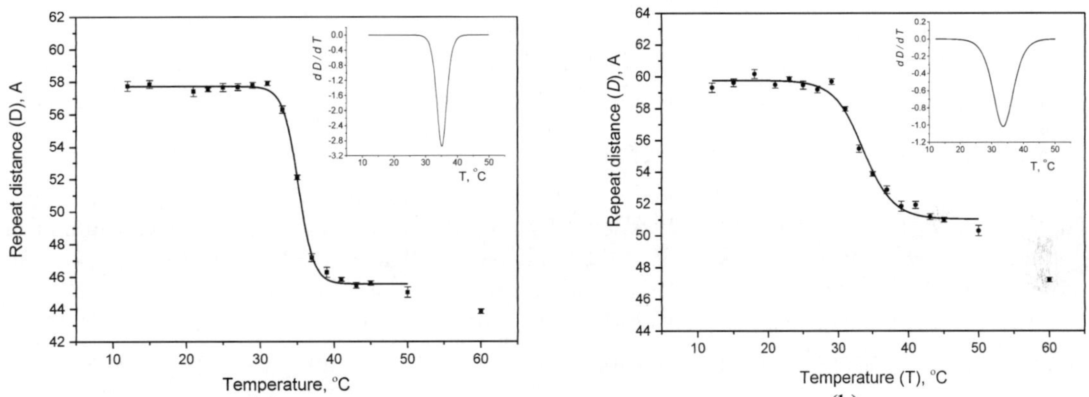

(a) (b)

Figure 5. Repeat distance vs. temperature for spontaneous MLVs DMPC (2 wt %) in DMSO/D_2O at X_{DMSO} = 0.2 (a) and in DESO/D_2O at X_{DESO} = 0.2 (b). Dots are the experimental data and lines are the fits by the Boltzmann function.

3.3. Distance between two planar layers

It is clear that the conditions of the sample preparation affect significantly the structure and phase behavior of the lipid membranes. The T_m is shifted by 4.6 °C and the repeat distance also depends on the type of the samples. Thus, the three regions in *D*-value can be selected for MLVs DMPC (2 wt %) in DESO/D_2O solution in the range of the investigated temperatures as presented in Figure 6. The region I (yellow) corresponds to the gel phase for both systems. The repeat distances are constant for prepared and spontaneously formed MLVs DMPC in this case. Nevertheless, the difference of 4 Å in *D*-values was calculated for these samples. The second region (II) represents the transition area between $L_{\beta'}$ and L_α phases. Here, the structure of the lipid molecules dramatically changes. The melting of the hydrocarbon tails led to reduction of the repeat distance of the DMPC membranes from 59.5 ± 0.2 Å to 53.1 ± 0.1 Å and from 55.5 ± 0.1 Å to 47.6 ± 0.1 Å for spontaneous formed and prepared MLVs, respectively. Finally, the region III (green) matches the liquid-crystalline phase. All experimental points are parallel shifted by 5.6 Å as it is clearly seen in Figure 6. Thus, we obtain the $D = 41.6$ Å for prepared MLVs using extrapolation (blue line) at $T = 55$ °C.

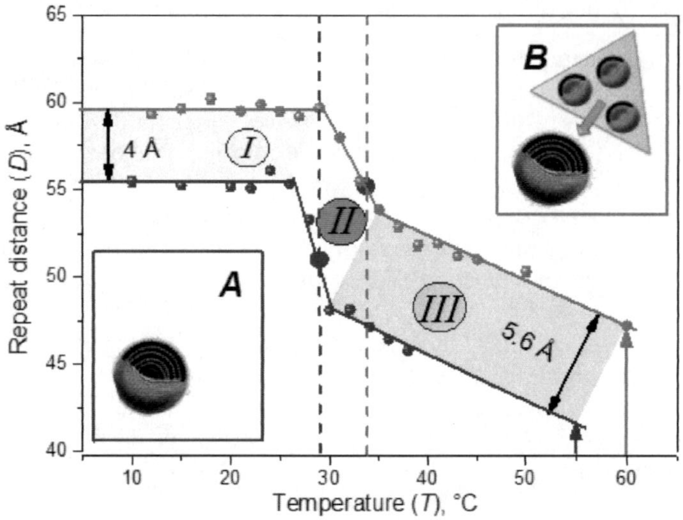

Figure 6. Repeat distance vs. temperature for MLVs DMPC (2 wt %) in DESO/D_2O at $X_{DMSO} = 0.2$. The blue dots correspond to prepared MLVs (schematically presented in inset A) and red dots correspond to spontaneously formed MLVs from ULVs (schematically presented in inset B). The red and blue vertical dashed lines indicate the phase transition temperatures for the corresponding samples.

Taking into account that the bilayer thicknesses (D_b) of the DMPC (2 wt %) are the same for prepared and spontaneously formed MLVs and equal to 37.7 Å in $L_{\beta'}$ phase at $T = 12$ °C and 26.5 Å in L_α phase at $T = 55$ °C [17], we calculated the distance between two planar layers (thickness of the solvent D_s) using relationship $D_s = D - D_b$. The results are presented in Table 2.

The data analysis has shown that the thickness of the solvent layer decreases with increasing of the temperature. However, it should be noted, that the reduction of the D_s-value is significant for the prepared MLVs. The distance between two planar layers decreases by 2.7 Å, while the changing of this value equal to 1.1 Å for spontaneously formed MLVs. It is important to note that the thickness of the solvent layer for the prepared MLVs is less by 4.0 Å in $L_{\beta'}$ phase and by 5.6 Å in L_α phase. This fact can be explained by the different interactions of the DESO molecules with polar heads of the lipid membranes. We can suppose that the number of the DESO molecules binding with D_2O molecules is higher in the case when the MLVs DMPC are spontaneously formed. However, this hypothesis should be proved in future investigations using contrast variation method.

Table 2. Structure parameters for the prepared and spontaneously formed MLVs DMPC (2 wt %) at $X_{DESO} = 0.2$ in gel $L_{\beta'}$ phase at T = 12 °C and in liquid-crystalline L_α phase at T = 55 °C.

	Structure parameters in $L_{\beta'}$ phase			Structure parameters in L_α phase		
	D(Å)	D_b (Å)	D_s (Å)	D (Å)	D_b (Å)	D_s (Å)
Prepared MLVs	55.5±0.1	37.7±0.3	17.8±0.2	41.6±0.1	26.5±0.3	15.1±0.2
Spontaneously formed MLVs	59.5±0.2	37.7±0.3	21.8±0.3	47.2±0.1	26.5±0.3	20.7±0.2

4. Conclusions

We can conclude that the intermembrane interaction of the PC membranes in the presence of DESO is similar as in case of the DMSO. First, the both polar aprotic solvents increase the temperature of the main phase transition. Nevertheless, the transition of the MLVs DMPC in the presence of DMSO from gel to liquid-crystalline phase occurs at lower temperature. Secondly, the dramatic changes in the structure of the PC membranes in excess solution take place at smaller DMSO and DESO mole fraction. The increasing of the temperature beyond T_m led to a linear decreasing of the repeat distance of the MLVs in investigated range of the temperatures. The study of the samples with different lipid concentration pointed out on non-dependent PC membrane – DESO interaction for the $C_l \leq 20$ wt %.

The method of the MLVs preparation directly affects the temperature of the main phase transition and its structure. The temperature of the main phase transition is higher with ~ 4.6 °C in case of the spontaneous forming MLVs from extruded ULVs. The thickness of the solvent layer for prepared MLVs is less by 4.0 Å in gel phase and by 5.6 Å in liquid-crystalline phase than the thickness of the solvent layer for spontaneously formed MLVs.

5. Acknowledgments

This research was supported by JINR-Romania Scientific Projects of 2017 year.

References

[1] Kharakoz D P 2001 *Uspekhi Biol. Khimii* **41** 333-364
[2] Winter R and Jeworrek C 2009 *Soft Matter* **5** (17) 3157-3173
[3] Kirchner S and Cevc G 1993 *Europhys. Lett.* **23** (3) 229-235
[4] Koynova R, Koumanov A and Tenchov B 1996 *BBA - Biomembranes* **1285** 101-108
[5] Ohline S M, Campbell M L, Turnbull M T and Kohler S J 2001 *Journal of Chemical Education* **78** (9) 1251
[6] Koyama T M, Stevens C R, Borda E J, Grobe K J and Cleary D A 1999 *Chem. Educator* **4** (1) 12-15
[7] Nagle J F 1973 *Proc. Nat. Acad. Sci. USA* **70** 12 (1) 3443-3444
[8] Nagle J F 1973 *The Journal of Chemical Physics* **58** (1) 252-264
[9] Shimshick E J and Mcconnell H M 1973 *Biochemistry* **12** 2351-2360
[10] King M E and Spector A A 1978 *J. Biol. Chem.* **253** 6493-6501
[11] Papahadjopoulos D, Jacobson K, Nir S and Isac I 1973 *BBA - Biomembranes* **311** (3) 330-348
[12] Sackmann E 1995 *Handbook of Biological Physics* ed R Lipowsky and E Sackmann (North-Holland) pp. 213-304
[13] Kobayashi Y and Fukada K 1998 *Chemistry Letters*, **27** (11) 1105-1106
[14] Seddon J M and Cevc G 1993 *Phospholipids handbook* ed G Cevc (New York: Marcel Dekker Inc.) pp. 403-454
[15] Gordeliy V I, Cherezov V G and Teixeira J 1996 *Journal of Molecular Structure* **383** (1-3) 117-124
[16] Bonora S, Markarian S A, Trinchero A and Grigorian K R 2005 *Thermochimica Acta*, **433** (1-2) 19-26
[17] Gorshkova Yu E 2015 *JOAM* **17** (9-10) 1532-1537
[18] Kuklin A I, Islamov A K and Gordeliy V I 2005 *Neutron News* **16** 16-18
[19] Soloviev A G, Solovieva T M and Kuklin A I *http://wwwinfo.jinr.ru/programs/jinrlib/sas/indexe.html*
[20] Ostanevich Y M 1988 *Makromolekulare Chemie. Macromolecular Symposia* **15** (1) 91-103
[21] Gorshkova Y E, Ivankov O I, Kuklin A I and Gordeliy V I 2012 *Journal of Physics: Conference Series* **351** (1) 012006
[22] Gorshkova J and Gordeliy V 2007 *Crystallography Reports*, **52** (3) 535-539
[23] Matsuki H, Goto M, Tada K and Tamai N 2013 *Int. J. Mol. Sci.* **14** (2), 2282-2302

IOP Conf. Series: Journal of Physics: Conf. Series **848** (2017) 012015 doi:10.1088/1742-6596/848/1/012015

Structure analysis of aqueous ferrofluids at interface with silicon: neutron reflectometry data

I V Gapon[1,2*], V I Petrenko[1,2], L A Bulavin[2], M Balasoiu[1,3], M Kubovcikova[4], V Zavisova[4], M Koneracka[4], P Kopcansky[4], H Chiriac[5], M V Avdeev[1]

[1] Frank Laboratory of Neutron Physics, Joint Institute for Nuclear Research, Dubna, Russia
[2] Taras Shevchenko National University of Kyiv, Kyiv, Ukraine
[3] National Institute of Physics and Nuclear Engineering, Bucharest, Romania
[4] Institute of Experimental Physics, SAS, Kosice, Slovakia
[5] National Institute of Research & Development for Technical Physics, Iasi, Romania

E-mail: gapon@jinr.ru

Abstract. Adsorption of nanoparticles from aqueous ferrofluids (FFs) on solid surface (crystalline silicon) was studied by neutron reflectometry (NR). Two kinds of FFs were considered. First kind was heavy water-based ferrofluids with magnetite nanoparticles coated by double layer of sodium oleate. Second one FF was cobalt ferrite nanoparticles stabilized by lauric acid/sodium n-dodecylsulphate layer and dispersed in water. It was obtained only a single adsorption layer for two types of ferrofluids. The impact of the magnetic nanoparticles concentration and geometry was considered in frame of the adsorption characteristic of FFs.

Introduction

Ferrofluids (FFs) or magnetic fluids (MFs) are unique systems combining magnetic properties and fluidity. They consist of magnetic nanoparticles (with a characteristic size of about 10 nm) placed in a liquid medium. To prevent aggregation, the particle surface is coated usually by a surfactant layer. Study of MFs is of great practical and fundamental importance (e.g, [1, 2]). Mainly papers devoted to the bulk structure of magnetic fluids, which is well observed by small-angle X-ray (SAXS) and neutron (SANS) scattering (e.g, see review in [3, 4]). However, ferrofluids have specific adsorption properties and the behavior of microstructure in MFs at interfaces with solids can be different from that in bulk, which is also important in respect of practical and fundamental point of view. are devoted The study structure organization of magnetic nanoparticles (MNPs) assembling into layer were done in recent few papers [5–7]. Using various methods it was shown that specific self-assembly of the nanoparticles occurs during the solvent evaporation, therefore interaction between the film and a substrate plays an important role in the ordering. Also an important point which should be considered is difference in the stability of MFs in bulk and at interfaces. It should be taken into account in the requirements for the stabilization of these systems with respect to their storage, namely to the interaction of the FFs particles with container walls under different conditions.

Thus, some kind of ordering into layered structures at the interface with silicon was reported [8] for a concentrated water-based MF (nanomagnetite coated by double layer of sodium oleate in heavy water) with magnetic volume fraction $\varphi_m \sim 10\,\%$. Specular neutron reflectivity curve was analyzed and layers formation was found at the silicon substrate. It was shown that this ordering is strongly dependent on the particle concentration and sensitive to the external magnetic field. In a similar way, less concentrated analogue of the water-based MF ($\varphi_m \sim 1\,\%$) was compared [9] with the highly stable MF based on organic non-polar liquid carrier (nanomagnetite coated with oleic acid in deuterated benzene). In both cases, the formation of only one well-defined adsorption layer of nanoparticles on silicon surface was concluded. An interesting question is at what extent the bulk structure of MFs influences their adsorption properties. To clarify this point, in the given work we have performed similar kind of NR experiments applied to the various kinds water-based MFs. First kind of MFs was based on heavy water (D_2O) (nanomagnetite coated by double layer of sodium oleate) with concentration $\varphi_m = 2.1\ vol.\ \%$ and $\varphi_m = 3.9\ vol.\ \%$. Second kind FF was prepared on H2O with cobalt ferrite nanoparticles (stabilization layer: lauric acid/sodium n-dodecylsulphate (DDS-Na)) with concentration of nanoparticles $\varphi_m \sim 3\ vol.\ \%$. Heavy water was added to initial FF of the second kind to satisfy the condition of the total specular reflection of the neutron beam in the NR experiments.

Experimental

Preparation of first type aqueous magnetic fluid was based on the co-precipitation reaction to produce nanomagnetite (Fe_3O_4). MF was synthesized at the Institute of Experimental Physics, Slovak Academy of Sciences (IEP SAS), Slovakia, in accordance with the procedure described in [10]. To avoid aggregation between magnetite particles they were covered by a double layer of sodium oleate ($C_17H_33COONa$) with the following surfactant/magnetite mass ratio 0.73 to 1. Two ferrofluids with different volume of magnetite $\varphi_m = 2.1\ vol.\ \%$ and $\varphi_m = 3.9\ vol.\ \%$ were prepared. Magnetite concentration in the magnetic fluids was determined by the static magnetization analysis (MPMS SQUID magnetometer, IEP SAS).

Ferrofluid of the second kind was prepared at the Institute of Technical Chemistry (Perm) [11, 12] by the co-precipitation of $Fe(OH)_3$ and $Co(OH)_2$, ferritisation of hydroxide mixture in 1M alkali aqueous solution, adsorption of lauric acid on ferrite particles and peptisation of hydrophobic precipitate in aqueous solution with sodium n-dodecyl sulphate.

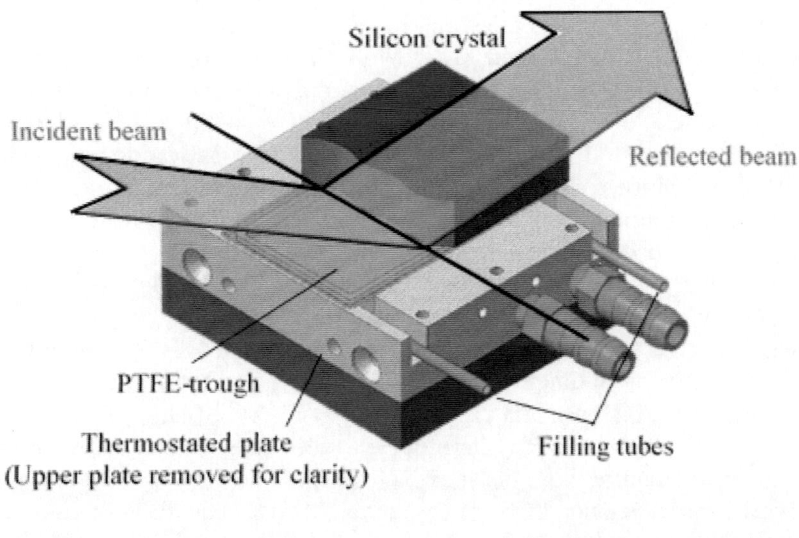

Figure 1. Schematic view of the neutron reflectivity cell for solid/liquid interface study

A principal scheme of the NR experiments with a solid-liquid interface was similar to that in our previous works [9, 13].The NR experiments were performed at the at the GRAINS reflectometer (Joint Institute for Nuclear Research, IBR2 reactor, Russia) with vertical scattering plane. Experiments at the GRAINS reflectometer were conducted in TOF (time-of-flight) mode with neutron wavelength range of $0.05 - 0.7$ nm. For experimental with liquid/solid interface special cell was build. Principal scheme is presented on Fig.1. It is some kind modification of the extensively used liquid/solid cell from V6, BERII [14]. Here the neutron beam passes the front facet of a monocrystalline 111 oriented silicon block (dimensions:80x50x15 mm^3). Reflection of the neutrons takes place at solid/liquid interface, a polished Si-surface with size80x50 mm^2 and roughness $\sigma_{Si} = 5$ A in contact with the ferrofluid. The required volume of liquid in a PTFE-trough is less than 10 ml. Holes in the edges of the trough allow changing the liquid volume bubble free. After leaving the backside of the Si-block the neutrons (reflected neutrons) are detected by the 2D position-sensitive detector. Before the measurements the polished Si-surface was cleaned and hydrophobized by applying the RCA-1 protocol. During the measurement a thermostat (Julabo F25-ME, JULABO GmbH, Germany) kept the sample under investigation at a constant temperature (20^oC).

Results and discussion

The experimental reflectivity, R, for magnetite/sodium oleate/heavy water MFs at the silicon interfaces are represented at Fig.2a as R Q_z^4 to enhance the differences in the data from the samples under study and demonstrate the quality of the fits. The deviations of the curves in Fig.2 from Fresnel law (reflection from two semi-infinite homogeneous mediums) indicate that there are some layer formations at the interface caused by adsorption of particles from MFs. The curves were treated in the frame of the Parrat formalism [15] by using Motofit program [16]. For this type MFs the best fits (solid lines in Fig.2a) were obtained using a two layer model with an effective SLD (corresponding to the momentum transfer vector values, seen in the curves) on the silicon substrate. The corresponding SLD profiles along the surface normal are shown in Fig.2b. These SLD profiles might be interpreted similar to previous works [9, 13]. The revealed two layers with significantly different SLDs can be considered as two sub-layers of one layer of colloidal core-shell particles adsorbed on the silicon substrate. The width (~ 5 nm) and SLD (~ 0 nm^{-2}) of the first sub-layer are close to the corresponding parameters of the surfactant shell (sodium oleate) around magnetite in MFs, while the width of the second sub-layer is well associated with the mean diameter of magnetite nanoparticles, D $\sim 6nm$. The mean SLD of this sub-layer is significantly lower than that of magnetite ($\rho = 6.9$ 10^{-4} nm^{-2}) because of the large fraction of surfactant in it. In addition, it is found concentration dependence of the filling of the adsorption layer. More concentrated magnetic fluid more clear and sharp border was observed. This result is comparable with the paper [8] where multilayer adsorption structure for concentrated MFs was observed. It should be noted that in this case we have to mention that important parameter is the concentration of single magnetic particles as that individual non-aggregated nanoparticles are preferably adsorbed [9].

The experimental curves together with best fits and Fresnel law for second type MFs are presented on Fig.3a. This type of magnetic fluid was prepared in light water with MNPs consist from cobalt ferrite/lauric acid/DDS-Na This liquid has a smaller SLD(($\rho \sim -0.52$ 10^{-4} nm^{-2}) than silicon ($\rho = 2.07$ 10^{-4} nm^{-2}) therefore we don't have total reflection. On Fig.3a. comparison of the experimental NR curve with Fresnel law are presented and additional structural organization as in previous case is concluded. But it was difficult to analyze the data in this case due to the high background from sample and small contrast between the solvent and the stabilizer shell of the particles. Obtained reflectivity data for such system after the addition of heavy water (about 50% D_2O in final MF) was used for analysis of the adsorption. With this contrast we have real critical momentum transfer vector (see Fig.3a) which was used to

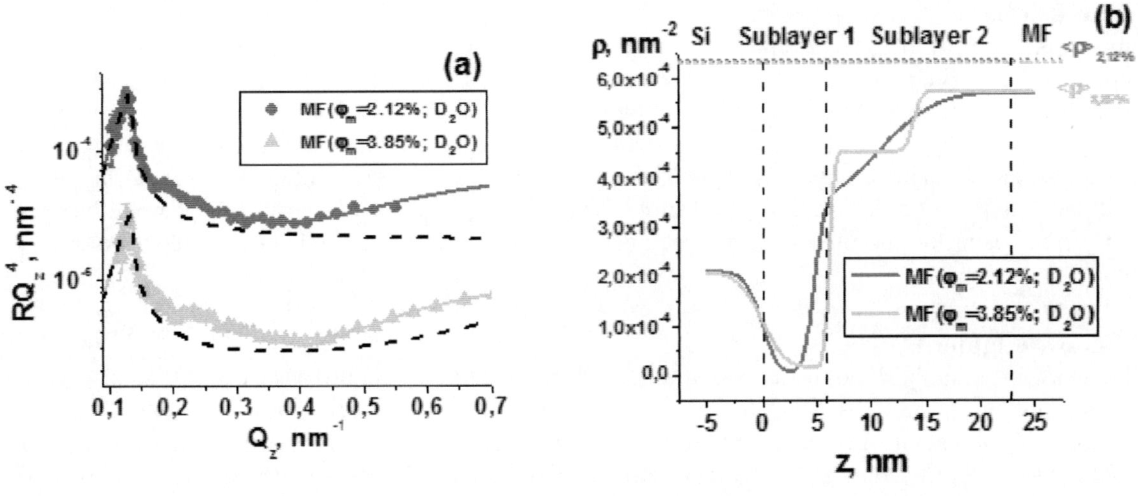

Figure 2. (a) Experimental reflectivity curves (points) together with fits (solid lines) for fist type MFs with concentration: 2.1 % (red circles), 3.9 % (green triangles); black curve (dashed line) is correspond with Fresnel law (the curves for the two concentrations of MFs are shifted over the reflectivity axis for the convenient representation); (b) found from experimental data SLD profiles for two studied kinds of solid/liquid interfaces.

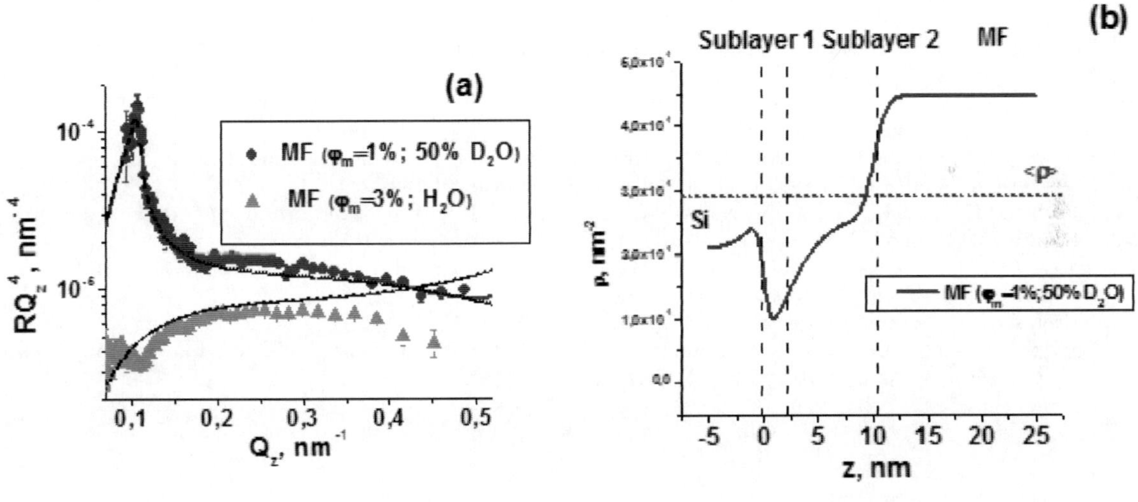

Figure 3. (a) Experimental reflectivity curves (points) together with fits (solid lines) for aqueous MF after dilution by D_2O (blue circles), initial aqueous magnetic fluids (magenta triangles); black dashed line is correspond with Fresnel law; (b) found from experimental data SLD profile for solid/liquid interface.

obtain bulk SLD. SLD profile which getting from best fit experimental data are show on Fig.3b. Model using to approximation is similar to previous case. The resulting fit was got first sublayer (consist of surfactant) has a size of about 2 *nm*. SLD of the second sublayer is 2.5 *nm* and the thickness of about 8 *nm* with large roughness (about 3 *nm*). This result is corresponding with bulk structure [17]. MNPs have ellipsoid forms with middle size about 8 *nm*. High roughness of second sublayer and low thickness of first sublayer correspondingly might be explained by disoriented ellipsoids of the MNPs.

Conclusions

Similar structural organizations of MFs at interface with silicon were concluded from NR experiments for two aqueous MFs with various bulk structures and chemical composition of magnetic nanoparticles and stabilization agents. It was obtained that individual non-aggregated nanoparticles are preferably adsorbed on the oxidized silicon surface from the aqueous MF to form a single adsorption layer. Values of the structural parameters of the adsorption layer are changed due to the different bulk structure of ferrofluids. Denser adsorption layer was observed at the increasing of the monomer concentration of magnetic nanoparticles in bulk of magnetic fluids.

Acknowledgments

The work is partially done in the frame of the RFBR grant # 14-22-01054 ofi_m. Also this work was supported by the VEGA 0045, Radiomag COST TD 1402, MERANET MACOSYS and by Ministry of Education Agency for Structural Funds of EU in frame of Projects 26220120005 and 26220120033. Authors acknowledge the grant of Romanian Plenipotentiary Representative No. 94/2016 item 16.

References

[1] Odenbach S (ed) 2009 *Colloidal Magnetic Fluids* vol 763 (Berlin: Springer-Verlag)
[2] Torres-Daz I and Rinaldi C 2014 *Soft Matter* **10** 8584–8602
[3] Balasoiu M, Avdeev M and Aksenov V 2007 *Cryst. Rep.* **52** 505–511
[4] Avdeev M V and Aksenov V 2010 *Phys. Uspekhi* **53** 971
[5] Sun S, Murray C, Weller D, Folks L and Moser A 2000 *Science* **287** 1989–1992
[6] Bodnarchuk M I, Kovalenko M, Pichler S *et al.* 2009 *ACS nano* **4** 423–431
[7] Mishra D, Benitez M, Petracic O *et al.* 2012 *Nanotechnology* **23** 055707
[8] Vorobiev A, Major J, Dosch H, Gordeev G and Orlova D 2004 *Physical Review Letters* **93** 267203
[9] Avdeev M V, Petrenko V I, Gapon I V *et al.* 2015 *Appl. Surf. Sci.* **352** 49–53
[10] Zavisova V, Koneracka M, Muckova M *et al.* 2011 *J. Magn. Magn. Mater.* **323** 14081412
[11] Kaur-Rajput J and Kaur G 2013 *Chinese Journal of Catalysis* **34** 1697
[12] Creanga D E, Iacob G, Ursache M, Nadejde C and Racuciu M 2008 *J. Optoelectron. Adv. Mater.* **10** 628
[13] Gapon I V, Petrenko V I, Avdeev M *et al.* 2014 *J. Surf. Investigations.* **9** 320–325
[14] Howse J H, Manzanares-Papayanopoulos E, McLure A *et al.* 2002 *Journal of Chemical Physics* **116** 7177
[15] Parratt L G 1954 *Phys. Rev.* **95** 359–369
[16] Nelson A 2006 *Journal of Applied Crystallography* **39** 273–276
[17] Balasoiu M, Ivankov O I, Soloviov D, Lysenko S, Yakushev R, Balasoiu-Gaina A M and Lupu N 2015 *Journal of Optoelectronics and Advanced Materials* **17** 1114 1121

Silicone rubber based magnetorheological elastomer: magnetic structure tested by means of neutron depolarization and magnetic force microscopy methods

M Balasoiu[1,2*], S V Kozhevnikov[1], Yu V Nikitenko[1], G E Iacobescu[3], M Bunoiu[4], I Bica[4]

[1]Joint Institute for Nuclear Research, Dubna, Russia
[2] "Horia Hulubei" National Institute for Physics and Nuclear Engineering, Bucharest, Romania
[3]Department of Physics, University of Craiova, 200585 Craiova, Romania
[4]West University of Timisoara, Faculty of Physics, 300223 Timisoara, Romania

E-mail: balas@jinr.ru

Abstract. Magnetorheological elastomer samples (MREs) prepared using silicone rubber and spherical Fe microparticles have been investigated by means of neutron depolarization (ND) and magnetic force microscopy (MFM). From MFM analysis it was found the average diameter of the microparticles in the case of the isotropic samples to be 4.85 μm, while for the anisotropic ones, 5.1 μm. It was applied and examined the neutron depolarization process in function the microparticle concentration variation and the sample anisotropy.

1. Introduction

It is well known that by introducing magnetic nano or microparticles in an elastic matrix, materials with tunable properties in the presence of external magnetic fields can be produced [1–5]. Based on the deformations and stresses induced in magnetorheological elastomers (MRE) [6-15], different application can be developed: giant magnetoresistors [7], quadrupolar magnetoresistors [7,8], Hall probes [8], deformation sensors [9,10], etc. Magneto-elastic properties of MRE are of high importance in defining their application areas [10]. The combination of different kinds of elastomers, magnetic particles and processes of manufacture may result in the obtaining of materials with a wide variety of properties [11-17].

The properties of these composites are the result of several effects, observable at different length scales and by different techniques. From small angle neutron scattering (SANS) data it was obtained earlier that doping with Fe_3O_4 nanoparticles and applying of a magnetic field during the polymerisation process led to a significant change in the local structure of the elastomer [18-23]. For an elastomer filled with a large amount of Fe microparticles a texture effect is observed by means of SANS, and this effect is larger for the samples polymerised in a magnetic field [19].

In the present paper results on the magnetic structure of a MRE investigated using magnetic force microscopy (MFM) and the neutron depolarization methods are reported.

* To whom any correspondence should be addressed.

2. Experimental

2.1. Samples preparation

The materials used in this study are silicone rubber (RTV3325, Rhône-Poulenc), catalyst (60R, Rhône-Poulenc), silicone oil (Merck), and stearic acid (S4641773, Merck). Soft magnetic carbonyl iron (CI) from Sigma with average diameter 5.0 μm was used as a dispersible microparticles. Using these materials, three different types of anisotropic MR elastomers were prepared. At first, the mixture of CI and silicone oil was heated and maintained at the temperature of 573 K (±10 % error range) for about 10 min, while the mixture was being homogenized. Stearic acid was then introduced in the mixture when its temperature dropped to 370 K (±10 % error range) with stirring. The as-obtained MR suspension contained 80 % CI, 15 % silicone oil, and 5 % stearic acid in weight. Pre-established amounts of MR suspension were introduced in silicone rubber and then mixed in the presence of the catalyst. A homogenous paste was then achieved containing 75 % silicone rubber, 20 % MR suspension, and 5 % catalyst for the S01 sample, 55 % silicone rubber, 40 % MR suspension and 5 % catalyst for the S02 sample, and 35 % silicone rubber, 60 % MR suspension and 5 % catalyst for the S03 sample with all in a weight fraction.

The following step was the introduction of the samples between two parallel glass plates; the distance between the plates was fixed at the 1 mm (±10 % error range). Further, the formed ensemble was introduced for 24 hours between the magnetic poles of a Weiss electromagnet (Phylatex type, Germany) of 500 kA/m fixed magnetic field intensity. The direction of the magnetic field intensity was perpendicular to the surface of the samples (SM1, SM2, SM3). At the end of the polymerization, MR elastomers stripes were obtained. For the investigations disks were cut out from the stripes. Each disk has a diameter of 20 mm and thickness of 1 mm [6].

2.2. Magnetic force microscopy

Magnetic force microscopy (MFM) is based on the magnetostatic interaction between the magnetic sample and a magnetic tip placed over the sample surface. The method has high spatial resolution (routinely better than 100 nm) and high surface sensitivity. MFM measurements are carried out using two-pass method. For each raster line, the first pass is made very close to the sample surface (AFM - atomic force microscopy) and yields knowledge of the surface topography. The second pass then follows the recorded topography but at an increased scan height. Obviously, the tip-sample distance must be large enough (usually 50-100 nm) to eliminate (or minimize) the short-range van der Waals' forces that provide the topographic contrast. Then in the second pass the tip is affected only (mainly) by long-range magnetic forces and corresponding MFM image of the sample surface is obtained.

For imaging the surface topographies of samples S01, S02, S03, SM1, SM2, SM3 the noncontact operating mode AFM (NC-AFM) it was used. The cantilever had a nominal length of 125 mm, a nominal force constant of 40 N/m, and oscillation frequencies in the range of 275–373 kHz. A horizontal line-by-line flattening planarization method was applied. The MFM images permitted to determine the average ferromagnetic particles diameter, using the line profile scan mode of the AFM data acquisition program applied for 20 different regions of the samples over an area of 500 μm [24].

2.3. Neutron depolarization

Polarized neutrons are useful to extract the magnetic contribution from the total result of the neutron interaction with magnetic media. A neutron technique for direct measurement of the magnetic induction from experimental data is the Larmor precession method. The polarization degree of a neutron beam is depicted by the following relationship

$$P = \frac{N^+ - N^-}{N^+ + N^-} \quad (1)$$

where, N^+ and N^- represent the neutrons number with the parallel spin, respectively antiparallel to the external applied magnetic field. The beam polarization degree reducing process is named as *depolarization*. The theory of depolarization of the neutron beams transmitted through ferromagnetic media was developed by Halpern and Holstein [25]. First experimental application of the neutron depolarization was done by Burgy et al. [26] and Drabkin et al. [27]. This method was developed and applied by Rekveldt [28,29] for the investigation of ferromagnetic materials. At present, this method is exploited at a few neutron centres in the world [30-38]. When the neutron passes through a magnetic cluster, the neutron spin is subjected to the Larmor precession around the local magnetic induction vector. Therefore, from the exponential depolarization curve, the dispersion of the magnetic induction inside the sample may be extracted.

In the present paper a time of flight (TOF) depolarization at Larmor [31,35,38,39] precession experiment, accomplished for testing the magnetic structure of silicone-rubber MRE's samples, is reported.

The neutron depolarization measurements were carried out on the polarized neutron spectrometer REMUR (see figure 1) installed at the pulsed reactor IBR-2 operated by the Frank Laboratory of Neutron Physics of JINR.

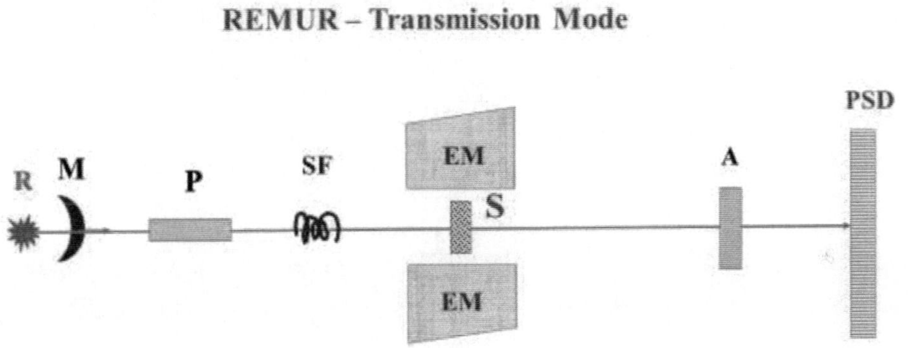

Figure 1. The REMUR spectrometer scheme for the polarized neutrons transmission mode experiment.

In figure 2 the geometrical features of the Larmor precession in a basic experiment are presented. A rectangular sample of a soft magnetic plate is placed under an angle β to the external applied magnetic field $\vec{H}$. The angle between the external applied magnetic field $\vec{H}$ and the average magnetic induction $<\vec{B}>$ in the sample is α. The neutron path through the ferromagnetic plate is d. The initial polarization of the neutron spin, $\vec{P_0}$ is directed along the applied magnetic field $\vec{H}$ and the angle between the vectors $\vec{P_0}$ and $<\vec{B}>$ is α. The existence of perpendicular polarization component of $<\vec{B}>$ results in the Larmor precession of the neutron spin in the local mean field $<B>$. The Larmor precession frequency is defined by the absolute value of magnetic induction through the relationship:

$$\omega_L = \gamma_n < B > \quad (2)$$

where, $\gamma_n = 1.83 \cdot 10^8 s^{-1} \cdot T^{-1}$ is the gyromagnetic ratio of the neutron. The Larmor precession angle in the sample is given by the expression:

$$\phi = \omega_L t = \gamma_n < B > \frac{d}{v}$$

where, t is the time that the neutrons of velocity v cross the distance d inside the sample. Using the expression of the momentum, $p = \frac{h}{\lambda}$, would mean that

$$\frac{1}{v} = \frac{m}{h}\lambda$$

where, m is the mass of the neutron, h is the Plank's constant and λ is the neutron wavelength [38]. Therefore, the angle of precession can be expressed as $\phi = \omega_\lambda \lambda$, where $\omega_\lambda = 0.04633 \cdot < B > \cdot d$ (ω_λ in [Å^{-1}], $< B >$ in [T], d in [μm]).

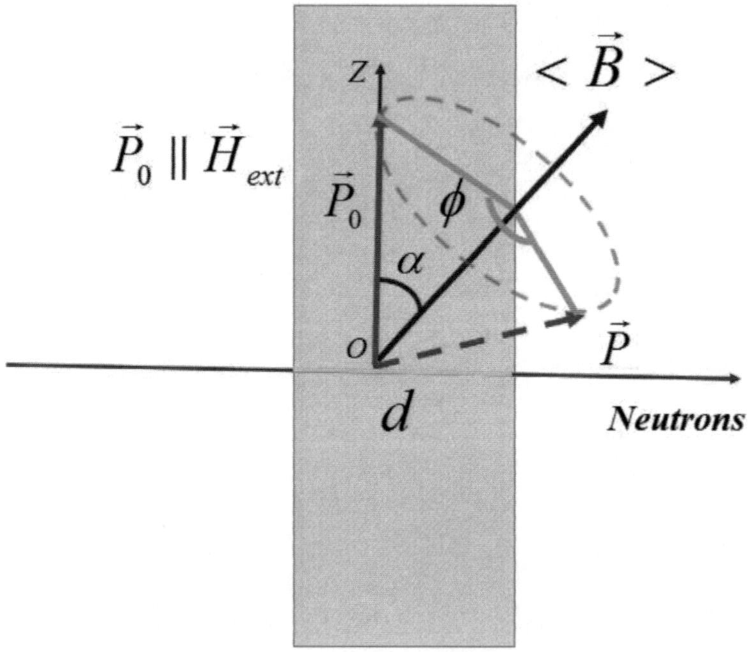

Figure 2. The geometrical features of the Larmor precession in a basic experiment are presented. A soft magnetic plate is placed in an external applied magnetic field $\vec{H}$. The angle between the external applied magnetic field $\vec{H}$ and the average magnetic induction $<\vec{B}>$ in the sample is α. The neutron path through the ferromagnetic plate is d. The initial polarization of the neutron spin, $\vec{P_0}$ is directed along the applied magnetic field $\vec{H}$ and the angle between the vectors $\vec{P_0}$ and $<\vec{B}>$ is α. The perpendicular polarization component of $<\vec{B}>$ determines the Larmor precession of the neutron spin in the local mean field $< B >$.

In the case of magnetic fluctuations arising along the neutron path through a magnetorheological membrane (see Fig.3) the following relation is obtained using the depolarization coefficient expression of Halpern and Holstein [25]:

$$\frac{P}{P_0}(\lambda) = \exp(-A\lambda^2) \quad (3)$$

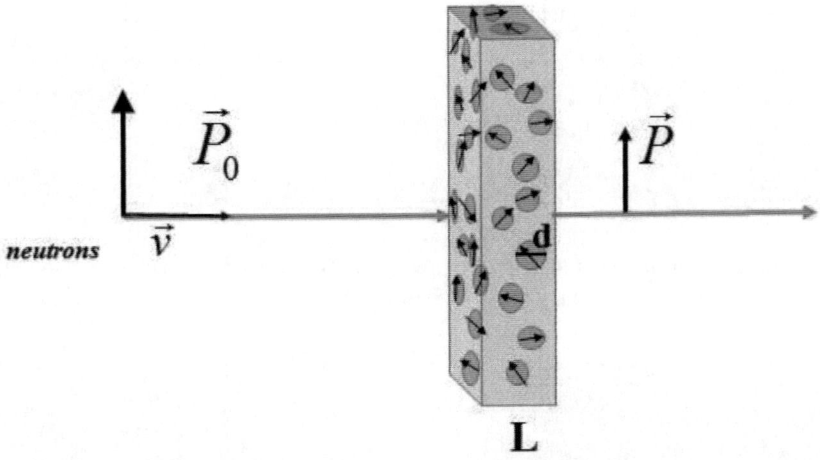

Figure 3. The geometry of the depolarization experiment in the isotropic magnetorheological elastomer sample S01 (P_0 and P denotes the initial and respectively final neutron spin polarization ; d is the particle diameter; L is the thickness of the sample).

3. Results and Discussions

In figure 4 MFM images from the surface of the isotropic MRE samples with different concentration of carbonyl iron (CI) suspension: (a) 20%; (b) 40%; (c) 60%, are presented. MFM method, provide images of the differences between the regions with different magnetic susceptibility.

In figure 5 MFM images from the surface of isotropic MRE samples, S01 (a) and anisotropic, SM1 (b) samples with 20% particle concentration and respectively 60% (S03, SM3) are given. The images show the tendency of magnetic particles agglomeration in the case of anisotropic MREs with high particle concentration (40 wt%, 60 wt%).

The MFM images can provide the average diameters of the ferromagnetic particles.

It was found the average diameter of the microparticles in the case of the isotropic samples to be 4.85 μm, while for the anisotropic ones, 5.1 μm.

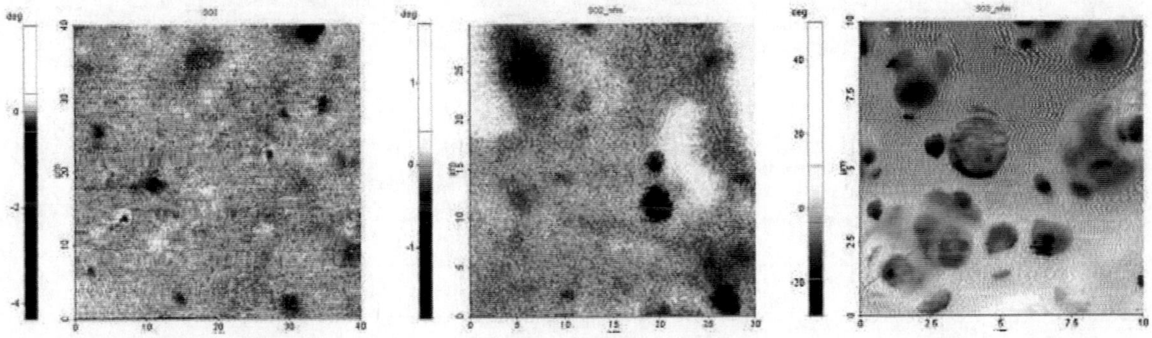

a b c

Figure 4. MFM images from the surface of the isotropic MRE samples with different concentration of carbonyl iron (CI) suspension: (a) 20% (S01); (b) 40% (S02); (c) 60% (S03).

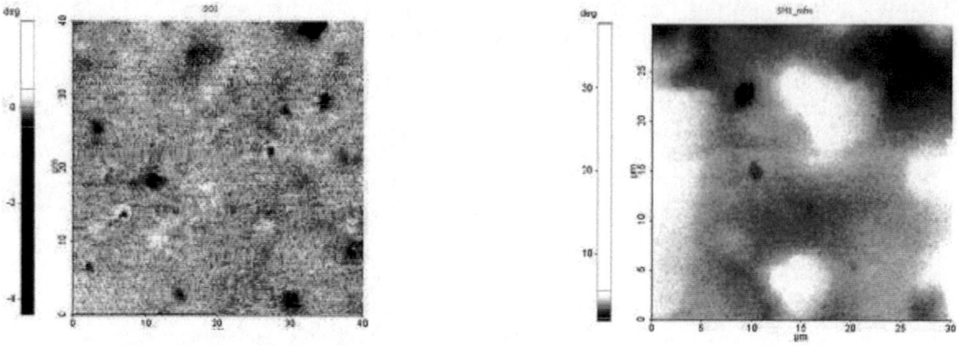

a b

Figure 5. MFM images from the surface of isotropic MRE samples S01 (a) and anisotropic SM1 (b) (75% silicone rubber, 20 % MR suspension, and 5 % catalyst).

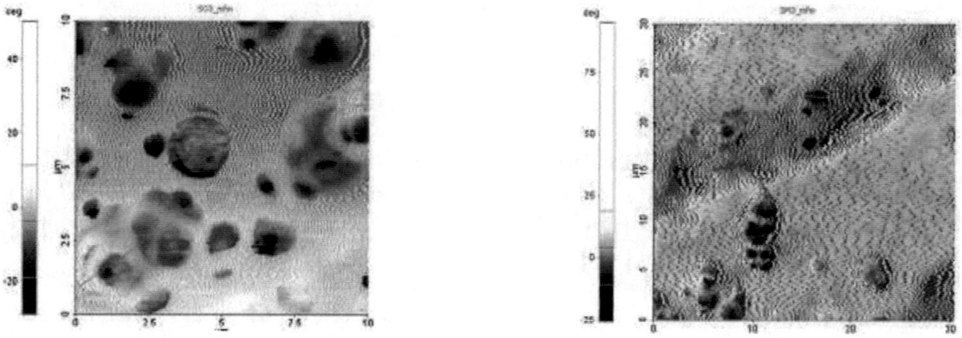

a b

Figure 6. MFM images of the surface of isotropic MRE sample SO3 (a) and anisotropic SM3 (b) (35% silicone rubber, 60 % MR suspension and 5 % catalyst).

The neutron depolarization experimental data and the fitting curves of the normalized polarization versus the wavelength for the isotropic samples S01, S02, S03 and the elastomeric matrix are depicted in figure 7.

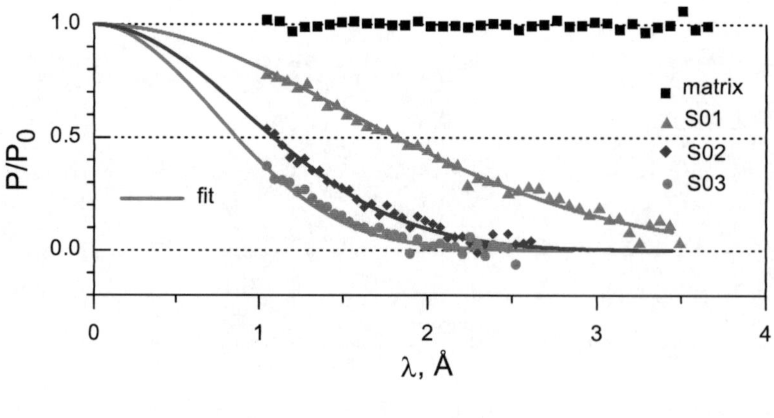

Figure 7. Experimental data (symbols) and the fitting curves (lines) of the normalized polarization versus the wavelength for the isotropic samples S01, S02, S03 and the elastomeric matrix.

It can be seen that there is no depolarization effect through the elastomeric matrix. For the isotropic samples S01, S02, S03 the depolarization enhancement with the particle concentration increase is observed.

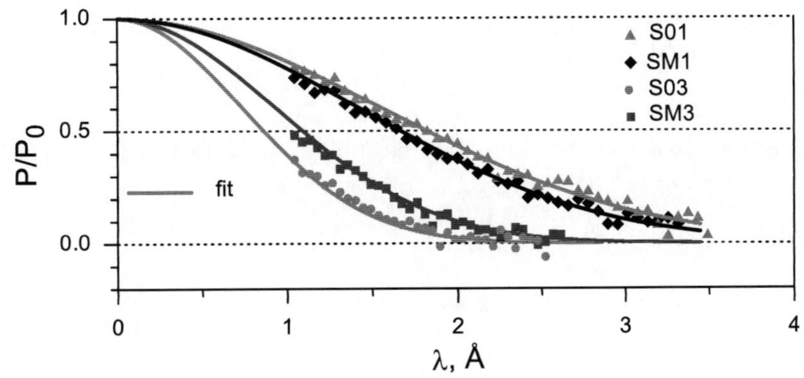

Figure 8. Experimental data and fitting curves of the normalized polarization as a function of the neutron wavelength from isotropic and anisotropic samples S01, SM1 and S03, SM3.

The comparison of the experimental data and fitting curves of the normalized polarization as a function of the neutron wavelength from isotropic and anisotropic samples S01, SM1 and S03, SM3 is presented in figure 8. For low particle concentration, the depolarization in the case of anisotropic sample overcomes the depolarization in isotropic one. On the contrary, for the highest particle concentration the depolarization for the anisotropic sample is lower than in the anisotropic one.
In table 1 the results of the neutron depolarization curve fits are given. The parameter A describes the magnitude of depolarization process.

Table 1. Values of parameter A obtained from neutron depolarization curve fits.

No.	Sample	A
1	S01	0.21
2	S02	0.59
3	S03	0.93
4	SM1	0.25
5	SM3	0.58

The depolarization process depends on the number of the magnetic fluctuations arising along the neutron path (through the mean square induction fluctuations) and the mean size of the magnetic inhomogeneites.

Modelling the depolarization process through the magnetorheological isotropic and anisotropic membrane details on the magnetic particles distribution and particle mean size can be obtained.

4. Conclusions

Magnetorheological elastomer samples (MREs) prepared using silicone rubber and spherical Fe microparticles have been investigated by means of neutron depolarization (N D) and magnetic force microscopy (MFM). From MFM analysis it was found the average diameter of the microparticles in the case of the isotropic samples to be 4.85 μm, while for the anisotropic ones, 5.1 μm.

Further it was examined the depolarization process with the microparticle concentration variation and with the sample anisotropy. The development of a model for processing de neutron depolarization data is in progress.

Acknowledgements
The authors acknowledge the RO-JINR Grant No.218/10.04.2017 item 14 and RO-JINR Projects Nos. 219/10.04.2017 and 220/10.04.2017 item 61.

References

[1] Abramchuk S, Kramarenko E, Stepanov G, Nikitin L, Filipcsei G, Khokhlov A, Zrinyi M 2007 *Polym. Adv. Technol.* **18** 883

[2] Stepanov GV, Chertovich AV, Kramarenko EYu 2012 *J. Magn. Magn. Mat.* **324** 3448

[3] Lopez-Lopez MT, Durán Juan DG, Iskakova, LYu, Zubarev AYu 2016 *Journal of Nanofluids* **5**(4) 479

[4] Stolbov OV, Yu L. Raikher YuL, Balasoiu M 2011 Soft Matter **7** 8484

[5] Andriushchenko P, Nefedev K & Stepanov G 2014 *Eur. Phys. J. B* 87 11 doi:10.1140/epjb/e2013-31097-1

[6] Bica I, Liu Y D, Choi H J 2012 *Colloid. Polym. Sci.* **290** 1115

[7] Bica I 2010 *Mater. Sci. Eng. B* **166** 94

[8] Bica I 2012 *J. Ind. Eng. Chem.* **18** 483

[9] Bica I 2012 *J. Ind. Eng. Chem.* **18** 1666

[10] Bica I, Balasoiu M, Kuklin A I 2012 *Solid State Phenom.* **190** 645

[11] Chertovich AV, Stepanov GV, Kramarenko EYu, Khokhlov AR, 2010 Macromol. Mater. Eng. **295** 336

[12] Philippova O, Barabanova A, Molchanov V, Khokhlov A 2011 Eur. Polym. J. **47**(4) 542

[13] Bica I, Anitas EM, Bunoiu M, Vatzulik B, Juganaru I 2014 *J. Ind. Eng. Chem.* **20** 3994

[14] Bica I, Balasoiu M, Bunoiu M, Iordaconiu L 2016 *Rom. Journ. Phys.* **61**(5–6) 926

[15] Balasoiu M and Bica I 2016 *Results in Phys.* **6** 199

[16] Bunoiu M, Bica I 2016 *J. Ind. Eng. Chem.* **37** 312

[17] Bica I, Balasoiu M, Bunoiu M, Iordaconiu L, Cirtina G 2015 *J. Optoelectron. Adv. Mater.* **17** (9-10) 1379

[18] Balasoiu M, Craus ML, Kuklin AI, et al 2008 *J. Optoelectron. Adv. Mater.* **10**(11) 2932

[19] Balasoiu M, Craus ML, Anitas EM, Bica I, Plestil J, Kuklin AI 2010 *Phys. Solid State* **52**(5) 917

[20] Balasoiu M, Lebedev VT, Orlova DN, Bica I 2011 *Crystallogr. Rep.* **56**(7) 93

[21] Balasoiu M, Bica I, Raikher Yu L, Dokukin E B, Almasy L, Vatzulik B, Kuklin A I 2011 *Optoelectron. Adv. Mater.* **5**(5) 514

[22] Balasoiu M, Lebedev VT, Orlova DN, Bica I, Raikher YL 2012 *JPCS* **351**(1) 012014

[23] Balasoiu M, Lebedev VT, Raikher YuL, Bica I, Bunoiu M 2017 *Journ Magn. Magn. Mater.* **431** 126-129

[24] Iacobescu GE, Balasoiu M, Bica I 2013 *J Supercond Nov Magn* **26** 785

[25] Halpern O, Holstein T 1941 *Phys. Rev.* **59** 960

[26] Burgy M, Hughes DJ, Wallace RJ, Heller RB, Woolf WE 1950 *Phys Rev. B* **80** 953

[27] Drabkin GH et al. 1965 *JETP* **20** 1548

[28] Rekveldt M Th 1973 Z. Phys. **259** 391

[29] Kraan W, Rekveldt M Th 1978 *J. Magn. Magn. Mater.* **8** 168

[30] Maleyev SV 1982 *J. de Physique* **43** (C7) 23

[31] Dokukin E, Korneev D, Loebner W, Pasjuk V, Petrenko V, Rzany H, 1988 *J. Phys. Coll.* **49** (C8) 2073

[32] Taketomi S, Itoh S, Endoh Y, Ogawa S, Miyajima H, Chikazumi S 1988 *J. Appl.Phys.* **64**(10) 5849

[33] Mitsuda S, Yoshizawa H, Endoh F 1992 *Phys. Rev. B* 45(17) 9788

[34] Rekveldt M. Th. in Studies of Magnetic Properties of Fine Particles and their Relevance to Material Science (J. L. Dorman and D. Fiorani, eds.). 1992. Amsterdam: Elsevier Science Publishers B.V. pp. 445

[35] Krezhov K, Lilkov V, Konstantinov P, Korneev D 1993 *J. Phys.: Condens. Matter* 5 9277

[36] Thibaudeau P, Ott F, Thiaville A, Dubuget V, Duverger F, *EuroPhys. Lett.* 93 (2011) 37003.

[37] Rekveldt M Th, Kraan WH 2013 *J. Magn. Magn. Mater.* 329 105

[38] Kozhevnikov SV, Ott F, Radu F 2016 *Journ Magn. Magn. Mater.* 402 83

[39] Balasoiu M, Dokukin E B, Kozhevnikov S V, Nikitenko Yu V 1999 *Proceedings of the international school and symposium on small angle scattering (Book series:*KFKI Preprint/Report Series vol 1999 issue 2) part E pp 90-94

Morphological structure of *Gluconacetobacter xylinus* cellulose and cellulose-based organic-inorganic composite materials

R Yu Smyslov[1, 2, *], K V Ezdakova[2], G P Kopitsa[2, 3], A K Khripunov[1], A N Bugrov[1, 4], A A Tkachenko[5], B Angelov[6], V Pipich[7], N K Szekely[7], A E Baranchikov[8], E Latysheva[9], Yu O Chetverikov[2] and V Haramus[10]

[1] Institute of Macromolecular Compounds RAS, St Petersburg, 199004, Russia
[2] B P Konstantinov Petersburg Nuclear Physics Institute (National Research Center "Kurchatov Institute"), Gatchina, Russia
[3] I V Grebenshchikov Institute of Silicate Chemistry RAS, St Petersburg, Russia
[4] Saint Petersburg Electrotechnical University "LETI", 197376, St Petersburg, Russia
[5] Department of Microbiology, S Petersburg State University, St Petersburg, 199178, Russia
[6] Institute of Physics ASCR, ELI Beamlines, 18000 Prague, Czech Republic
[7] JCNS, Forschungszentrum Juelich GmbH, Outstation at MLZ, 85747, Garching, Germany
[8] N S Kurnakov Institute of General and Inorganic Chemistry RAS, Moscow, Russia
[9] RC of SPbSU "Centre for diagnosis of functional materials in medicine, pharmacology and nanoelectronics", S Petersburg, 198504, Russia
[10] Helmholtz- Zentrum Geesthacht Outstation at EMBL/DESY, Notkestrasse 85, Hamburg, Germany

[*] Author to whom any correspondence should be addressed

E-mail: urs@mail.macro.ru

Abstract. Scanning electron microscopy, ultra-small-angle neutron scattering (USANS), small-angle neutron and X-ray scattering (SANS and SAXS), as well as low-temperature nitrogen adsorption, were used in the studies of micro- and mesostructure of polymer matrix prepared from air-dry preliminarily disintegrated cellulose nano-gel film (synthesized by *Gluconacetobacter xylinus*) and the composites based on this bacterial cellulose. The composites included ZrO_2 nanoparticles, Tb^{3+} in the form of low molecular weight salt and of metal-polymer complex with poly(vinylpyrrolydone)-poly(methacryloyl-*o*-aminobenzoic acid) copolymer. The combined analysis of the data obtained allowed revealing three levels of fractal organization in mesostructure of *G. xylinus* cellulose and its composites. It was shown that both the composition and an aggregation state of dopants have a significant impact on the structural characteristics of the organic-inorganic composites. The composites containing Tb^{3+} ions demonstrate efficient luminescence; its intensity is an order of magnitude higher in the case of the composites with the metal-polymer complex. It was found that there is the optimal content of ZrO_2 nanoparticles in composites resulting in increased Tb^{3+} luminescence.

1. Introduction

In recent years, due to new requirements for quality of materials, great interest has been generated in organic-inorganic composites which are promising materials in various fields including optoelectronics, medicine, cosmetology, textile industry etc. [1–3]. Cellulose-based composites possess natural biodegradability and biocompatibility; they consist of cellulose nanocrystals (**CNs**), which warrants attention because of CNs' unsurpassed physical and chemical properties [4]. Cellulose is one of the most widespread natural polymers on Earth. It is known that cellulose can be obtained from various evolutionary sources; it can be synthesized by plants, some animals (e.g., members of the subphylum *Tunicata*), or bacteria (e.g., *Gluconacetobacter xylinus*). It should be noted that biosynthesis of cellulose performed by various organisms yields products with varying morphological structures which differ significantly in supramolecular organization [5]. At the same time, cellulose biosynthesis by *G. xylinus* is environmentally friendly.

Unlike plants, *G. xylinus* accomplish producing cellulose (**GXC**) on the nano-sized level [6]. GXC synthesized in stationary conditions on the surface of nutrient medium is pellicle, which is able to retain water in the approximate ratio of 1:100 (dry substance : H_2O) [7]. Due to certain physicochemical properties and the unique structure, this nano-gel film (**NGF**) is widely used as a matrix in preparation of various organic-inorganic composite materials (after preliminary treatment). The GXC-based composites containing gold, silver, selenium nanoparticles (**NPs**), TiO_2, SiO_2, CdSe and calcium phosphate NPs have been described [8]. GXC can be used as a scaffold (rigid supramolecular structure) for regeneration of a wide variety of biological tissues. Thus, the authors of [3, 9] demonstrated that GXC can serve as a matrix for mechanical reinforcement of hydrogels from poly(acrylamide), poly(vinyl alcohol) and similar polymers. The presence of hydrophilic synthetic polymers in a material imparts elasticity and the necessary swelling degree, while the rigid-chain base (cellulose) provides strength to a composite. Composite hydrogels reinforced by GXC are similar to different types of cartilage in their water content and mechanical properties; therefore, these materials can be used as substitutes for damaged cartilage fragments [10]. Thus, GXC may become excellent material for medical nanotechnologies [3, 11].

Introducing lanthanides into hybrid polymer matrix allows developing materials with tunable or modulated optical properties [1]. Lanthanide ions possess the unique photophysical properties, which are important for applications in photonics [12–15], for generation and enhancement of laser light [16] or in producing optical amplifiers, *e.g.* erbium-doped fiber amplifiers [17, 18]. Lanthanide ions are widely used in development of light-emitting materials for active light guides in visible and near-infrared spectral regions, active coatings, biomedical devices and chemosensors. Thus, lanthanide ions can be used in numerous areas of material science and technology related to integration, miniaturization and functionality of devices. For example, the membranes based on composites of GXC with ITO and SiO_2 can be used as flexible substrates in producing organic light-emitting diodes [19].

2. Materials and methods

2.1. Biosynthesis of bacterial cellulose by Gluconacetobacter xylinus

Biosynthesis of cellulose was performed by *G. xylinus* (bacterium strain № 1629 CALU deposited at the Department of Microbiology of S. Petersburg State University). The nutrient medium contained aqueous solutions of yeastrel, glucose, peptone, ethanol and beer wort; concentrations of these nutrients were found after optimization of the medium [20, 21]. The pH of the medium was 5.9—6.0. The *G. xylinus* strain cultivated for 48 hrs. in the nutrient medium containing mixture of yeastrel and beer wort (6 degrees Balling) in the 1:1 ratio to the mixture of glucose (2 wt.%) and ethanol (1 vol.%) was used as an inoculum. Biosynthesis of GXC was carried out at 30.0 ± 0.5 °C for 6–7 days; then, cellulose was isolated and washed periodically with renewed boiling aqueous solution of NaOH (concentration 0.5-1%) until all *G. xylinus* cells were removed. Then, in order to remove NaOH, GXC was washed with distilled water, acetic acid (0.5% solution) and again with distilled water until the neutral pH was achieved. The purified bacterial cellulose (BC) was stored either in the form of NGF in distilled water at 5°C or in autoclave (after sterilization) at room temperature as well as in the dry form (after freeze-drying).

2.2. Synthesis of hybrid organic-inorganic composites

The GXC nano-gel film intended for preparation of composite material was preliminarily disintegrated in laboratory blender at 2000 rpm. Then aqueous suspension of disintegrated NGF was obtained; the necessary amount of ZrO_2 NPs (obtained by hydrothermal method) in the form of powder with an approximate diameter of 200 ± 50 Å [22] was added to cellulose suspension under continuous stirring and ultrasound treatment (35 kHz, 100 W) (*note* K3, K5, K6, K8 on table 1).

Luminescent material was obtained using Tb^{3+}. These ions were introduced into cellulose either in the form of low molecular weight luminophore ($TbCl_3$) or in the form of metal-polymer complex (**MPC**) [Tb(*polyLig*)](III) in aqueous solution. Copolymer of vinyl pyrrolidone (**VP**) with 12 mol.% of methacryloyl-*o*-aminobenzoic acid (**M-*o*-ABA**) or **P(VP-*co*-MABA)** was used as a polymeric ligand (***polyLig***) in the MPC (*note* the copolymer formula in the footnote of table 1) [23].

Sorption of aqueous solution of terbium chloride (III) by suspension of disintegrated GXC was carried out for 24 hrs. In this procedure, we used terbium chloride hexahydrate ($TbCl_3{\cdot}6H2O$) (99.9% purity) purchased from Sigma-Aldrich (CAS 13798-24-8) without further purification. Aqueous solution of this reactant at a concentration of 0.72 mg/mL was prepared, and the necessary amount of this solution was added to water suspension of GXC (table 1).

Table 1. The composites based on GXC with added Tb^{3+} (in the form of salt or MPC) and ZrO_2 NPs. The values of luminescence efficiency θ are given at an excitation wavelength (λ_{ex}) of 300 nm and at an observation wavelength of 543 nm (equation 4).

System	GXC, *wt.%*	$c_{Tb^{3+}}$, *wt. %*	*Additives*		θ
			MPC*)	ZrO2	
K0	100	0	–	–	0
K1	99.98	0.02	–	–	1
K2	99.8	0.21	–	–	1.40
K3	98.8	0.21	–	1	1.73
K4	97.5	2.50	–	–	0.33
K5	90.7	0.19	–	9.1	0.94
K6	66.6	0.14	–	33.3	1.16
K7	63.6	1.60	34.7	-	7.74
K8	59.8	1.50	32.6	6	11.86

*) the MPC was formed on the base of P(VP-*co*-MABA):

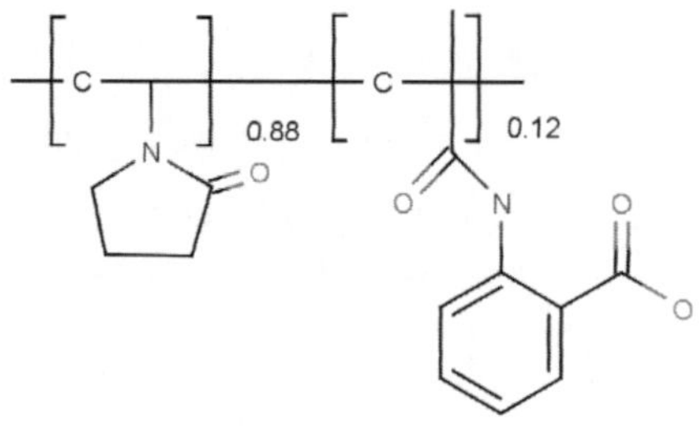

VP M-*o*-ABA

The MPC was prepared beforehand in the following way. Copolymer was dissolved, and side groups of M-*o*-ABA were ionized; then, terbium (III) ions were added in the ratio of [M-*o*-ABA] to

[Tb^{3+}] equal to 4 :1. The obtained complex [Tb(*polyLig*)](III) was added in the necessary amount to aqueous suspension of disintegrated GXC nano-gel film (*note* K7, K8 on table 1).

The prepared aqueous suspensions of disintegrated GXC nano-gel film containing additives were dried in thermostat at 50° C until constant weight was reached, and thus air-dry films of composites were obtained. Composition of materials based on the polymer matrix (air-dry previously disintegrated native GXC) with added ZrO_2 NPs and terbium ions in the form of $TbCl_3$ or [Tb(*polyLig*)](III) is given on table 1.

The density of GXC is 1.5 g/cm^3, the density of the P(VP-*co*-MABA) copolymer is 1.4 g/cm^3, the density of ZrO_2 is approximately 6 g/cm^3. Thus, volume fractions of components in air-dry composite films (organic-inorganic hybrid material) varied as follows: from 99.98 to 59.8 wt.% for GXC, from 0 to 33.3 wt.% for ZrO_2 NPs, and from 0.02 to 2.5 wt.% for Tb^{3+}. The portion of MPC was 32-35 wt.% (table 1). The studied objects were rectangular films with the following dimensions: $l \times w \times h$ = 25 mm × 25 mm × (70÷150) μm.

2.3. Luminescence studies

Luminescence emission and excitation spectra of composites containing Tb^{3+} and other additives (table 1) were recorded using a LS-100 BASE luminescence spectrophotometer (PTI Lasers Inc., Canada) equipped with a solid sample holder. Luminescence intensity, I_{lum}, was registered in the direction of incident exciting light in the phosphorescence mode with a signal integration window of 100÷2000 μs. The glancing angle between exciting light beam and a sample was ~ 30°. The wavelength range of 460÷700 nm was used for emission spectra (the excitation wavelength λ_{ex} = 299 nm); the wavelength range of 210÷400 nm was used in observation of luminescence (the observation wavelength λ_{obs} = 543 nm). The used slit width of excitation and luminescence monochromators was 4 nm. Photomultiplier tube gain values were 200 and 500. For correct comparison, the measured values of I_{lum} were normalized by the internal laboratory standard.

Evaluation of luminescence efficiency of the composites, θ, was performed using the I_{lum} value of the luminescence band at λ_{max} = 543 nm; this maximum corresponds to the $^5D_4$ — $^7F_5$ transition of Tb^{3+} from the emission resonance level to the ground state, which demonstrates the maximum intensity.

2.4. Neutron measurements

The small-angle and ultra-small-angle neutron scattering measurements were performed on the KWS-2 and KWS-3 scattering facilities (FRM-II reactor, Garching, Germany). The KWS-2 facility is a classical SANS diffractometer operating in near point geometry. The measurements were performed at the neutron wavelength λ = 4.55 Å with $\Delta\lambda/\lambda$=0.1. The range of momentum transfer $3\cdot10^{-3} < q < 4.3\cdot10^{-1}$ Å^{-1} was obtained using three sample-to-detector distances S_D (2, 8 and 20 m). The scattered neutrons were detected by a two-dimensional position-sensitive detector (128×128 cells with a spatial resolution of 5 mm × 5 mm) based on ^{6}Li glass scintillators.

The KWS-3 is a high-resolution SANS diffractometer running on the focusing mirror principle; it allows performing scattering experiments with high momentum transfer resolution up to 10^{-4} Å^{-1} [24, 25]. The measurements were performed at the neutron wavelength λ = 12 Å with $\Delta\lambda/\lambda$=0.2. The range of momentum transfer $2.5\cdot10^{-4} < q < 1.4\cdot10^{-2}$ Å^{-1} was obtained using two sample-to-detector distances *SD* (1 and 10 m). The scattered neutrons were detected by a two-dimensional position-sensitive detector (diameter of the active zone was 8.7 cm, spatial resolution was 0.36 × 0.39 mm2) based on ^{6}Li glass scintillators.

In both cases, films of preliminarily disintegrated air-dry GXC and GXC-based organic-inorganic composites were clamped between two quartz glass plates. The raw data were corrected using the standard procedures [26] taking into account the scattering from the setup equipment and quartz glasses plates. The resulting 2D isotropic spectra were averaged azimuthally and obtained 2D spectra of SANS and USANS were merged. Their absolute values were determined by normalization to the incoherent scattering cross section of plexiglas with inclusion of the detector efficiency [26] and thickness L_s for each sample. All the measurements were done at room temperature. The QtiKWS software [27] was used for data processing.

The intensity of SANS and USANS analysed in this work ($I_s(q)$) was determined as:

$$I_s(q) = I(q) - T \cdot I_0(q) \tag{1}$$

where $I(q)$ and $I0(q)$ are the q-distributions of scattered neutrons after sample and neutron beam without sample, respectively; T is the transmission coefficient of the neutrons passed through a sample:

$$T = \frac{I(0)}{I_0(0)} = \exp\left(-\Sigma_{tot} \cdot L\right) \tag{2}$$

where $\Sigma = \sigma_s + \sigma_a$ is the integral cross-section including nuclear scattering σ_s and absorption σ_a; L is the sample thickness.

The measured values of SANS and USANS intensity are related to the scattering law $S(q)$ by the following equation:

$$I_s(q) = I_0 \cdot L \cdot \int F(q - q_1) S(q) dq_1 \tag{3}$$

where $F(q)$ is the function of instrument resolution which was approximated by the Gaussian function [28].

Thus, the use of these two methods (SANS and USANS) allowed us to obtain the complete picture of scattering by the polymer matrix prepared from preliminarily disintegrated air-dry GXC and GXC-based organic-inorganic composites in the q range from $2.5 \cdot 10^{-4}$ to 0.43 $Å^{-1}$, which corresponded to the analysing structure of the studied objects in the range of characteristic dimensions from 10 Å to several micrometers.

2.5. Measurements of Small-angle X-ray scattering

The measurements of small-angle X-ray scattering (**SAXS**) were carried out at the BioSAXS facility situated on the line 12P of the synchrotron PETRA III (DESY, Hamburg) and running in the axial geometry. Using photons of wavelength $\lambda = 0.124$ nm and the distance L = 3 m for the sample detector allowed to measure the scattering intensity in the range of momentum transferred $7 \cdot 10^{-3} \leq q \leq 1$ $Å^{-1}$. The installation is implemented with the gas-filled detector Pilatus 2M (Dectris) with the active area size equal to 253.7×288.8 mm and the pixel size equal to 172×172 μm^2.

Source spectra were corrected with taking into account the scattering by accessories, adhesive tape, and the background of the hall. The two-dimensional isotropic spectra obtained were azimuthally averaged with due regard for the efficiency of the detector. All measurements were performed at ambient temperature. For pre-processing of data, one used the ATSAS software package [29].

It should be noted that use of small-angle X-ray scattering, which is complementary to SANS, gave us opportunity to obtain additional information about the structure of the polymer matrix prepared from preliminarily disintegrated air-dry GXC and GXC-based organic-inorganic composites in the range of typical sizes not exceeding *ca.* 500 Å.

2.6. Scanning electron microscopy

In parallel with SANS and X-ray scattering experiments, microstructure of films was studied with the use of a Carl Zeiss N Vision 40 scanning electron microscope. Films were not subjected to any special treatment (e.g., coating with metal). Microphotographs were taken at an accelerating voltage of 1 kV.

2.7. The low-temperature nitrogen adsorption

The determination of specific surface area S_{sp} for films of organic-inorganic composites based on the polymer matrix of the air-dry pre-disintegrated GXC was performed by low-temperature nitrogen adsorption using a commercial QUADRASORB SI instrumentation. Before measuring, the samples were degassed at 80° C under vacuum for 5 ÷ 7 hours, depending on a sample. Based on the data obtained, one calculated specific surface area S_{BET} using the Brunauer–Emmett–Teller (BET) equation on six points.

3. Results and discussion

3.1. Luminescence studies

Luminescence of hybrid polymer-inorganic GXC-based nanocomposites is provided by the presence of Tb^{3+} in the form of $TbCl_3$ and [Tb(*polyLig*)](III). The content of Tb^{3+} in composites was varied in order to find the optimal luminescence behaviour which reflects interaction between ZrO_2 NPs and the matrix in composites (compare samples K1, K2 and K4 in table 1). So, one of our tasks was to determine whether NPs are aggregated in the matrix, i.e. whether the GXC matrix is capable of dispersing ZrO_2 over the whole volume at the expense of possible hydrogen bonds. Since terbium (III) ions are able to adsorb on both the surface of NPs and the crystalline nanofibrils of GXC, it was necessary to reveal whether possible aggregation of NPs affect luminescence properties of the films. Figure 1 presents luminescence spectra of the composites containing varied amount of Tb^{3+} in the form of salt or MPC (with and without ZrO_2 NPs).

Analysis of luminescence spectra demonstrates the presence of characteristic bands at 490, 543, 580 and 620 nm within the luminescence band; these peaks can be compared with resonance transitions in luminescence spectra of Tb^{3+} corresponding to spectral terms $^5D_4 - {}^7F_J$, where J = 6—0 [12, 23, 30].

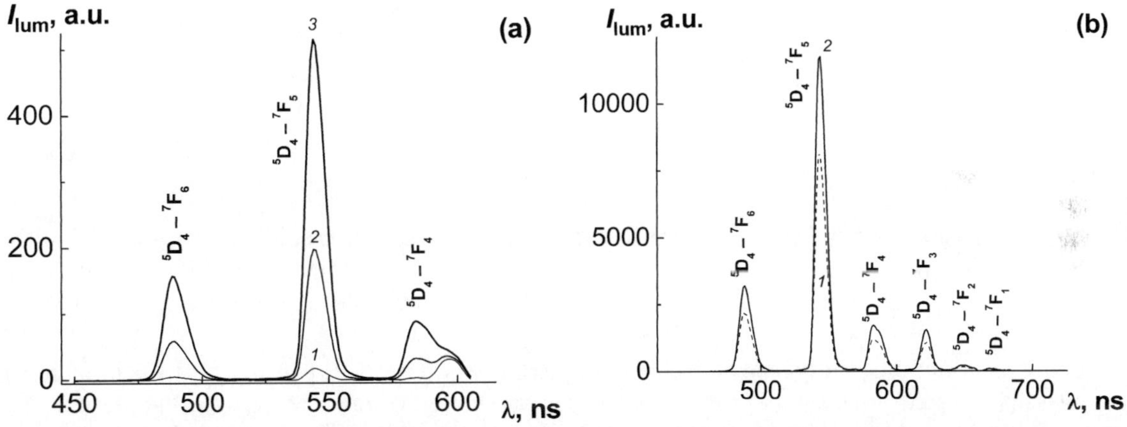

Figure 1. Luminescence spectra of Tb^{3+} in the form of $TbCl_3$ in the GXC matrix (a) for K1 (*1*), K2 (*2*), and K4 (*3*), or in the form of MPC in the composites based on GXC (b) for K7 (*1*) and K8 with added ZrO_2 NPs (6 wt.%) (*2*). The excitation wavelength was 300 nm. PMT gain was 500.

We have characterized luminescence efficiency of the studied composites with the value, θ, estimated from luminescence spectra using the following equation:

$$\theta = \frac{I_{300} / c_{Tb(III)}}{I_{300}^{KI} / c_{Tb(III)}^{KI}}, \quad (4)$$

where I_{300} is the luminescence intensity of a composite during excitation by light with a wavelength of 300 nm; $c_{Tb(III)}$ is the content of terbium (III) ions in a composite (wt.%); K1 is the superscript which indicates that the value applies to the composite K1 (table 1).

As seen from table 1, the efficiency value of θ varies little within the wide range of luminophore (*viz* Tb^{3+}) concentration from 0.02 to 1 wt.% (*note* K1 and K2 in table 1). However, increase in luminophore concentration up to 2.5 wt.% results in sharp decrease in the θ value from 1.43 to 0.33

(*note* K2 and K4 in table 1). At that, the total luminescence intensity I_{lum} increases (figure 1 (a)), though. These facts indicates the mode of concentration quenching has not set in, yet.

Assessment of luminescence efficiency through the value θ estimated demonstrated that the optimal content of Tb^{3+} is 0.21 wt.% (0.54 μmol per 40 mg of dry GXC); this value corresponds to terbium content in the composite K2. We have also obtained luminescence spectra of the composites containing [Tb(*polyLig*)](III) (figure 1 (b)). It was established that luminescence efficiency characterized by θ for the composites K7 and K8 are significantly higher than that in the case of low molecular weight luminophore $TbCl_3$ (compare with the values of θ for K1 and K2 on table 1). This result can be explained by antenna effect arising due to the formation of complex between Tb^{3+} and ligand fragment of monomer unit (*viz* 2-*o*-aminobenzoic acid) [23]. The suggested structure of [Tb(*polyLig*)](III) is shown in figure 2. The energy of exciting light (300 nm) is efficiently transferred to excited resonance nuclear level of terbium (III) ion according to the Förster resonance energy transfer mechanism (FRET) and then is transformed into visible radiation with large pseudo-Stokes shift [12].

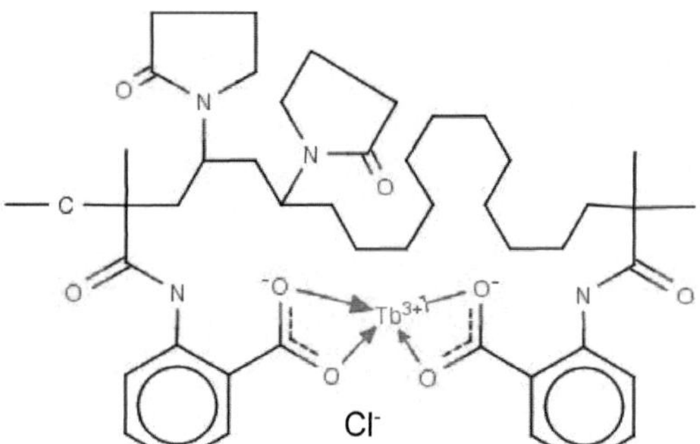

Figure 2. Scheme of [Tb(*polyLig*)](III) that includes P(VP-*co*-MABA) (88:12%).

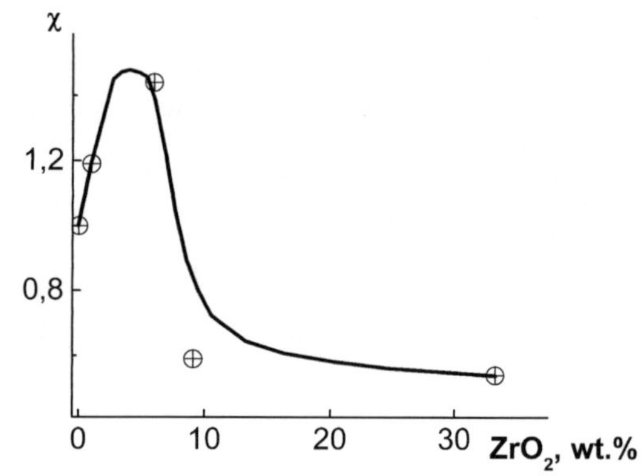

Figure 3. Enhanced luminescence of Tb^{3+} ions in GXC-based composites with adding ZrO_2 NPs. The observation wavelength is 543 nm.

Luminescence of Tb^{3+} in composites containing ZrO_2 NPs (*note* K3, K5, K6, and K8 in table 1) was also investigated. Analysis of luminescence spectra of these composites showed the presence of bands attributed to Tb^{3+} emission (*note* curve 2 in figure 1 (b)). Calculation of the values of θ characterizing luminescence efficiency in this case demonstrated increase in the value of I_{lum} in composites containing MPC with small amount of ZrO_2 NPs (not exceeding 6 wt.%).

The χ parameter characterizing enhancing luminescence upon addition of NPs was defined in compliance with:

$$\chi = \frac{\theta_{ZrO_2}}{\theta_{bm}}, \tag{5}$$

where θ_{ZrO_2}, θ_{bm} are the luminescence efficiency values of ZrO_2-containing composites and the corresponding composites without ZrO_2 as *benchmarks*, respectively. Variation of this parameter is shown in figure 3. It turned out that maximum luminescence is observed in the ZrO_2 concentration range of 1 to 6 wt.%. This fact may be related to change in the refractive index and dielectric constant of the filler which facilitates enhancing internal reflectance of light in a composite and more complete energy transfer to Tb^{3+} [31]. Further increase in ZrO_2 concentration in the polymeric matrix leads to luminescence quenching caused by beginning the aggregation of NPs.

3.2. SANS and USANS

Figure 4 shows experimental dependences of neutron scattering cross-section $d\Sigma(q)/d\Omega$ on momentum transfer q on a double logarithmic scale obtained in the experiments with the polymer matrix prepared from air-dry preliminarily disintegrated GXC nano-gel film and GXC-based organic-inorganic composites containing Tb^{3+} in the form of $TbCl_3$ (figure 4 (a)) or of [Tb(*polyLig*)](III) (figure 4 (b)) and ZrO_2 NPs. The samples of organic-inorganic composites for small-angle neutron scattering were selected on the basis of the data of luminescence measurements (*note* K4, K5, K7 and K8 on table 1).

It can be seen from the data presented in figure 4 that scattering curves corresponding to the polymer matrix and composites containing Tb^{3+} in the form of $TbCl_3$ or [Tb(*polyLig*)](III) have qualitatively similar character. The common feature of scattering curves for these samples is the presence of two ranges; one is located in the area of high ($> q_c$) momentum transfers, and the other is located in the area of low ($< q_c$) momentum transfers q. The point of crossover from one scattering regime to another lies at $q_c \approx 6.5 \cdot 10^{-3}$ Å^{-1}. At the same time, behaviour of scattering cross-section function $d\Sigma(q)/d\Omega$ before and after the crossover point obeys power laws $q^{-\Delta}$ with different exponents $\Delta = n_1$ and n_2, respectively.

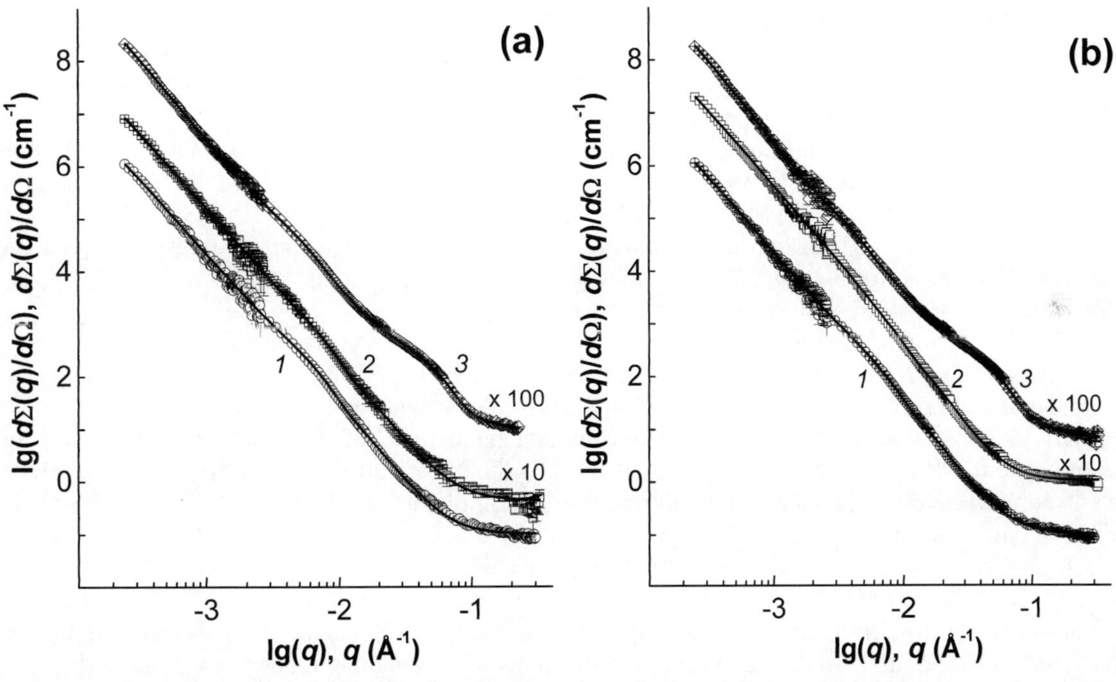

Figure 4. USANS and SANS differential cross sections $d\Sigma(q)/d\Omega$ in log-log scale for the matrix of GXC (*1*), and for the composites with Tb^{3+} in the form of $TbCl_3$ (a), and of MPC (b) without (*2*) and with ZrO2 NPs (*3*). Fits of experimental data by equation 7 are shown as solid lines. For the sake of clarity, cross section values for some samples were multiplied by 10 and 100 (corresponding factors are given next to the curves).

The scattering pattern observed for the studied films is typical of two-level hierarchical structures [30-33] with different characteristic scale and aggregation type for every level. The scattering from the first structural level (with lower scale) is observed in the area of high momentum transfers ($q > q_c$) and is described by the power dependence q^{-n_1}. The values of exponent n_1 found from the slope of linear parts of experimental $d\Sigma(q)/d\Omega$ curves plotted on a double logarithmic scale are equal to 3.04 ÷ 3.05 ± 0.02; these values correspond to the scattering by inhomogeneities (clusters) possessing highly developed fractal surface [34]. Dimensionality of this surface can be defined as $D_{s1} = 6 - n_1 = (2.95 \div 2.96) \pm 0.02$. In this connection, in the subsequent analysis of scattering at $q > 6.5\cdot10^{-3}$ Å^{-1} we used the two-phase structure model (where pore is a solid phase) with fractal phase interface [35]. According to this model, the object consists of inhomogeneities (pores) with developed surface. Thus, if the total surface area of an inhomogeneity (pore) measured on a scale of its dimension R is proportional to R^2, measurement of its surface on a scale $r << R$ gives surface area of the order of $R^2(R/r)^{\Delta}$, where $0 < \Delta < 1$, and $n = 4 - \Delta$. Fractal dimensionality of surface ($D_s = 2 + \Delta$) is higher than 2.

Deviations of scattering cross-section curve $d\Sigma(q)/d\Omega$ from the q^{-n_1} power law are observed both in the vicinity of the crossover point q_c and in the area of high momentum transfers $q > 10^{-1}$ Å^{-1} for all samples. Near the crossover point q_c, this deviation is caused by the beginning of the Guinier regime [36], where scattering is determined by the characteristic size R_{c1} (in the case of fractal system, this is the upper boundary of self-similarity) independently of scattering fractal objects. Analysis of scattering in the Guinier regime (the slope of the $\ln(d\Sigma(q)/d\Omega)$ *vs.* q^2 plots) can give estimated radius of gyration R_{g1} of fractal objects and, correspondingly, their characteristic size R_{c1}. For example, in the case of spherical fractal objects, these values have the relation [37]:

$$R_{c1} = \left[(D_{s1} + 2)/D_{s1}\right]^{1/2} \cdot R_{g1} .$$

At $q > 10^{-1}$ Å^{-1}, scattering cross-section $d\Sigma(q)/d\Omega$ no longer depends on q (*i.e.*, becomes constant) and is apparently related to incoherent scattering by hydrogen atoms which are present in GXC in abundance. In this regard, analysis of the observed scattering and estimation of the lower boundary of self-similarity of fractal objects do not seem possible.

The scattering from the second (larger) structural level observed in the $q < q_c$ area is described by the q^{-n_2} power dependence. The values of exponent n_2 found from the slope of linear parts of experimental $d\Sigma(q)/d\Omega$ curves vary from 2.80 ± 0.07 (for the polymer matrix) to 2.87 ± 0.04 (for the composite containing Tb^{3+} in the form of MPC). It is known [34] that these values are typical of scattering from mass fractals (aggregates) with fractal dimensionality $2.80 \pm 0.07 \leq D_{m2} = n_2 \leq 2.87 \pm 0.04$.

The lower boundary of self-similarity of mass fractals on this structural level is determined by characteristic size R_{c1} of surface fractals (clusters) of the first structural level. In turn, absence of deviation of scattering curves at low q (figure 4) from power dependence (*i.e.*, transition to the Guinier regime) indicates that characteristic size of inhomogeneities R_{c2} exceeds the maximum size of inhomogeneities $R_{max} \approx 3.5/q_{min}$ [35]. The latter value is the maximum size of scattering objects which can be experimentally registered at a given resolution of an instrument. In our case, $q_{min} = 2.5\cdot10^{-4}$ Å^{-1} and, correspondingly, $R_{c2} > R_{max} \approx 14000$ Å.

As seen from figure 4, addition of ZrO_2 NPs together with Tb^{3+} (both in the form of low molecular weight salt $TbCl_3$ and MPC) leads to the appearance of an additional shoulder on the corresponding scattering curves (in the Guinier area) at $q > 1.8\cdot10^{-2}$ Å^{-1}. Here, scattering is determined by the characteristic size of independent scattering inhomogeneities R_{c0} and the related power dependence q^{-n_0}. Analysis of this dependence can give information about local structure of these inhomogeneities. The values of exponent n_0 found from the slope of linear parts of experimental $d\Sigma(q)/d\Omega$ curves are equal to 4.14 ± 0.06 and 4.30 ± 0.10 for the sample containing MPC and the sample without MPC, respectively. This scattering law is typical of porous systems with the so-called diffuse surface for which the exponent n is equal to $4 + 2\beta > 4$, where $0 \leq \beta \leq 1$ is the exponent characterizing the law of change in nuclear density ρ in the surface layer of inhomogeneities [38].

Let us suppose that inhomogeneities have virtually smooth surface, then nuclear density ρ would depend only on the distance x from a point on the surface. Therefore, nuclear density $\rho(x)$ can be presented as:

$$\rho(x)=\begin{cases} 0, & x<0 \\ \rho_0 (x/\alpha)^\beta, & 0\le x\le\alpha \\ \rho_0, & x>\alpha \end{cases} \quad (6)$$

The parameter, α, is the width of the intermediate ("diffuse") layer where nuclear density ρ increases from zero to ρ_0. In this case, the appearance of the "diffuse" layer apparently results from modification of surface of ZrO_2 NPs with accompanying components (Tb^{3+} in the form of low molecular weight salt $TbCl_3$ or of MPC) which are used in the synthesis. From this assumption, in further analysis of scattering in the q area above $1.8\cdot10^{-2}$ Å^{-1} by the samples containing ZrO_2 NPs, the two-phase model (solid phase and pore) with diffuse phase interface [38] was used.

From the analysis of scattering in the Guinier regime (slope of dependences between $\ln(d\Sigma(q)/d\Omega)$ and q^2), we made the following estimates: R_{g0} = 87 ± 3 and 77 ± 8 Å and estimated characteristic dimensions ($R_{c0}=\sqrt{5/3}\,R_{g0}$ = 112.2 ± 3.9 Å) for the sample with and without MPC, respectively. The obtained estimated values indicate that "additional" scattering observed in the case of composites containing ZrO_2 NPs can be attributed to scattering by almost spherical ZrO_2 NPs with "diffuse" surface and diameter d_{c0} ~ 200 Å. Besides, correlation between the values of d_{c0} and dimensions of NPs (d_{ZrO2} = 200 ± 50 Å) used in the synthesis confirms the absence of aggregation of these particles. Difference in the values of d_{c0} and the β exponent for composites obtained without and with MPC is apparently related to different types of interaction between modifying additives and the surface of ZrO_2 NPs. In the first case, low molecular weight complexes containing Tb^{3+} are adsorbed on the surface of NPs as a thin layer, and in the second case, a thick layer of MPC (not less than 13 Å) covers the surface of ZrO_2 NPs.

It should be noted that in the region of low q (less than $1.8\cdot10^{-2}$ Å^{-1}), the shape of these scattering curves is similar to those observed for both the matrix of air-dry disintegrated GXC and its composites without ZrO_2, *i.e.*, these curves also include two regions with power dependence $d\Sigma(q)/d\Omega \sim q^{-\Delta}$ and different values of exponent ($\Delta = n_1$ and n_2). The first region is attributed to scattering by clusters with developed fractal surface and dimensionality D_{s1} = 2.85 ± 0.02 (independent of the presence of MPC in a composite). At the same time, use of MPC in the synthesis of hybrid organic-inorganic composites containing ZrO_2 NPs has a considerable effect on morphology of a higher structural level, which is described by power dependence q^{-n_2}, and this assumption is confirmed by estimation of values of the $n2$ exponent. Thus, for composite with ZrO2 NPs without MPC, n_2 = 2.96 ± 0.04, which corresponds to scattering by mass fractal aggregates with dimensionality D_{m2} = 2.96 ± 0.04 (similarly to the case of polymer matrix prepared from air-dry disintegrated GXC). On the other hand, introducing ZrO_2 NPs together with MPC into this GXC matrix leads to changes in the structure of the formed large scale aggregates. From the estimated value n_2 = 3.04 ± 0.03, we can conclude that surface fractals are formed instead of mass fractals. In other words, in this case, the second structural level of a composite is comprised of aggregates with highly developed fractal surface and dimensionality D_{s2} = 2.96 ± 0.02.

Thus, the SANS pattern observed for film of air-dry previously disintegrated GXC and GXC-based hybrid organic-inorganic composites containing Tb^{3+} in the form of low molecular weight salt or MPC as well as ZrO_2 NPs clearly indicates that these films contain two types of scattering inhomogeneities which differ strongly in their characteristic scale. Most likely, these samples consist of large-scale (R_{c2} > 14000 Å) mass fractal aggregates; in the case of composites containing ZrO_2 NPs together with MPC, these aggregates are constructed from surface fractal clusters with characteristic dimensions:

$$R_{c1} = \left[(D_{s1}+2)/D_{s1}\right]^{1/2}\cdot R_{g1}.$$

Addition of ZrO_2 NPs into composites leads to uniform distribution of these particles in the polymer matrix and absence of aggregation. Based on the above observations, in the detailed analysis of the scattering pattern observed for films of air-dry previously disintegrated GXC and GXC-based hybrid organic-inorganic composites containing Tb^{3+} in the form of low molecular weight salt or MPC as well as ZrO_2 NPs, we used unified exponential/power law expression which takes into account the presence of several structural levels in the scattering system [39]:

$$\frac{d\Sigma(q)}{d\Omega} = \sum_{i=0}^{m} (G_i \cdot \exp\left(-\frac{q^2 R_{gi}^2}{3}\right) + B_i \exp\left(-\frac{q^2 R_{g(i-1)}^2}{3}\right)\left[\frac{(erf(qR_{gi}/\sqrt{6}))^3}{q}\right]^{n_i}) + I_{inc} \quad . \qquad (7)$$

Summation in equation 7 is taken with respect to all structural levels. In the general case, equation 7 provides for the presence of 4 free parameters for every structural level, *i.e.* G_i (the Guinier prefactor), R_{gi} (radius of gyration), B_i (power-law prefactor) and n_i (the exponent). The parameter, I_{inc}, is a certain constant which does not depend on q and results from incoherent scattering by hydrogen atoms.

For obtaining final results, equation 7 was convoluted with the resolution function of the instrument. Experimental dependences of differential scattering cross-section $d\Sigma(q)/d\Omega$ were processed using the least squares method over the whole studied range. The results of this analysis are presented in figure 4 and in table 2.

Table 2. Structural parameters of the polymer matrix (air-dry pre-disintegrated GXC nano-gel film) and of hybrid organic-inorganic GXC-based composites, obtained from the SANS and USANS data.

System	R_{c0}, Å	$n_0 = 4 + 2\beta$	R_{c1}, Å	$n_1 = 6 - D_{s1}$	R_{c2}, Å	$n_2 = D_{m2}$; $6 - D_{s2}$
K0{Polymer matrix}	–	–	516 ± 26	3.05 ± 0.01		2.80 ± 0.07
K4{GXC / Tb^{3+}}	–	–	517 ± 25	3.04 ± 0.02		2.84 ± 0.04
K5{GXC / ZrO_2 / Tb^{3+}}	99.3 ± 10.3	4.30 ± 0.10	522 ± 19	3.15 ± 0.02	> 14000	2.96 ± 0.04
K7{GXC / Tb^{3+}+(MPC)}	–	–	548 ± 29	3.04 ± 0.02		2.87 ± 0.04
K8{GXC / ZrO_2 / Tb^{3+}(MPC)}	112.2 ± 9.5	4.14 ± 0.06	552 ± 20	3.15 ± 0.02		3.04 ± 0.03

3.3. SAXS

It is known [6] that GXC nano-gel film consists of water virtually to the extent of 90% and, correspondingly, the products based on this film contain rather high amounts of water. Thus, analysis of experimental SANS data for films of air-dry previously disintegrated GXC and GXC-based hybrid organic-inorganic composites in the area of $q > 0.15$ Å^{-1} was practically impossible due to the presence of strong background incoherent scattering by hydrogen atoms. At the same time, it is known [36] that amplitude of SAXS by an atom with atomic number Z (containing Z electrons) is directly proportional to Z. Thus, contribution of scattering by hydrogen atoms into small-angle scattering cross-section $d\Sigma(q)/d\Omega$ is negligible in comparison with contribution of atoms of other elements which are present in the studied samples (carbon, oxygen, terbium and zirconium). It was also necessary to increase the studied q range in order to obtain information about mesostructure on larger scale. In this connection, we carried out additional studies of the polymer matrix and cellulose-based hybrid organic-inorganic composites by SAXS. The corresponding SAXS curves are presented on a double logarithmic scale in figure 5.

It can be seen that behaviours of SAXS curves $I_S(q)$ measured for films of air-dry previously disintegrated GXC and composites containing Tb^{3+} in the form of $TbCl_3$ or of MPC are qualitatively similar. Scattering curves include two regions: high values of momentum transfer $q > q_c$ and low values of momentum transfer $q < q_c$, where q_c is equal to *ca.* $8.0 \cdot 10^{-1}$Å^{-1}. Behaviour of scattering intensity curve $I_S(q)$ obeys power laws $q^{-\Delta}$ with different values of exponent $\Delta = n_1$ and n_0, respectively. As was stated above, this pattern is typical of scattering by hierarchical two-level structures with different characteristic scale and type of aggregation for each level.

The scattering by the first of these structural levels (small-scale) is observed in the region of high momentum transfer ($q > q_c$) and described by power law dependence q^{-n_0}. The values of the n_0 exponent vary from 3.56 to 3.63, this corresponding to scattering by inhomogeneities (particles) with developed fractal surface and dimensionality $2.44 \le D_{s0} = 6 - n_0 = 2.37$. Deviations of the $I_S(q)$ scattering curve from the power law q^{-n_0} are observed in the vicinity of the crossover point $q_c \approx 8.0 \cdot 10^{-1}$ Å^{-1}; position of this point allows estimating characteristic dimension (upper limit of self-similarity) $R_{c0} = [(D_{s0}+2)/D_{s0}]^{1/2} \cdot R_{g0}$ of particles with fractal surface. Deviations are also observed for all samples in the region of high momentum transfers, where scattering intensity $I_S(q)$ does not depend on q and is related to incoherent scattering by inhomogeneities with scattering density ρ and dimensions of the order of wavelength of the used irradiation λ.

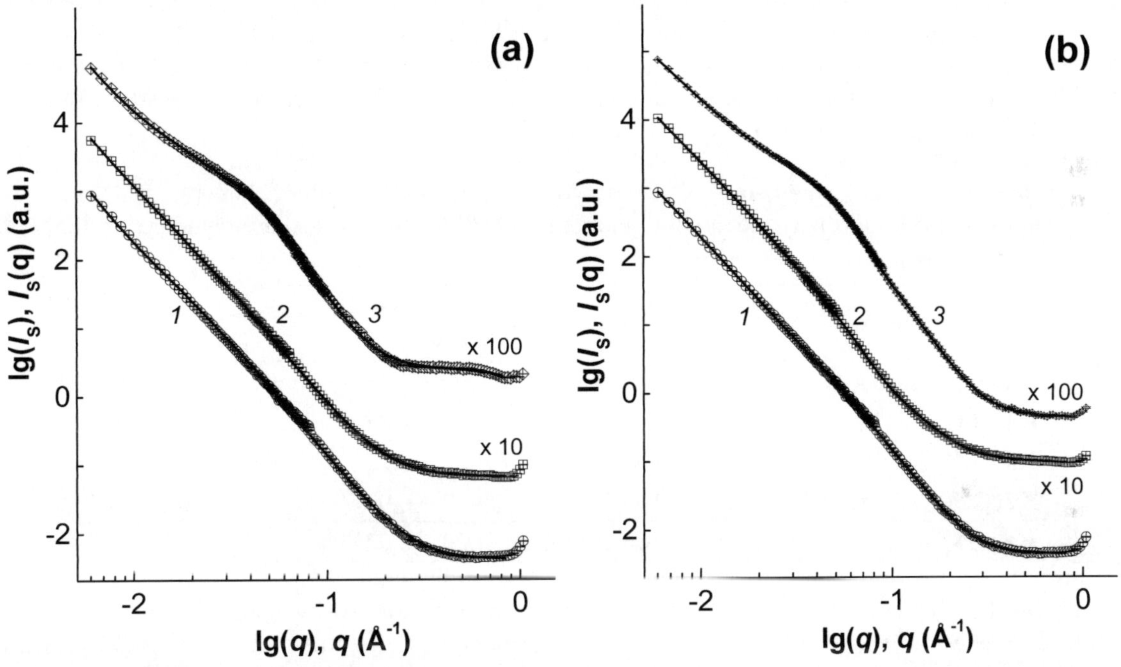

Figure 5. SAXS intensity in log-log scale for the matrix of GXC (*1*), and for composites with Tb^{3+} in the form of $TbCl_3$ (a), and of MPC (b) without (*2*) and with ZrO2 NPs (*3*). Fits of experimental data by equation 7 are shown as solid lines. For the sake of clarity, cross section values for some samples were multiplied by 10 and 100 (corresponding factors are given next to the curves).

Apparently, in the region of low q ($q < q_c$), SAXS observed for films of air-dry previously disintegrated GXC composites containing Tb^{3+} in the form of $TbCl_3$ or of MPC corresponds (similarly to SANS, *note* figure 4) to scattering by inhomogeneities (clusters) with highly developed fractal surface and dimensionality $2 \le D_{s1} = 6 - n_1 < 3$. This conclusion is confirmed by estimations of the $n1$ exponent obtained from the slopes of linear parts of experimental $I_S(q)$ curves; values of this exponent vary from 3.02 to 3.14. The lower limit of self-similarity of surface fractal clusters of this structural level is determined correspondingly by characteristic size R_{c0} of primary surface-fractal particles (inhomogeneities), and the upper limit of self-similarity R_{c1} was estimated earlier using SANS (table 2). The difference in values of fractal dimensionality D_{s1} obtained from the SANS and SAXS data is apparently related to difference in contrasts $\Delta\rho$ (solid phase-pore) of scattering densities for X-rays and neutrons.

The shape of SAXS curves $IS(q)$ obtained for hybrid organic-inorganic composites containing ZrO_2 NPs (figure 5) is similar to that of SANS curves in the region of momentum transfer values $q > 7\cdot10^{-3}$ Å^{-1}. Therefore, it can be concluded that the observed SAXS is also related mainly to scattering by almost spherical ZrO_2 NPs and nanoparticles with "diffuse" surface ($n_0 = 4 + 2\beta \approx 4.3$). It is virtually impossible to isolate contribution of the polymer matrix into scattering because of considerable difference between scattering density of GXC ($\rho_{GXC} = 1.34\cdot10^{11}$ cm^{-2}) and scattering density of ZrO_2 ($\rho_{ZrO_2} = 4.21\cdot10^{11}$ cm^{-2}). It should be noted that similar difference in the values of scattering density ρ is observed in SANS experiments; $\rho_{GXC} = 1.76\cdot10^{10}$ cm^{-2} and $\rho_{ZrO_2} = 5.04\cdot10^{10}$ cm^{-2}, respectively.

Thus, the use of SAXS allowed us to reveal another (primary) level in structural organization of the polymer matrix and GXC-based hybrid organic-inorganic composites.

In view of the above, in the detailed analysis of the small-angle X-ray scattering pattern observed for films of air-dry previously disintegrated GXC and GXC-based hybrid organic-inorganic composites, we also used the unified exponential/power law expression 7. For obtaining final results, equation 7 was convoluted with the resolution function of the instrument. Experimental dependences of SAXS intensity $I_S(q)$ were processed using the least squares method over the whole studied range. The results of this analysis are presented in figure 5 and in table 3.

Table 3. Structural parameters of the polymer matrix (air-dry, pre-disintegrated GXC nano-gel film) and of hybrid organic-inorganic GXC-based composites, obtained from the SAXS data.

System	R_{c0}, Å	$n_0 = 6 - D_{s0}$; $4 + 2\beta$	R_{c1}, Å	$n_1 = 6 - D_{s1}$
K0{Polymer matrix}	42.0 ± 3.2	3.63 ± 0.03		3.07 ± 0.01
K4{GXC / Tb^{3+}}	50.7 ± 4.1	3.56 ±0 .04		3.14 ± 0.01
K5{GXC / ZrO_2 / Tb^{3+}}	105.5 ± 5.6	4.32 ± 0.02	> 400	3.15 ± 0.01
K7{GXC / Tb^{3+}(MPC)}	104.5 ± 6.0	3.62 ± 0.02		3.02 ± 0.01
K8{GXC / ZrO_2 /Tb^{3+}(MPC)}	104.5 ± 5.5	4.29 ± 0.02		3.54 ± 0.03

The results obtained in the studies of the initial polymer matrix by complementary methods (USANS, SANS and SAXS, *note* tables 2 and 3) clearly indicate that this matrix is a porous system with three-level fractal structural organization. The lower structural level of the polymer matrix consists of almost spherical primary inhomogeneities with developed fractal surface, dimensionality $D_{s0} = 2.37 \pm 0.03$ and upper limit of self-similarity $R_{c0} = 42.0 \pm 3.2$ Å. These inhomogeneities form secondary surface fractal clusters with dimensionality $D_{s1} = 2.95 \pm 0.01$ and upper limit of self-similarity $R_{c1} = 516 \pm 26$ Å. It can be assumed that during biosynthesis of GXC nano-gel film, surface fractals of the second structural level aggregate to form large-scale ($R_{c2} > 1.4$ μm) mass fractals with dimensionality $D_{m2} = 2.80 \pm 0.07$.

Structural organization of hybrid organic-inorganic composites is similar to a large extent to that of the initial polymer matrix. All synthesized composites also demonstrate three-level fractal structure which is mainly formed by the initial polymer matrix. At the same time (tables 2 and 3) both composition and state of matter of the additives used in the synthesis exert considerable influence on morphology of the obtained composites.

Thus, introducing Tb^{3+} in the form of low molecular weight salt into the polymer matrix influences only structural parameters of the primary surface fractal inhomogeneities (*i.e.*, leads to increase in their fractal dimensionality D_{s0} from 2.37 to 2.44 and increase in their characteristic size R_{c0} from 42 to 50.7 Å). The structure of surface fractal clusters of the secondary structural level and large-scale mass fractal aggregates remains virtually unchanged. Therefore, it can be assumed that molecular complexes containing Tb^{3+} are adsorbed on primary inhomogeneities, loosen their surface and lead to increase in their characteristic size R_{c0}.

At the same time, introducing Tb^{3+} in the form of MPC into the polymer matrix has virtually no influence on surface fractal dimensionality of the primary inhomogeneities D_{s0} and dimensionality of clusters D_{s1}, but leads to considerable increase in their characteristic sizes R_{c0} and R_{c1}. Thus, the size of the primary inhomogeneities R_{c0} increases almost by a factor of 2.5 (up to $\approx$ 105 Å), and the size of surface fractals of the secondary structural level increases from 516 to 548 Å, respectively. Apparently, MPC uniformly envelops both the primary inhomogeneities and surface fractal clusters and does not change their surface morphology. It should be noted that use of MPC as an additive also results in compacting of mass fractal aggregates (because voids between clusters are partially filled) and thus leads to increase in fractal dimensionality D_{m2} from 2.80 to 2.87.

Doping with ZrO_2 NPs together with Tb^{3+} (both in the form of low molecular weight salt and in the form of MPC) exerts the most pronounced influence on morphological structure of the initial polymer matrix. As it has been established previously, ZrO_2 NPs modified during synthesis are uniformly distributed in pores of surface fractal clusters and voids of large-scale mass fractals which are comparable in size to these particles. This process leads, first, to slight smoothing of cluster surface (i.e., decrease in their fractal dimensionality D_{s1} from 2.95 to 2.85), while their characteristic sizes R_{c1} remain practically unchanged within the uncertainty of measurements. Secondly, when low molecular weight salt is used as a second modifying additive, compacting of aggregates occurs (which displays itself in increase in their fractal dimensionality D_{m2} from 2.8 to 2.96). When ZrO_2 NPs are introduced into the polymer matrix together with MPC, type of fractals changes from mass fractals (D_{m2} = 2.95) to surface fractals (D_{s2} = 2.96). This structural transition is apparently related to complete filling voids between clusters which form spatial network of mass fractal aggregate by ZrO_2 NPs and MPC; thus, “dense” aggregate is formed. Surface of this aggregate consists of clusters with highly developed fractal surface.

Analysis of morphology of the first structural level of hybrid organic-inorganic composites containing ZrO_2 NPs is impossible due to significant difference between scattering density ρ_{ZrO_2} and scattering density of GXC (ρ_{GXC}). However, we can assume that mesostructure of the primary particles of bacterial cellulose NGF remains unchanged due to considerable difference in their sizes: $d_{c0}(ZrO_2) \sim$ 200 Å, and $d_{c0}(GXC) \sim$ 80 Å.

The above-proposed model of the organization of the micro- and mesostructure of hybrid organic-inorganic composites based on polymer matrix of air-dry pre-disintegrated nano-gel GXC film is confirmed by the analysis of the experimental data obtained by low-temperature nitrogen adsorption and SEM. As can be seen from table 4, the obtained values of specific surface area, S_{BET}, clearly indicate that these composites have a porous structure with the sufficiently developed system of open pores.

Table 4. Specific surface area, S_{BET}, for hybrid organo-inorganic composites based on the polymeric matrix of air-dryed, pre-disintegrated nano-gel-film of GXC by low-temperature nitrogen adsorption

System	S_{BET}, m^2/g
K4{GXC / Tb^{3+}}	141.5
K5{GXC / ZrO_2 / Tb^{3+}}	69.6
K7{GXC / Tb^{3+}(MPC)}	112.8
K8{GXC / ZrO_2 / Tb^{3+}(MPC)}	329.3

The composite being the closest in structural parameters to a polymeric matrix of GXC is received by doping with Tb^{3+} in the form of $TbCl_3$ (tables 2 and 3), with the value of S_{BET} making 141.5 m^2/g for it. At the same time, the porosity of the composites under investigation explicitly depends on the chemical nature of the additives in the process of their synthesis. Thus, the use of [Tb(*polyLig*)](III), or of ZrO_2 NPs together with Tb^{3+} in the form of $TbCl_3$ significantly reduces the specific surface area, S_{BET}, for composites to 112.8 and 69.6 m^2/g, respectively. In the first case, it is apparently related to the increase in the characteristic dimensions of the primary inhomogeneities, R_{c0}, and of fractal clusters of the 2nd structural level, R_{c1} (tables 2 and 3). In the second one it is due to filling of the pores of the cellulose matrix with ZrO_2 NPs. The highest value of S_{BET} equal to 329.3 m^2/g was observed for the composite containing ZrO_2 NPs together with [Tb(*polyLig*)](III) (table 4). Obviously, the growth of S_{BET} for this composite is associated with a different nature of the fractal structure organization for the large aggregates of the third structural level, comparing to the initial polymer matrix of GXC and other composites based on it: That is with changing the fractal type from mass to the surface one.

Analysis of SEM images presented in figure 6 demonstrated that morphology of the polymer matrix obtained from air-dry previously disintegrated GXC nano-gel film changes in various composites depending on chemical nature of components. It is well-known [40] that morphology of GXC nano-gel film can be presented as a polymer network, which remains virtually unchanged in composites and after previous disintegration of GXC with subsequent drying this dispersion in air. In our case, SEM images illustrate the process of filling this network with various additives. Composites with [Tb(*polyLig*)](III) demonstrate more compact surface structures. This phenomenon is caused by the presence of flexible polymer in MPC. These microphotographs clearly demonstrate the third structural level which was detected by USANS.

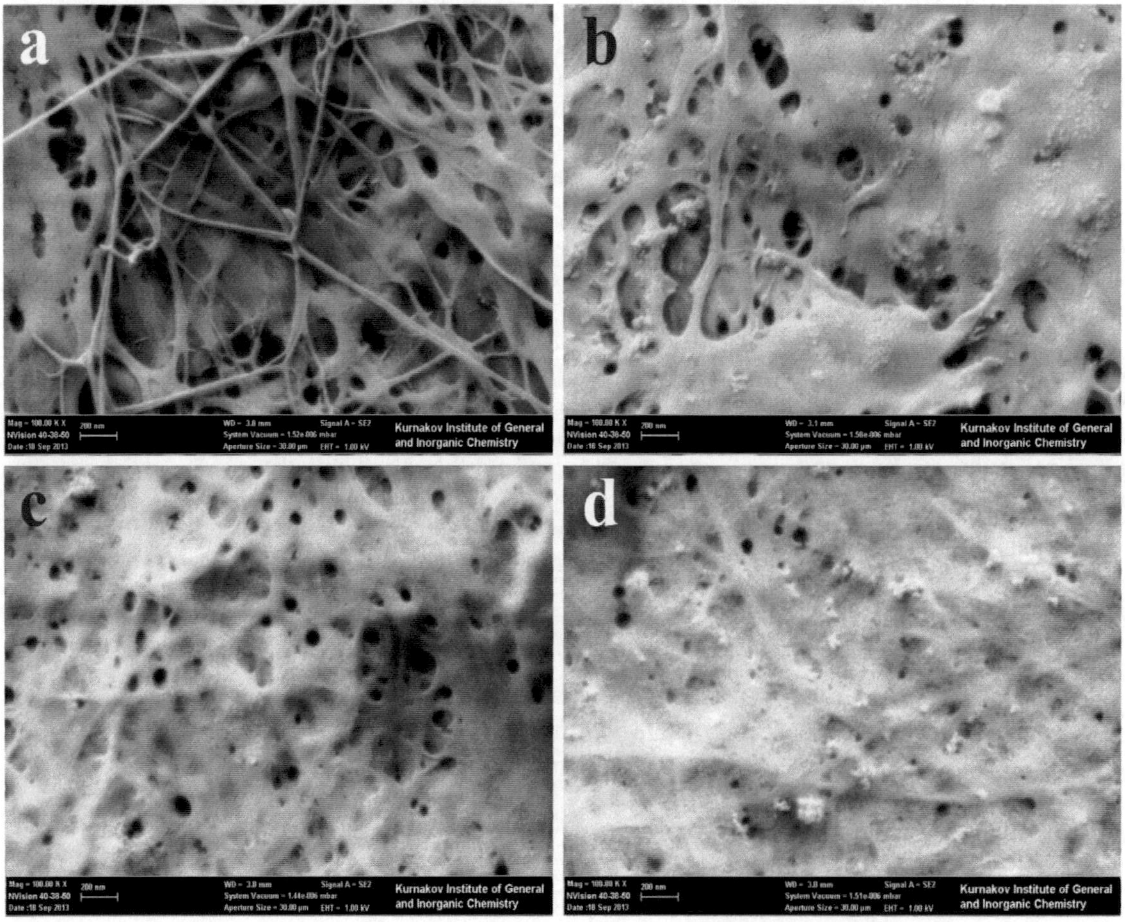

Figure 6. Microphotographs of GXC composites with Tb^{3+} in the form of $TbCl_3$ (a), and of MPC (b);

the analogous composites but with ZrO_2 NPs, respectfully (c, d). The scale bar is 200 nm.

4. Conclusions

Micro- and mesostructure, including fractal one, of the polymeric matrix prepared from air-dry, pre-disintegrated GXC, and of composites on its base with the addition of ZrO_2 NPs and Tb^{3+} in the form of $TbCl_3$ and in [Tb(*polyLig*)](III) was investigated by SEM techniques, USANS, SANS, and SAXS, as well as by low-temperature nitrogen adsorption.

Comprehensive analysis of the experimental data showed that the initial polymeric matrix is a porous system with a complex – three-level structure of the fractal organization. The lower structural level of the polymeric matrix consists of initial inhomogeneities being near-spherical and possessing a developed fractal surface with $D_{s0} = 2.37 \pm 0.03$ and an upper limit of self-similarity $R_{c0} = 42.0 \pm 3.2$ Å. Then these inhomogeneities formed the secondary surface fractal clusters with $D_{s1} = 2.95 \pm 0.01$ and an upper limit of self-similarity $R_{c1} = 516 \pm 26$ Å. On the third structural level, surface fractal clusters aggregate into large-scale ($R_{c2} > 1.4$ microns) mass-fractal aggregates with $D_{m2} = 2.80 \pm 0.07$.

It is found that the structural organization of the hybrid organic-inorganic composites in many ways, and in the case of doping by Tb^{3+} almost entirely maintains the original character of the structural organization of the polymer matrix. For all of the synthesized composites, they are also characterized by a three-level fractal structure, in the formation of which the leading role is clearly belongs to the original polymer matrix.

At the same time, it was revealed that the composition and aggregation state of dopants used in the synthesis have a significant impact on the structural characteristics of the organic-inorganic composites. Thus, the introduction of Tb^{3+} in the form of $TbCl_3$, and as the part of [Tb(*polyLig*)](III) leads to increase of the characteristic size R_{c0} of initial inhomogeneities. Upon using MPC, the growth of the characteristic size R_{c1} of clusters of the *2nd* structural level was also observed, with their surface fractal dimension D_{s1} decreasing. In turn, it was found that the presence of doping ZrO_2 NPs in the composition has a significant impact only on the morphology of large-scale units of the *3rd* structural level as a consequence of the nanoparticle characteristic size R_{ZrO_2} equal to *ca.* 100 Å. This effect is resulting in the significant increase in the mass-fractal dimension D_{m2} of aggregates from 2.80 to 2.96 for the composite K5 {GXC / ZrO_2 / Tb^{3+}}, or in the change of a fractal type from the mass to the surface one for the composite K8 {GXC / ZrO_2 / [Tb(*polyLig*)](III)}. It clearly indicates more compact filling of the cellulose matrix by doping components.

Luminescence measurements have shown that upon using MPC with Tb^{3+} based on P(VP-*co*-MABA) as the luminophore, the efficiently luminescent composite materials were produced. Adding MPC in a polymer matrix allowed to overcome the concentration quenching in the matrix and made it possible to realize an efficient transfer of electronic excitation energy by the Förster mechanism from an *o*-aminobenzoic moieties to the central lanthanide atom due to the antenna effect. It is also found that the use of ZrO_2 NPs as doping ingredients at a concentration of from 1 to 6 wt.% can significantly enhance the luminescence of Tb^{3+} in the composite material due to a possible change in the refractive index of the composite.

Certainly, the results obtained can be used in the field of regenerative medicine, as well as in optoelectronics.

Acknowledgements

The work was financially supported by Program of Fundamental Research of Presidium of RAS # 1 "Nanostructures: physics, chemistry, biology, the basics of technologies" and Russian Foundation for Basic Research (grant № 16-02-00987A). We would like to express special thanks to Dr. Zhendong Fu (JCNS, Forschungszentrum Juelich GmbH, Outstation at MLZ, Garching, Germany) for support in performing the USANS experiments.

References

[1] Binnemans K 2009 Lanthanide-based luminescent hybrid materials *Chem. Rev.* **109** 4283–374

DOI 10.1021/cr8003983

Carlos L D, Ferreira R A S, Bermudez V de Z and Ribeiro S J L 2009 Lanthanide-containing light-emitting organic-inorganic hybrids: a bet on the future *Adv. Mater.* **21** 509–34 DOI 10.1002/adma.200801635

[2] Liao C, Zhang M, Yao M Yu, Hua T, Li L and Yan F 2015 Flexible organic electronics in biology: materials and devices *Adv. Mater.* **27** 7493–527 DOI 10.1002/adma.201402625

Nogi M and Yano H 2008 Transparent nanocomposites based on cellulose produced by bacteria offer potential innovation in the electronics device industry *Adv. Mater.* **20** 1849–52 DOI 10.1002/adma.200702559

Nogi M, Iwamoto S, Nakagaito A N and Yano H. 2009 Optically transparent nanofiber paper *Adv. Mater.* **21** 1595–98 DOI 10.1002/adma.200803174

[3] Czaja W K, Young D J, Kawecki M and Brown R M Jr 2007 The Future Prospects of Microbial Cellulose in Biomedical Applications *BioMacromolecules* **8** 1–12 DOI 10.1021/bm060620d

[4] Habibi You, Lucia L A and Rojas O J 2010 Cellulose Nanocrystals: Chemistry, Self-Assembly, and Applications *Chem. Rev.* **110** 3479–500 DOI 10.1021/cr900339w

Klemm D, Kramer F, Moritz S, Lindström T, Ankerfors M, Gray D and Dorris A 2011 Nanocelluloses: A New Family of Nature-Based Materials *Angew. Chem. Int. Ed.* **50** 5438–66 DOI 10.1002/anie.201001273

Eichhorn S J, *et al.* 2010 Review: current international research into cellulose nanofibres and nanocomposites *J Mater Sci* **45** 1–33 DOI 10.1007/s10853-009-3874-0

Chen P, Cho S Y and Jin H-J 2010 Modification and Applications of Bacterial Celluloses in Polymer Science *Macromolecular Research* **18** 309-20 DOI 10.1007/s13233-010-0404-5

Moon R J, Martini A, Nairn J, Simonsen J and Youngblood J 2011 Cellulose nanomaterials review: structure, properties and nanocomposites *Chem. Soc. Rev.* **40** 3941–3994 DOI: 10.1039/c0cs00108b

[5] Brown R M Jr, Saxena I M and Kudlicka 1996 Cellulose biosynthesis in higher plants *Trends in Plant Sciences* **1** 149–156 DOI:S1360-1385(96)10017-0

Baklagina Y G, Khripunov A K, Romanov D P, Klechkovskaya V V, Lavrent'ev V K, Smyslov R Y, Bugrov A N, Tkachenko A A, Romanov V N and Vinogradova K L 2012 Phase transitions of native celluloses from evolutionarily different sources into polymorph IV *Russian Journal of Applied Chemistry* **85** 1923-9 DOI 10.1134/S1070427212120018

[6] Brown R M Jr 1996 The biosynthesis of cellulose *J. M. S. – Pure Appl. Chem.* A **33** 1345–73

Huang Y, Zhu C, Yang J, Nie Y, Chen C and Sun D 2014 Recent advances in bacterial cellulose *Cellulose* **21** 1–30 DOI 10.1007/s10570-013-0088-z

[7] Klechkovskaya V V, Baklagina Yu G, Stepina N D, Khripunov A K, Buffat P A, Suvorova E I, Zanaveskina I S, Tkachenko A A and Gladchenko S V 2003 Structure of cellulose Acetobacter xylinum *Crystallography Reports* **48** 755–62

[8] Maria L C S, Santos A L C, Oliveira P C et al 2009 *Materials Letters* **63** 797-99

Baklagina Y G, *et al.* 2006 Interaction of Se^0 nanoparticles stabilized by poly(vinylpyrrolidone) with gel films of cellulose Acetobacter xylinum *Crystallography Reports* 51 619-26 DOI 10.1134/S1063774506040134

Khripunov A K, *et al.* 2007 Formation of a composite from Se^0 nanoparticles stabilized with polyvinylpyrrolidone and Acetobacter xylinum cellulose gel films *Russian Journal of Applied Chemistry* **80** 1549–57 DOI 10.1134/S1070427207090200

Volkov V V, Klechkovskaya V V, Shtykova E V, Dembo K A, Arkharova N A, Ivakin G I and Smyslov R Yu 2009 Determination of the Size and Phase Composition of Silver Nanoparticles in a Gel Film of Bacterial Cellulose by Small-Angle X-Ray Scattering, Electron Diffraction, and Electron Microscopy *Crystallography Reports* **54** 169–73 DOI 10.1134/S1063774509020011

Wang W, Zhang T J, Zhang D W et al 2011 *Talanta* **84** 71–7

Gutierrez J, Tercjak A, Algar I et al 2012 *Journal of Colloid and Interface Science* **377** 88–93

Yang Z, Chen S, Hu W et al. 2012 *Carbohydrate Polymers* **88** 173–78
[9] Buyanov A L, Gofman I V, Revelskaya L G, Khripunov A K and Tkachenko A A 2010 Anisotropic swelling and mechanical behavior of composite bacterial cellulose–poly(acrylamide or acrylamide–sodium acrylate) hydrogels *Journal of the mechanical behaviour of biomedical materials* **3** 102–11 DOI 10.1016/j.jmbbm.2009.06.001
[10] Buyanov A L, Gofman I V, Bozhkova S A, Saprykina N N, Kochish A Yu, Netyl'ko G I, Khripunov A K, Smyslov R Yu, Afanas'ev A V and Panarin E F 2016 *Russian Journal of Applied Chemistry* **89** 772-9 DOI 10.1134/S1070427216050141
Velichko E V, Buyanov A L, Chetverikov Yu O, Duif C P, Bouwman W G and Smyslov R Yu 2016 Mesostructure anisotropy of bacterial cellulose—polyacrylamide hydrogels as studied by spin-echo small-angle neutron scattering *Preprint* arXiv 1608.00544
[11] Lina F, Yue Z, Jin Zh and Guang Y 2011 Bacterial Cellulose for Skin Repair Materials *Biomedical Engineering – Frontiers and Challenges* ed R Fazel (Rijeka, Shanghai: InTech) chapter 3 pp 251-74 http://www.intechopen.com/books/biomedical-engineering-frontiers-and-challenges/bacterialcellulose-for-skin-repair-materials
[12] Bünzli J.-C. G. 2010 Lanthanide luminescence for biomedical analyses and imaging Chem. Rev. 110 2729–55 DOI 10.1021/cr900362e
[13] Werts M H V 2005 Making sense of lanthanide luminescence *Sci. Prog.* **88** 101–31
[14] Blasse G and Grabmaier B C 1994 *Luminescent Materials* (Berlin: Springer)
Justel T, Nikol H and Ronda C 1998 *Angew. Chem. Int. Ed.* **37** 3085
[15] Bugrov A N, Rodionov I A, Zvereva I A, Smyslov R Y and Almjasheva O V 2016 Photocatalytic activity and luminescent properties of Y, Eu, Tb, Sm and Er-doped ZrO_2 nanoparticles obtained by hydrothermal method International *Journal of Nanotechnology* **13** 147–57
[16] Marling J, 1978 1.05-1.44 micron tunability and performance of the CW Nd/3+/:YAG laser *IEEE J. Quantum Electron.* **14**, 56–62 DOI 10.1109/JQE.1978.1069667
[17] Kuriki K, Koike Y, Okamoto Y, 2002 *Chem. Rev.* **102** 2347
[18] Polman A, van Veggel F C J M, 2004 *J. Opt. Soc. Am.* B **21** 871
[19] Legnani C. *et al.* 2008 *Thin Solid Films* **517** 1016–1020
[20] Patent RU №2141530
Patent RU №2189394
[21] Baklagina Yu G, *et al.* 2003 *Russian Journal of Applied Chemistry* **76** 989-96
[22] Bugrov A N and Almjasheva O V 2013 Effect of hydrothermal synthesis conditions on the morphology of ZrO_2 nanoparticles. *Nanosystems: Physics, Chemistry, Mathematics* **4** 1–6
[23] Yakimanskii A V, Goikhman M Y, Podeshvo I V, Anan'eva T D, Nekrasova T N and Smyslov R Y 2012 Luminescence of Ln^{3+} Lanthanide Complexes in Polymer Matrices *Polymer Science* A **54** 921-41 DOI 10.1134/S0965545X1209009X
[24] Radulescu A, *et al.* 2005 *Neutron News* 2005 **16** 7
[25] Goerigk G and Varga Z. 2011 *J. Appl. Cryst.* **44** 337
[26] Wignall G D and Bates F S 1987 *J. Appl. Cryst.* **20** 28
[27] http://iffwww.iff.kfa-juelich.de/~pipich/dokuwiki/doku.php/qtikws
[28] Schmatz W, Springer T, Schelten J and Ibel K. 1974 *J. Appl. Cryst* **7** 96
[29] https://www.embl-hamburg.de/biosaxs/download.html
Petoukhov M V, Franke D, Shkumatov A V, Tria G, Kikhney A G, Gajda M, Gorba C, Mertens H D T, Konarev P V and Svergun D I 2012 New developments in the ATSAS program package for small-angle scattering data analysis *J. Appl. Cryst.* **45** 342–50 DOI 10.1107/S0021889812007662
[30] Carnall W T 1979 *Handbook on the Physics and Chemistry of Rare Earths* vol 3, ed K A Gschneidner Jr and L Eyring (Amsterdam: North Holland Publ. Co.) p 172
[31] Reisfeld R, Zelner M and Patra A 2000 Fluorescence study of zirconia films doped by Eu, Tb and Sm and their comparison with silica films *Journal of Alloys and Compounds* **300–1** 147–

IOP Conf. Series: Journal of Physics: Conf. Series **848** (2017) 012017 doi:10.1088/1742-6596/848/1/012017

51
Fox M 2010 *Optical Properties of Solids (Oxford Master Series in Physics) 2nd ed* (Oxford: Oxford University Press)
[32] Beaucage G, Ulibarri T A, Black E P and Schaefer D W 1995 *Hybrid Organic-Inorganic Composites (ACS Symposium Series)* ed J Mark *et al.* (Washington, DC: American Chemical Society) p 97
[33] Štěpánek M, Matějíček P, Procházka K, Filippov S K, Angelov B, Šlouf M, Mountrichas G and Pispas S 2011 *Langmuir* **27** 5275–81
Gubanova N N, Kopitsa G P, Ezdakova K V, Baranchikov A E, Angelov B, Feoktystov A, Pipich V, Ryukhtin V and Ivanov V K 2014 Structure of zirconium dioxide based porous glasses *Journal of Surface Investigation. X-ray, Synchrotron and Neutron Techniques* **8** 963–71 DOI 10.1134/S1027451014050309
Khamova T V, Shilova O A, Kopitsa G P, Angelov B and Zhigunov A 2016 Effect of Biocidal Additives on the Mesostructure of Epoxy–Siloxane Bioactive Coatings *Journal of Surface Investigation. X-ray, Synchrotron and Neutron Techniques* **10** 113–22 DOI 10.1134/S1027451015060312
[34] Teixera J 1986 *On Growth and Form-Fractal and Non-Fractal Pattern in Physics* ed H E Stanley, N Ostrovsky (Boston: Martinus Nijloff)
[35] Bale H D and Schmidt P W 1984 *Phys. Rev. Lett.* **38** 596
[36] Guinier A 1939 *Ann. Phys.* **12** 161
[37] Guinier A, Frournet G, Walker C B and Yudowitch K L 1955 *Small-Angle Scattering of X-rays* (New York: Wiley) p 17
[38] Schmidt P W, Avnir D and Levy D *et al.* 1991 *J. Chem. Phys.* **94** 1474
[39] Beaucage G and Schaefer D W 1994 *J. Non-Cryst. Solids* **172–74** 797
[40] Svensson A, Nicklasson E, Harrah T, Panilaitis B, Kaplan D L, Brittberg M and Gatenholm P 2005 *Biomaterials* **26** 419–31

The effect of thermal treatment on the atomic structure of PtCu core-shell nanoparticles in PtCu/C electrocatalysts

D B Shemet, V V Pryadchenko, V V Srabionyan, S B Belenov, A S Mikheykin, L A Avakyan, V E Guterman, L A Bugaev
Southern Federal University, Rostov-on-Don, Russia

E-mail: vvpryadchenko@sfedu.ru

Abstract. Nanocatalysts PtCu/C with core-shell structure of PtCu nanoparticles were synthesized by the method of sequential chemical reduction of Cu^{2+} and Pt(IV) in carbon suspension, prepared on the basis of ethylene glycol–water solvent. The characterization of atomic structure of "as prepared" PtCu nanoparticles and obtained after thermal treatment at 350 °C was performed by Pt L_3- and Cu K-edge extended X-ray absorption fine structure (EXAFS), complemented with TEM and XRD studies. The processing of EXAFS was performed by the technique that enables to reduce the effect of correlations among fitting parameters on the determined values of local structure parameters of the absorbing atoms, which have the nearest surrounding consisting both Pt and Cu atoms.

1. Introduction

Among various important applications, the platinum nanoparticles (NPs) on carbon supports are a basic component of catalytic layers in low temperature fuel cells. The main disadvantages of using these materials are i) insufficient corrosive and morphological stability of NPs, as well as of the metal-carbon composite itself during the operation and ii) relatively high cost. One of the key factors affecting the proton exchange membrane fuel cells (PEMFC) performance is the slow kinetics of the oxygen reduction reaction (ORR) under PEMFC operating conditions. An efficient way to overcome these problems and to obtain promising electrocatalysts for ORR is the synthesis of catalytic layer containing bimetallic NPs of composition PtM (where M – transition metal, such as Cu, Co, Ni, Ag etc.) with a core-shell structure (where shell is formed by Pt atoms and core - by M atoms) deposited onto highly developed surface of carbon supports [1,2].

In this work the structural study of PtCu/C electrocatalysts containing bimetallic PtCu NPs synthesized by sequential chemical reduction of components before and after thermal treatment was performed. The post-treatment is a quite simple method that enables to estimate the morphological stability of prepared materials. Moreover, in some cases the thermal treatment can improve the electrocatalytic activity of such materials [3]. In order to study the stability of the core-shell structure of PtCu NPs during the thermal treatment, the analysis of Pt L_3- and Cu K-edge XAS for "as prepared" and heated at 350 °C PtCu/C electrocatalysts was performed, complemented with TEM and XRD studies. The information on atomic structure of bimetallic PtCu NPs was derived from XAS spectra using the suggested fitting technique [4,2], which enables to reduce the effect of correlations between fitting parameters on the obtained values of structural parameters. Another fitting technique [5], which does not use Fourier transformation, was used to confirm the obtained values of structural parameters.

2. Experimental results.

Carbon supported PtCu NPs were synthesized by the chemical reduction of Pt (IV) (H_2PtCl_6) and Cu^{2+} ($CuSO_4$) in a suspension of carbon powder (Vulcan XC72, Cabot), prepared on the basis of water-ethylene glycol solvent at pH = 10 (an excess of NH_3). The volume ratio of (water):(ethylene glycol) in the mixed solvent was 1:1. A 0.5 M solution of $NaBH_4$ was used as a reducing agent. In order to synthesize core-shell PtCu NPs with the internal core of Cu atoms and the outer shell of Pt atoms, a method of sequential chemical reduction was used. First, nanostructured material Cu/C was obtained by the reduction of copper from the carbon slurry, prepared on the basis of $CuSO_4$ solution. After that, the resulting Cu/C material was removed from the solution, washed and dried. Then, the obtained Cu/C composite was added to the aqueous solution of H_2PtCl_6 containing an excess of NH_3 (pH = 10). After stirring, an excess of $NaBH_4$ was added as a 0.5 M solution. The final material PtCu/C was separated by filtration, washed and dried for several hours in an oven at 100 °C. A part of the resulting catalyst was heated at 350 °C for 1 hour in an atmosphere of Ar.

The composition of the synthesized PtCu/C nanocatalysts corresponds to mass fraction of the metallic component of ~30% by weight. The chemical composition was determined by x-ray fluorescence analysis using the x-ray spectrometer ARL OPTIM'X to be $Pt_{0.8}Cu$ and remain quite stable during heating.

Microphotographs of the synthesized materials were obtained using a JEM-2100 (JEOL, Japan) microscope operated at an accelerating voltage of 200 kV and resolution of 0.2 nm. To conduct electron-microscopic studies a drop of specially prepared catalyst ink (about 0.5 mg of the catalyst was dispersed ultrasonically in 1 ml of isopropanol for 5 min) was applied on a copper grid covered with a thin layer of amorphous carbon film (to fix the microparticles on the surface) and dried for about 20 minutes in an air atmosphere at room temperature. TEM microphotographs and the histograms of particles' size distributions for "as prepared' catalysts and obtained after treatment are shown on Figures 1 and 2.

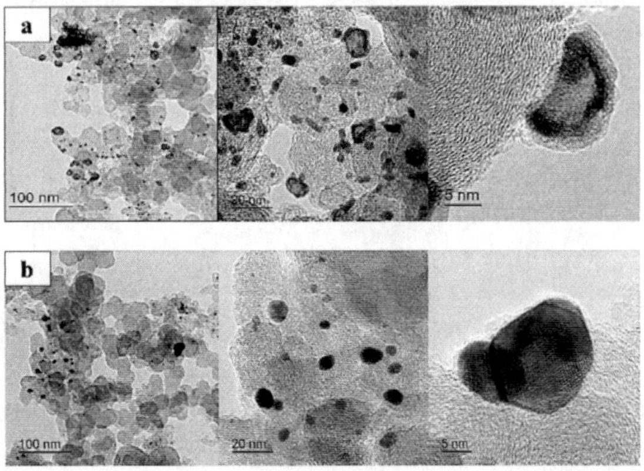

Figure 1. TEM microphotographs of PtCu/C electrocatalysts synthesized by sequential chemical reduction: "as prepared" (a) and after heating (b).

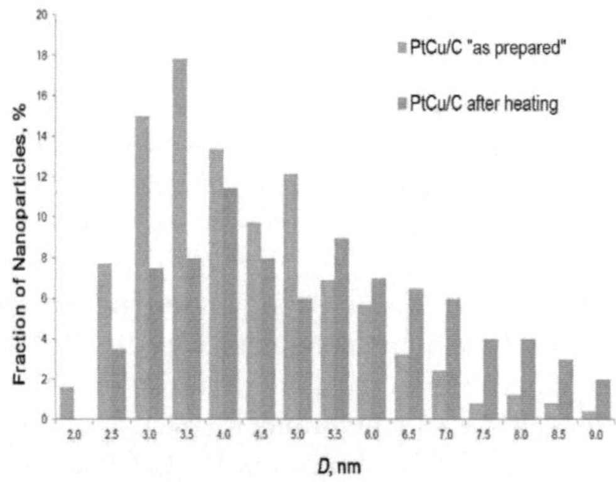

Figure 2. Hystograms of NPs' size distributions in PtCu/C electrocatalysts.

Pt L_3- and Cu K-edge EXAFS spectra for the studied materials were measured at the μSpot beamline of the BESSY-II Synchrotron Radiation Facilities (Berlin, Germany). Mean electron current of a storage ring was maintained during experiment at 300 mA in Top-Up mode. The measurements were performed in the transmission mode utilizing a Si (111) channel-cut monochromator and two ionization chambers and diode for a reference channel. The photon energy scanning steps were adjusted to δE = 1.0 eV and δk = 0.05 Å^{-1} in the XANES and EXAFS regions respectively, where E is the energy of the incident radiation and k is the photoelectron wavenumber.

XRD Synchrotron X-ray powder diffraction patterns were collected at the BM01 beamline [6] of the Swiss-Norwegian Beamlines at European Synchrotron Radiation Facility (SNBL at ESRF, Grenoble, France). The monochromatic beam with a wavelength λ = 0.69489 Å was used. The sample-to-detector distance, the tilt angles of the detector and the wavelength were calibrated using LaB_6 NIST standard sample.

3. Discussion and conclusions

TEM microphotographs of "as prepared' electrocatalysts and obtained after thermal treatment are shown on Figure 1. As can be seen, there are NPs of different structures and sizes in "as prepared" materials. On the central image of Figure 1a (scale 20 nm) one can distinguish small homogeneous NPs (dark and light small spots), as well as a larger inhomogeneous NPs that could be attributed to the core-shell structures (light inner region with darker surface). One of such NPs is represented on right image of Figure 1a. After the thermal treatment, all NPs probably have similar structure, and besides, there are no evidences of the core-shell NPs on TEM images (homogeneous grey spots, Figure 1b). So the qualitative analysis of TEM images suggests that thermal treatment leads to the alloying of initially inhomogeneous NPs .

The histograms of particles' size distributions derived from TEM micrographs are presented on Figure 2. To obtain good statistics, sizes of about ~1000 particles were measured for each sample. It can be seen, that the enlargement of NPs occurs as the result of thermal treatment at 350 °C. Observable bimodal character of size distributions can be explained by the presence of differently structured NPs in "as prepared" catalyst (see above), which are characterized by different size distributions.

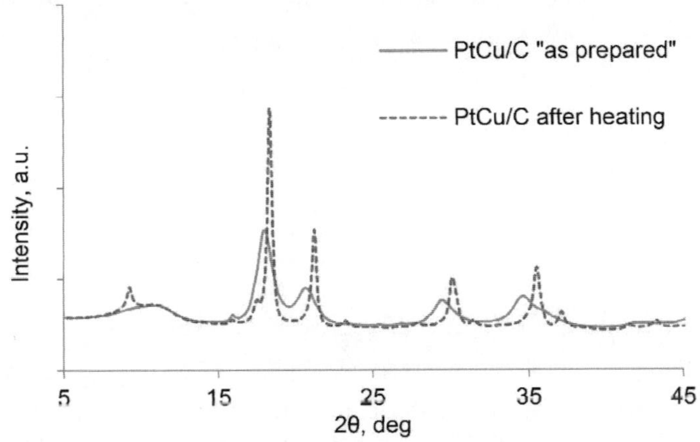

Figure 3. XRD patterns of "as prepared" (blue solid line) and heated at 350 °C (red dashed line) PtCu/C electrocatalysts, synthesized by the method of sequential reduction of metals on a carbon support.

The increase in the average size of NPs as the result of heating is confirmed by the decrease of FWHM of XRD reflections, presented in Figure 3. Furthermore, Figure 3 shows that positions of the reflections are shifted toward higher values of 2θ, which indicates the decrease of the mean value of lattice parameter. Appearance of additional super-lattice reflections on XRD patterns at ~ 9°, 17°, 23°, 28° etc. indicates the presence of the long-range order in alloyed NPs [3,7].

In Figure 4 the magnitudes of the Fourier-transforms $|F(R)|$ of experimental and theoretical Cu *K*- and Pt L_3-edge EXAFS are compared. The fit of the Fourier-transforms $F(R)$ of EXAFS spectra was performed using the following expression for oscillatory parts $\chi(k)$:

$$\chi_{Pt}(k) = S_0^2(N_{Pt\text{-}Pt} \cdot \chi_{Pt\text{-}Pt} + N_{Pt\text{-}Cu} \cdot \chi_{Pt\text{-}Cu}) \quad (1)$$

$$\chi_{Cu}(k) = S_0^2(N_{Cu\text{-}Cu} \cdot \chi_{Cu\text{-}Cu} + N_{Cu\text{-}Pt} \cdot \chi_{Cu\text{-}Pt}) \quad (2)$$

where χ_{Pt-Pt}, $\chi_{Pt\text{-}Cu}$, χ_{Cu-Cu}, $\chi_{Cu\text{-}Pt}$ – the contributions in $\chi(k)$, which correspond to photoelectron scattering processes on a one nearest atom of a specific type (the first label in subscript indicates the absorbing atom and the second label indicates the scattering one). Determination of structural parameters for the nearest neighbors of the absorbing atoms was performed by Fourier-transform analysis of EXAFS using the shortened *k*-range with k_{min} ~ 6 Å^{-1}, which is sufficient for the used number of number of fitting parameters [8] and enabled to neglect the contribution of photoelectron multiple scattering processes and backscattering on the light atoms (carbon atoms of the support and oxygen atoms) [9,4].

The fit gave the average values of structural and non-structural parameters: partial coordination numbers N_{Pt-Pt}, N_{Pt-Cu}, N_{Cu-Pt}, N_{Cu-Cu} and corresponding Debye-Waller parameters σ^2_{Pt-Pt}, σ^2_{Pt-Cu}, σ^2_{Cu-Pt}, σ^2_{Cu-Cu}, interatomic distances R_{Pt-Pt}, R_{Pt-Cu}, R_{Cu-Pt}, R_{Cu-Cu} and energy-shift parameters e_0(Pt), e_0(Cu) [10]. One of the criteria of correctness of the fit results was the coincidence of parameters σ^2_{Pt-Cu} with σ^2_{Cu-Pt} and R_{Pt-Cu} with R_{Cu-Pt}.

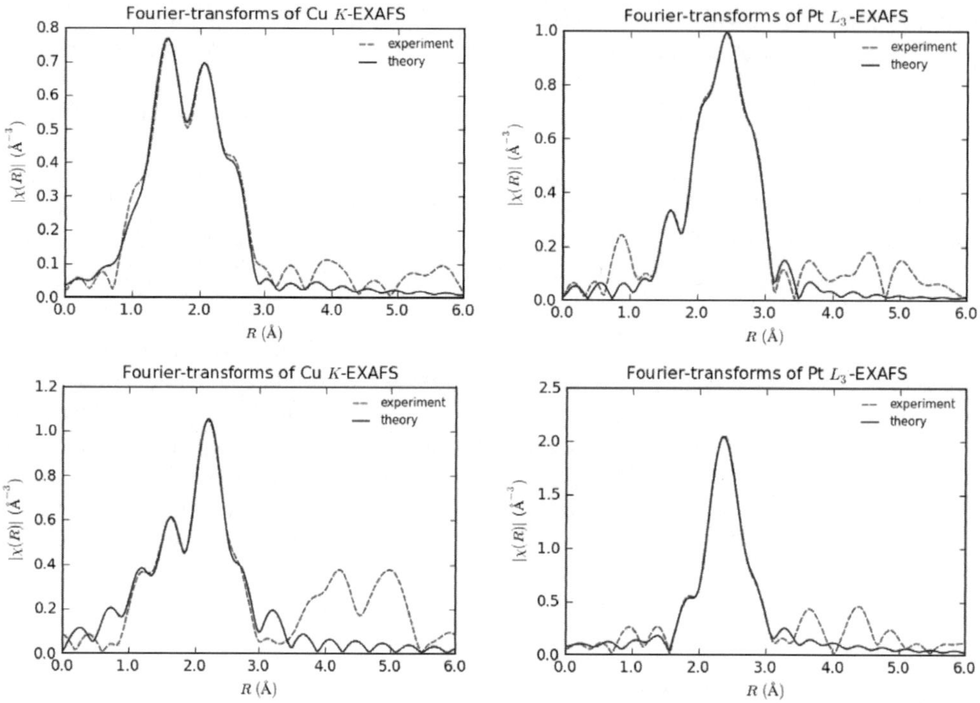

Figure 4. Experimental (red dashed lines) and theoretical (blue solid lines) magnitudes of the Fourier-transforms of Cu *K*- and Pt L_3-edge EXAFS for "as prepared" (top panel) and heated at 350 °C (bottom panel) PtCu/C electrocatalysts.

The obtained values of local structure parameters for Pt and Cu atoms in bimetallic PtCu NPs are shown in Table 1. Small values of partial coordination numbers of Cu indicate that a significant fraction of Cu atoms are presented in oxide form in PtCu/C, which is, however, amorphous at both temperatures and therefore cannot be seen in XRD. Increase of the total coordination number of Cu is probably due to the reduction of Cu ions during thermal treatment [11].

One more source of structural information are the interatomic distances R_{Pt-Pt}, R_{Pt-Cu} и R_{Cu-Cu}. In "as prepared" NPs their values correspond to those obtained recently for the core-shell NPs [2], whereas after heating the values of interatomic distances become almost equal. The latter is not typical neither for NPs with homogeneous solid solution structure nor with core-shell structure, but is an evidence in favor of the above assumption on the presence of local order in NPs, which is typical for intermetallic compounds [7].

Table 1. Values of local structure parameters of Pt and Cu atoms in bimetallic PtCu nanoparticles

Sample	***R*, Å (±0.01)**			**σ^2, Å^2 (±0.001)**			**Partial coord. numbers**			
	R_{Pt-Pt}	R_{Pt-Cu}	R_{Cu-Cu}	σ^2_{Pt-Pt}	σ^2_{Pt-Cu}	σ^2_{Cu-Cu}	N_{Pt-Pt}	N_{Pt-Cu}	N_{Cu-Pt}	N_{Cu-Cu}
"As prepared" PtCu/C	2.74	2.64	2.54	0.007	0.009	0.007	8.4	2.2	1.9	1.9
Heated at 350 °C	2.66	2.64	2.64	0.005	0.005	0.009	8.3	3.4	3.0	3.6

For detailed understanding of significant structural changes in PtCu NPs between room temperature and 350 °C, further structural investigations of PtCu/C electrocatalysts at different temperatures from these interval required.

4. Acknowledgements

The work was supported by Grant of Southern Federal University (№ 213.01-07.2014/10ПЧВГ). We thank SNBL at ESRF for providing beamtime for XRD measurements and my-Spot beamline at BESSY II for providing beamtime for EXAFS measurements.

5. References

[1] Salgado J R C, Antolini E and Gonzalez E R 2005 Carbon supported Pt70Co30 electrocatalyst prepared by the formic acid method for the oxygen reduction reaction in polymer electrolyte fuel cells *J. Power Sources* **141** 13–8

[2] Pryadchenko V V, Srabionyan V V, Mikheykina E B, Avakyan L A, Murzin V Y, Zubavichus Y V., Zizak I, Guterman V E and Bugaev L A 2015 Atomic Structure of Bimetallic Nanoparticles in PtAg/C Catalysts: Determination of Components Distribution in the Range from Disordered Alloys to "Core–Shell" Structures *J. Phys. Chem. C* **119** 3217–27

[3] Xiong L and Manthiram A 2005 Effect of atomic ordering on the catalytic activity of carbon supported PtM (M = Fe, Co, Ni, and Cu) alloys for oxygen reduction in PEMFCs *J. Electrochem. Soc.* **152** A697–703

[4] Srabionyan V V., Bugaev A L, Pryadchenko V V., Avakyan L A, van Bokhoven J A and Bugaev L A 2014 EXAFS study of size dependence of atomic structure in palladium nanoparticles *J. Phys. Chem. Solids* **75** 470–6

[5] Pryadchenko V V., Galustov A D, Srabionyan V V. and Bugaev L A 2014 Determination of the local atomic structure of material from X-ray absorption spectroscopy data without fourier analysis of experimental spectra *Opt. Spectrosc.* **117** 187–93

[6] Dyadkin V, Pattison P, Dmitriev V and Chernyshov D 2016 A new multipurpose diffractometer PILATUS@SNBL *J. Synchrotron Radiat.* **23** 825–9

[7] Oezaslan M, Hasché F and Strasser P 2012 PtCu3, PtCu and Pt3Cu Alloy Nanoparticle Electrocatalysts for Oxygen Reduction Reaction in Alkaline and Acidic Media *J. Electrochem. Soc.* **159** B444–54

[8] Bugaev L a., Avakyan L a., Srabionyan V V. and Bugaev a. L 2010 Resolution of interatomic distances in the study of local atomic structure distortions by energy-restricted x-ray absorption spectra *Phys. Rev. B - Condens. Matter Mater. Phys.* **82** 1–9

[9] Srabionyan V V., Bugaev A L, Pryadchenko V V., Makhiboroda A V., Rusakova E B, Avakyan L A, Schneider R, Dubiel M and Bugaev L A 2013 EXAFS study of changes in atomic structure of silver nanoparticles in soda-lime glass caused by annealing *J. Non. Cryst. Solids* **382** 24–31

[10] Koningsberger D C, Mojet B L, van Dorssen G E and Ramaker D E 2000 XAFS spectroscopy; fundamental principles and data analysis *Top. Catal.* **10** 143–55

[11] Lee S Y, Mettlach N, Nguyen N, Sun Y M and White J M 2003 Copper oxide reduction through vacuum annealing *Appl. Surf. Sci.* **206** 102–9

Structure and oil responsiveness of viscoelastic fluids based on mixed anionic/cationic wormlike surfactant micelles

A V Shibaev[1], A V Makarov[1], A L Aleshina[1], A V Rogachev[2,3], A I Kuklin[2,3] and O E Philippova[1]

[1] Physics Department, Moscow State University, Moscow 119991, Russia
[2] Joint Institute for Nuclear Research, Dubna 141980, Russia
[3] Moscow Institute of Physics and Technology, Dolgoprudniy 141700, Russia

E-mail: shibaev@polly.phys.msu.ru

Abstract. In this work, a combination of small-angle neutron scattering, dynamic light scattering and rheometry was applied in order to investigate the structure and oil responsiveness of anionic/cationic wormlike surfactant micelles formed in a mixture of potassium oleate and *n*-octyltrimethylammonium bromide (C_8TAB). A new facile method of calculating the structure factor of charged interacting wormlike micelles was proposed. It was shown that the mean distance between the micelles decreases upon the increase of the amount of cationic co-surfactant and lowering of the net micellar charge. It was demonstrated that highly viscous fluids containing mixed anionic/cationic wormlike micelles are highly responsive to oil due to its solubilization inside the micellar cores, which leads to the disruption of micelles and formation of microemulsion droplets. Experimental data suggest that solubilization of oil proceeds differently in the case of mixed anionic/cationic micelles in the absence of salt, and anionic micelles of the same surfactant in the presence of KCl.

1. Introduction

Surfactant molecules can self-assemble due to the balance between hydrophobic interaction of their alkyl tails and repulsion of polar head groups, and form wormlike (or cylindrical) micelles in aqueous solutions [1,2]. If wormlike micelles are sufficiently long, they form a network of topological entanglements, which imparts high viscosity and strong viscoelastic properties to the solutions. These properties are the key factors for numerous applications of wormlike micellar solutions, e.g. in cosmetics, drag reduction in heating and cooling systems, and oil industry [3-5].

The formation of wormlike micelles was described in numerous works [6-9] when a low-molecular weight salt was added to a solution of an ionic surfactant. Salt molecules dissociate in aqueous solution, providing sufficient screening of electrostatic repulsions between the charged surfactant head groups. This results in closer packing of head groups and formation of elongated cylindrical aggregates which can attain several micrometers in length. For instance, the network of long wormlike micelles was observed for an anionic surfactant potassium oleate [6,7] or a cationic surfactant erucyl bis(hydrohyethyl)methylammonium chloride (EHAC) [8,9] in the presence of KCl.

Long wormlike micelles can also be formed with no added salt in the mixtures of an ionic surfactant and an oppositely charged co-surfactant or hydrotrope salt [10-15]. The molecules of co-surfactant (or hydroptrope) embed into surfactant micelles, and their charged groups provide very effective screening of electrostatic repulsions between the surfactant polar heads. At the same time, hydrophobic part of the co-surfactant (or hydrotrope) interacts with the alkyl surfactant tails and resides closer to the micellar core. This leads to the formation of long micellar aggregates and their assembly into networks at very low surfactant concentrations. For example, wormlike micelles were observed in the mixtures of cationic surfactants cetyltrimethylammonium chloride (CTAC) [10] or EHAC [11,12] with sodium salicylate; or anionic surfactant sodium dodecyl sulfate (SDS) with *p*-toluidine hydrochloride (PTHC) [13], potassium oleate with alkyltrimethylammonium bromides [14], or sodium erucate with benzyl trimethylammonium bromide (BTAB) and tetramethylammonium bromide (TMAB) [15].

In the paper [14], the mixtures of potassium oleate and alkyltrimethylammonium bromides (C_nTAB) with different alkyl tail length n = 6, 8, 10, 12 were investigated in detail, and it was shown that an optimal asymmetry of surfactant and co-surfactant tail lengths is necessary to obtain 1-phase solution which contains a network of wormlike micelles. The best results were found in the case of *n*-octyltrimethylammonium bromide (C_8TAB): the solutions showed strong viscoelastic behavior and viscosity up to 1800 Pa·s at the total concentration of surfactant of 3 wt%. Structural investigations by small-angle neutron scattering (SANS) were also performed, and they proved the formation of cylindrical surfactant aggregates at different surfactant concentrations and [C_8TAB]/[potassium oleate] molar ratios. However, the structure factor of interacting charged wormlike micelles was not analyzed.

In the present work, we propose a new facile method of calculating the structure factor of charged wormlike micelles by dividing the total intensity I(q) by the form factor of a cylinder P(q) obtained from fitting the scattering curve at high q. By using this method, we analyze the position of electrostatic structure peak at different molar ratios [C_8TAB]/[potassium oleate] and show that the mean distance between the micelles decreases when they become less charged.

Due to their self-assembled nature, wormlike micelles are highly responsive to external stimuli [16]. For example, the addition of oils (hydrocarbons) totally disrupts the micelles and drives the formation of small spherical microemulsion droplets [17-19]. This effect is explained by the solubilization of hydrophobic oil inside the micellar cores, which leads to the rearrangement of surfactant molecules packing, and is accompanied by a drastic drop of viscosity by several orders of magnitude. High responsiveness to oils is a key factor determining the use of viscoelastic micellar solutions as the main component of fracturing fluids in oil recovery [5].

However, to the best of our knowledge, the responsiveness to oils was proven only for wormlike micelles formed in the presence of low molecular weight salts. In this work, we first investigate the effect of oil (*n*-decane) on charged anionic/cationic micelles of potassium oleate and C_8TAB in the absence of salt, and we show that the addition of small amount of oil (2 wt%) leads to the complete disruption of wormlike micelles and formation of small microemulsion droplets. Surprisingly, the radius of microemulsion droplets (2 nm) is found to be much less than those for potassium oleate system in the presence of KCl [19] indicating different packing of oil and surfactant tails inside the droplet core.

2. Experimental section

2.1. Materials

Potassium oleate (TCI, purity > 98%), C_8TAB (ABCR, purity > 98%) and *n*-decane (Aldrich) were used without further purification. D_2O (99.9% isotopic purity) supplied by Deutero GmbH was used to prepare the solutions for SANS measurements. For rheological measurements, the solutions were prepared using distilled deionized water purified by the Millipore Milli-Q system.

2.2. Samples preparation

First, aqueous stock solutions of potassium oleate and C_8TAB were prepared, which were then mixed in appropriate quantities. The obtained samples were stirred for 1-3 days and then left for 1-2 weeks to remove air bubbles. To study the effect of oil, *n*-decane was added on top of the sample, which was then stirred for 1 day.

IOP Conf. Series: Journal of Physics: Conf. Series **848** (2017) 012019 doi:10.1088/1742-6596/848/1/012019

2.3. Small-angle neutron scattering

SANS experiments were conducted using the YuMO spectrometer at the IBR-2 reactor facility (Frank Laboratory of Neutron Physics, Joint Institute for Nuclear Research, Dubna, Russia). Specially made dismountable cells with parallel disk-like quartz plates and beam path of 2 mm were used. The temperature during measurements was controlled at $20.0 \pm 0.3\ ^0$C by a water-circulating thermostat. The data were recorded by two detectors covering the range of scattering vectors q of 0.007 - 0.5 Å^{-1}.

All the samples were prepared in D_2O. The scattering length densities of the surfactant tails and D_2O are equal to ca. $-0.30 \cdot 10^{-6}$ Å^{-2} and $6.36 \cdot 10^{-6}$ Å^{-2}, respectively. The intensity profiles were corrected for sample transmission, beam path length and electronic noise, and calibrated at the absolute scale by a vanadium standard. Incoherent scattering from blank solvent (D_2O) was subtracted from the scattering curves.

The program SasView was used to fit the scattering curves by form-factors of a cylinder or a sphere. Only a part of the scattering curve at q higher than the structure peak position q* (q > 0.06-0.07 Å^{-1}) was used for fitting, and then the fitted form-factor was calculated in the whole q-range. One fitting parameter (radius of a cylinder or a sphere) was used.

2.4. Rheometry

Rheological measurements were made on a rotational rheometer Anton Paar Physica MCR 301. The details of the measurements are described elsewhere [20,21]. Cone-plate geometry with a diameter of 50 mm and cone angle of 1^0 was used for investigation. Temperature was controlled by a water circulating thermostat and Peltier elements at $20.00 \pm 0.05\ ^0$C. A specially constructed vapor lock was used to prevent solvent evaporation from the sample.

2.5. Dynamic light scattering (DLS)

DLS experiments were performed with ALV/DLS/SLS-5022F compact goniometer system equipped with ALV-6010/EPP digital time correlator (ALV, Langen, Germany), helium-neon laser operating at a wavelength of 632.8 nm and computer controlled and stepping motor driven variable angle detection system. Prior to measurements the samples were filtered through 0.20 µm polytetrafluoroethylene Millex-LG filters (Millipore). The details of the experiments are described elsewhere [22].

In DLS experiments, the normalized time autocorrelation function $g^{(2)}(q,t)$ of the scattered intensity is measured. To obtain the decay times distribution $A(\tau)$ and extract translational diffusion coefficients, Contin analysis based on the inverse Laplace transformation was applied [23]. With Stokes-Einstein equation the diffusion coefficients were converted to hydrodynamic radii.

3. Results and discussion

In this work, we studied the structure and responsiveness to oils of surfactant solutions containing 190 mM potassium oleate and *n*-octyltrimethylammonium bromide (C_8TAB). The molar ratio [C_8TAB]/[potassium oleate] was varied in the range 0.2-0.4. According to the literature data [14], a network of long wormlike micelles is formed at these conditions. Since the amount of cationic co-surfactant is less than the amount of anionic surfactant, the micelles are negatively charged.

Scattering curves for the solutions at different molar ratios are presented in Figure 1. On all curves one can see a structure peak, which originates from electrostatic repulsion of similarly charged micelles [14]. In the intermediate and high q-range (above the structure peak position q*) the curves can be well fitted by a form-factor of a cylinder (Figure 1).

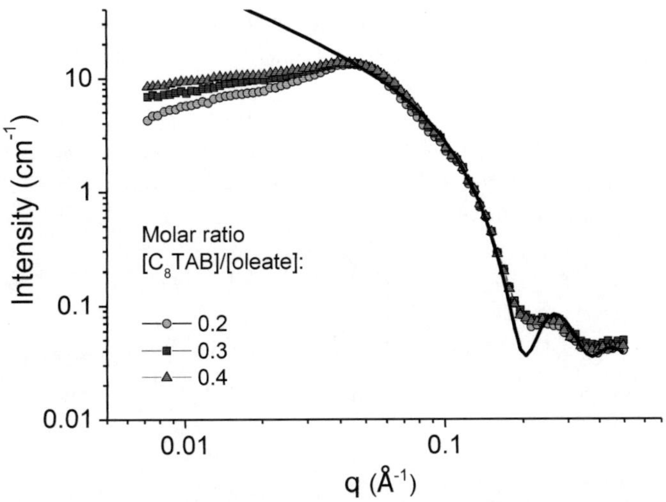

Figure 1. Scattering curves I(q) for aqueous solutions containing 190 mM potassium oleate and different concentrations of C_8TAB: 38 mM (circles), 57 mM (squares), 76 mM (triangles) in D_2O at 20^{0}C. Corresponding molar ratios [C_8TAB]/[potassium oleate] are shown in the picture. Line is a fit of the scattering curves by a form-factor of cylinder with radius R = 18.8 $\pm$ 0.2 Å.

The cylinder radii (R) for different molar ratios obtained from the fit are given in Table 1. It is seen that R does not change at the variation of the molar ratio [C_8TAB]/[potassium oleate] and remains close to the length of a fully extended alkyl tail of potassium oleate (1.9 nm). At the same time, R is much higher than the length of C_8TAB tail (0.8 nm). This fact indicates that the radius of mixed micelles at the excess of potassium oleate is governed only by its tail length, and not by the tail length of co-surfactant. Since the volume of C_8TAB tail is nearly 2 times smaller than those of potassium oleate, and the volume of the C_8TAB polar head is larger, one can conclude that the incorporation of more C_8TAB molecules into the micelles leads to closer packing of potassium oleate head groups, which is necessary to avoid the formation of free space inside the micellar hydrophobic cores.

Table 1. Structural parameters of 190 mM potassium oleate aqueous solutions in the presence of different concentrations of C_8TAB and oil obtained from SANS data

Molar ratio [C_8TAB]/ [potassium oleate]	Oil	Form of the objects	Radius of cylinder or sphere (R), nm	Structure peak position (q*), Å^{-1}	Average distance between the objects (d), nm
0.2	-	cylinder	1.86	0.053	11.9
0.3	-	cylinder	1.88	0.056	11.2
0.4	-	cylinder	1.88	0.059	10.6
0.4	140 mM *n*-decane	sphere	2.05	0.055	11.4

It can be seen from Figure 1 that the scattered intensity below the structure peak (q < q*) decreases with decreasing q and is lower than the form-factor of a cylinder, indicating the presence of strong repulsion between the micelles. In order to obtain the structure factor S(q), the following procedure was applied. First, the scattering curves were fitted by the form-factor of a cylinder at q > q*, where S(q) ~

1. Then, the structure factor was reconstructed in the whole q-range using the radii obtained from the fit. Such reconstruction is possible in our experimental range of scattering vectors (q > 0.007 Å^{-1}), since for charged micelles very high values of persistence length in the order of 70-90 nm are expected [1,2], and, therefore, the crossover to a persistent cylinder lies outside or near the low-q experimental limit. Finally, S(q) was determined by division of the total intensity I(q) by P(q).

The structure factors at different molar ratios [C_8TAB]/[potassium oleate] obtained using this procedure are presented in Figure 2. A well-defined structure peak can be seen at all curves. The expected position of the structure peak was calculated using the theoretical equation, which assumes locally hexagonal packing of wormlike micelles [24]

$$q^* = \frac{1}{R}\sqrt{\frac{8\pi\varphi}{\sqrt{3}}} \quad (1)$$

where φ is the volume fraction of surfactants and R is the radius of the micelles taken from Table 1. At the molar ratio 0.4 it is equal to 0.058 Å^{-1}, which coincides perfectly with q* extracted from S(q) (Table 1).

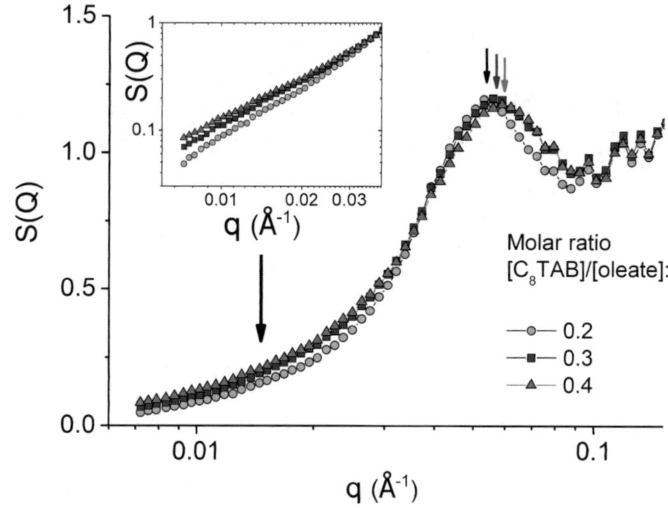

Figure 2. Calculated structure factors S(q) for aqueous solutions containing 190 mM potassium oleate and different concentrations of C_8TAB: 38 mM (circles), 57 mM (squares), 76 mM (triangles) in D_2O at 20^{0}C. Corresponding molar ratios [C_8TAB]/[potassium oleate] are shown in the picture. Inset: the low-q part of the same curves in log-log coordinates.

It can be seen from Figure 2 that the position of the structure peak q* slightly shifts to lower q upon decreasing content of C_8TAB (Table 1). The q* value is related to the mean distance between the micelles by the equation [14]

$$d = 2\pi/q^* \quad (2)$$

Therefore, the decrease of C_8TAB content in the micelles making their total charge higher results in a slight increase of average intermicellar distance. At low q (Figure 2, inset), the intensity drops with the decrease of C_8TAB content, which indicates strengthening of the repulsive electrostatic interactions between the micelles when less co-surfactant is added.

In the further study, the molar ratio [C_8TAB]/[potassium oleate] was fixed at 0.4, which corresponds to less charged micelles. At these conditions, the solutions possess very high zero-shear viscosity of 3400 Pa·s (Figure 3), which is more than 6 magnitudes higher than the viscosity of water. They also

IOP Conf. Series: Journal of Physics: Conf. Series **848** (2017) 012019 doi:10.1088/1742-6596/848/1/012019

show shear-thinning behavior and strong viscoelasticity. This is indicative of a formation of a highly entangled wormlike micellar network.

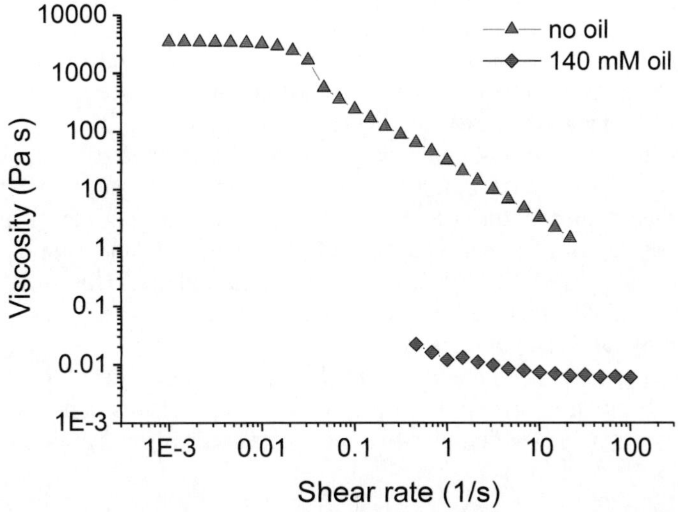

Figure 3. Flow curves for aqueous solutions containing 190 mM potassium oleate and 76 mM C_8TAB (molar ratio $[C_8TAB]/[\text{potassium oleate}] = 0.4$) in the absence (triangles) and in the presence (diamonds) of 140 mM *n*-decane at 20^{0}C.

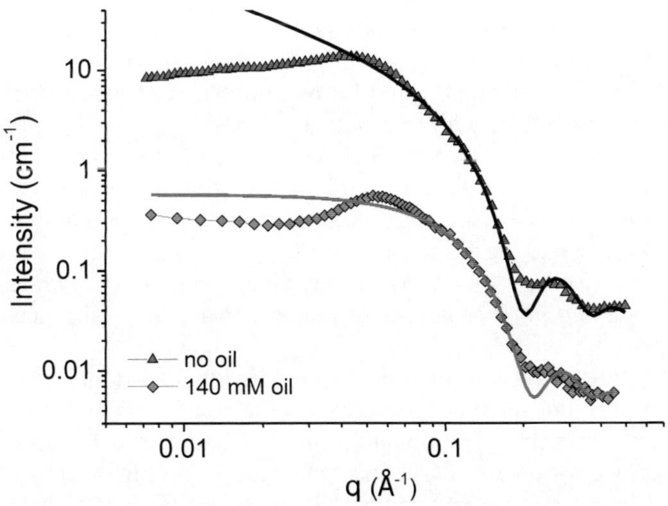

Figure 4. Scattering curves I(q) for aqueous solutions containing 190 mM potassium oleate and 76 mM C_8TAB (molar ratio $[C_8TAB]/[\text{potassium oleate}] = 0.4$) in the absence (triangles) and in the presence (diamonds) of 140 mM *n*-decane at 20^{0}C. The data in the presence of *n*-decane are offset by a factor of 5 for better representation. Black line is a fit of the scattering curve by a form-factor of cylinder with radius R = 18.8 $\pm$ 0.2 Å. Grey line is a fit of the scattering curve by a form-factor of sphere with radius R = 20.5 $\pm$ 0.2 Å.

We studied the effect of aliphatic oil (*n*-decane) on the rheological properties of such viscoelastic networks. It can be seen (Figure 3) that the addition of a small amount of oil (2 wt% = 140 mM) induces a dramatic drop of viscosity by 6 orders of magnitude, which is equal to 0.006 Pa·s. This effect is accompanied by a complete loss of viscoelastic properties of the solution, which becomes liquid-like and shows Newtonian behavior. It means that mixed anionic/cationic surfactant micelles formed in the absence of salts are highly responsive to oils, as their counterparts formed in the presence of salts [18,19], and incorporation of an oppositely charged co-surfactant into the micellar surface does not alter the ability of micelles to solubilize hydrocarbon in their cores and to break.

In order to get insight into the structural changes occurring in the micellar networks upon addition of oil, SANS experiments were performed. Fig. 4 shows the scattering intensity profiles of the micellar solutions before and after the addition of oil. It is seen that in the presence of oil the scattering curve can no longer be fitted by a form-factor of a cylinder, but in the intermediate and high q-range (above the structure peak position q*) it is well fitted by form-factor of a sphere. This means that spherical microemulsion droplets are formed as a result of solubilization of oil in the micelles.

This result is confirmed by DLS data, which show monomodal distribution of decay times τ (Figure 5a) with a narrow peak. The dependence of the decay rate $\Gamma = 1/\tau$ on q^2 is a direct proportionality (Figure 5b), which confirms that the mode is diffusive [23]. From the slope of this dependence $\Gamma = Dq^2$, the diffusion coefficient D was extracted, which was then converted to hydrodynamic radius R_h by using the Stokes-Einstein equation

$$D = \frac{kT}{6\pi\eta_s R_h} \tag{3}$$

where k is the Boltzmann constant, T is the absolute temperature, and η_s is the solvent viscosity. In this way, the value of R_h = 2.0 nm was obtained. It coincides perfectly with R calculated from SANS data (Table 1).

This result indicates that all cylindrical micelles are disrupted upon addition of oil, and only microemulsion droplets are present in the solution. The same effect was observed when *n*-dodecane was added to potassium oleate/KCl wormlike micelles [18,19], which were transformed into microemulsion droplets with a radius of 3.8 nm, consisting of an oil droplet in the core with radius of 1.9 nm, and a surrounding layer of surfactant molecules with thickness 1.9 nm.

Surprisingly, in the present system potassium oleate/C_8TAB the radius of the droplets is much less and is equal to 2 nm. This value is even close to the length of potassium oleate alkyl tail (1.9 nm), meaning that if the droplet had had the same core-shell structure as in potassium oleate/KCl system, then the inner oil core would have had the radius of only ~ 0.1 nm, which is far not enough to solubilize all added oil. It should be noted that 140 mM is the maximum amount of hydrocarbon that can be solubilized in the system containing 190 mM potassium oleate, and at higher oil concentrations its excess forms a second upper phase.

We suppose that in the present system microemulsion droplets of slightly different structure are formed due to the presence of a short-chain co-surfactant. Indeed, as mentioned above, the volume of C_8TAB alkyl tail is 2 times less than that of potassium oleate. It is more probable that C_8TAB polar head resides near the micellar surface (where its charge interacts with potassium oleate charge) and does not penetrate deeply into the micellar core. It means that the surfactant layer is formed not only by long potassium oleate tails (with length 1.9 nm), but also by shorter C_8TAB tails (note that at molar ratio 0.4 there are nearly 30% C_8TAB molecules and 70% potassium oleate molecules in the mixture). One can suggest that there is some space in the surfactant tails layer where oil can be solubilized in addition to the core, and such a small radius of microemulsion droplet can be explained by the penetration of oil into this layer.

Therefore, highly viscous solutions of mixed anionic/cationic wormlike micelles with no added salt keep responsiveness to oils inherent to the micellar chains formed in the presence of salts. It opens a new way of constructing oil-responsive systems for the applications in petroleum industry.

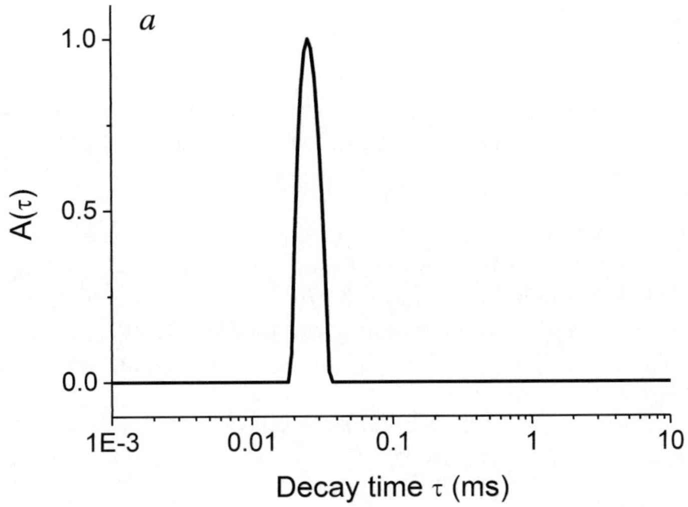

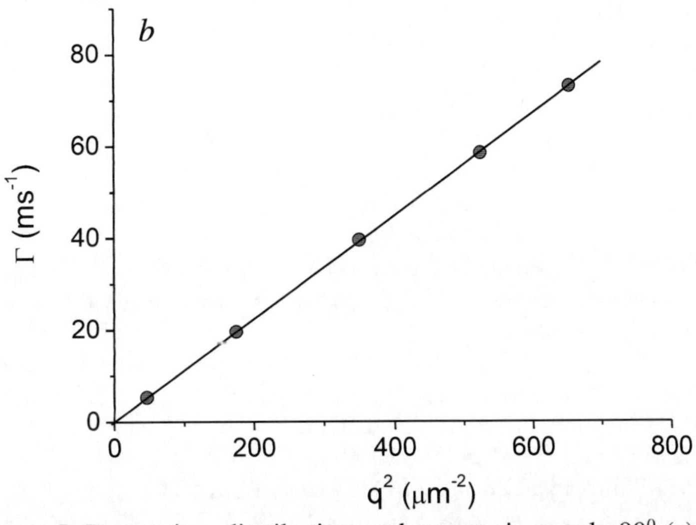

Figure 5. Decay time distribution at the scattering angle 90^0 (a) and the dependence of decay rate Γ on q^2 (b) for aqueous solutions containing 190 mM potassium oleate, 76 mM C_8TAB (molar ratio [C_8TAB]/[potassium oleate] = 0.4) and 140 mM *n*-decane at 20^0C.

4. Conclusions

In this work, the structure of mixed potassium oleate/C_8TAB wormlike micelles was investigated by SANS. It was shown that the radius of the micelles does not depend on the amount of cationic co-surfactant in the mixture, and is determined by the length of a fully extended anionic surfactant tail. Calculation of the structure factors revealed that the mean distance between the micelles slightly increases when they become more charged.

It was discovered that the networks formed by mixed anionic/cationic surfactant micelles are highly responsive to the addition of oil. The viscosity of the solution drops by 6 orders of magnitude when a small amount of oil is added, which is explained by the transformation of wormlike micelles into spherical microemulsion droplets. However, the radius of microemulsion droplets formed in potassium

IOP Conf. Series: Journal of Physics: Conf. Series **848** (2017) 012019 doi:10.1088/1742-6596/848/1/012019

oleate/C_8TAB/n-decane system is nearly 2 times less than that in potassium oleate/KCl/n-dodecane microemulsions, suggesting the solubilization of alkane in the surfactant layer in the presence of a short-chain cationic co-surfactant.

Acknowledgements
This work was funded by the Russian Foundation for Basic Research and by the National Intellectual Development Foundation (NIDF), according to the research project No. 16-33-80162 mol_ev_a.

References
[1] Dreiss C A 2007 *Soft Matter* **3** 956
[2] Magid L J 1998 *J. Phys. Chem.* B **102** 4064
[3] Yang J 2002 *Curr. Opin. Colloid Interface Sci.* **7** 276
[4] Rose G D and Foster K L 1989 *J. Non-Newtonian Fluid Mech.* **31** 59
[5] Chase B, Chmilowski W, Marcinew R, Mitchell C, Dang Y, Krauss D, Nelson E, Lantz T, Parham C and Plummer J 1997 *Oilfield Rev.* **9** 20
[6] Flood C, Dreiss C A, Croce V, Cosgrove T and Karlsson G 2005 *Langmuir* **21** 7646
[7] Molchanov V S, Philippova O E, Khokhlov A R, Kovalev Yu A and Kuklin A I 2007 *Langmuir* **23** 105
[8] Croce V, Cosrgove T, Maitland G, Hughes T and Karlsson G 2003 *Langmuir* **19** 8536
[9] Shashkina J A, Philippova O E, Zaroslov Yu D, Khokhlov A R, Pryakhina T A, Blagodatskikh I V 2005 *Langmuir* **21** 1524
[10] Kern F, Zana R and Candau S J 1991 *Langmuir* **7** 1344
[11] Raghavan S J and Kaler E W 2001 *Langmuir* **17** 300
[12] Raghavan S J, Edlund H and Kaler E W 2002 *Langmuir* **18** 1056
[13] Hassan P A, Raghavan S J and Kaler E W 2002 *Langmuir* **18** 2543
[14] Raghavan S J, Fritz G and Kaler E W 2002 *Langmuir* **18** 3797
[15] Han Y, Feng Y, Sun H, Li Z, Han Y and Wang H 2011 *J. Phys. Chem.* B **115** 6893
[16] Chu Z, Dreiss C A and Feng Y 2013 *Chem. Soc. Rev.* **42** 7174
[17] Hoffmann H and Ulbrich W 1989 *J. Colloid Interface Sci.* **129** 388
[18] Shibaev A V, Tamm M V, Molchanov V S, Rogachev A V, Kuklin A I, Dormidontova E E and Philippova O E 2014 *Langmuir* **30** 3705
[19] Shibaev A V, Molchanov V S and Philippova O E 2015 *J. Phys. Chem.* B **119** 15938
[20] Philippova O E, Shibaev A V, Muravlev D A and Mityuk D Yu 2016 *Macromolecules* **49** 6031
[21] Shibaev A V, Abrashitova K A, Kuklin A I, Orekhov A S, Vasiliev A L, Iliopoulos I and Philippova O E 2017 *Macromolecules* **50** 339
[22] Korchagina E V and Philippova O E 2010 *Biomacromolecules* **11** 3457
[23] Provencher S W 1979 *Macromol. Chem.* **180** 201
[24] Gamez-Corrales R, Berret J.-F, Walker L M and Oberdisse J 1999 *Langmuir* **15** 6755

IOP Conf. Series: Journal of Physics: Conf. Series **848** (2017) 012020 doi:10.1088/1742-6596/848/1/012020

SAS program for two-detector system: seamless curve from both detectors

A G Soloviev[1], T M Solovjeva[1], O I Ivankov[1,2,3], D V Soloviov[1,2,3], AV Rogachev[1,2], and A I Kuklin[1,2]

[1]JINR, Dubna 141980 Russia

[2]Moscow Institute for Physics and Technology, Dolgoprudny, Russia

[3]Institute for Safety Problems of NPP, Kyiv, Ukraine

E-mail: kuklin@nf.jinr.ru, solovjev@jinr.ru;

Abstract. The changes of the SAS program have been made as an attempt to adapt it to a two-detector system for the YuMO spectrometer. It is shown that applying weight algorithm technique for the raw data averaging procedure allows one to avoid a "sewing" problem resulting from using different detectors in rather wide q-range. The operating scheme and the background details of the experiment are discussed as well.

1. Introduction

The methodical specific features of experiment configuration at a small angle neutron scattering spectrometer are essential for the raw data treatment. The YuMO spectrometer located at the 4-th beam-line of the IBR-2 reactor (JINR, Dubna, Russia) has specific features, namely: two-detector system, vanadium standard in front of each detectors, high flux on a sample, a direct view of the reactor core, a central hole of the detectors and special geometry of the beam [1-4].

The methodological specifications of small-angle neutron scattering (SANS) experiments are always beyond the scope of the papers aimed at highlighting the features of measurements [5]. The main feature of YuMO spectrometer is a pulsed reactor. It allows using a wide range of wavelengths (from 0,7 Å to 10 Å) and applying time of flight method [1]. The second feature is two-detector system for small-angle scattered neutrons [3, 4]. Before having been mounted in Dubna in 2000, a similar system did never exist. Now there is at least a couple of known SANS instruments which have two-detector system [6-8]. The other point is centered detectors without using a neutron guide or bender. Such geometry allows obtaining a high flux on the sample comparable to the flux of ILL in terms of SANS spectrometer [9], although the power of ILL reactor is much higher. One more point is the usage of a special geometry of scaling rings of gas ^{3}He detectors, i.e. a hole in the center of the detectors. This allows applying a direct-beam detector and decreasing background scattering. Finally, a specific feature of the installation is vanadium standards in front of each detector. They periodically overlap the direct beam of neutrons that have already come through the sample.

In the paper, we investigate one of the problems connected with the experimental data treatment: sewing SAS curves obtained from different detectors. In this case, accurate background calculation and direct-beam detector usage as well as other spectrometer features mentioned above sound crucial.

IOP Conf. Series: Journal of Physics: Conf. Series **848** (2017) 012020 doi:10.1088/1742-6596/848/1/012020

Detailed answers to these questions are not the goals of the paper. Here we just illustrate the possibilities of the SAS program [4, 13], adapted to the two-detector system, i.e. the seamless curves obtained by two different detectors.

2. Brief description of specific YuMO setup features

The time-of-flight SANS YuMO spectrometer at the high flux pulse IBR-2 reactor [1, 3, 11] was modernized in 2000. One of the goals of the improvement was to increase the dynamic Q-range of the instrument. It was done by using two-detector system with central holes on the YuMO instrument. For this purpose, "OLD' and "NEW' detectors were used. The main view of the setup of the modernized YuMO instrument as well as examples of the SANS experiments are presented in figure 1.

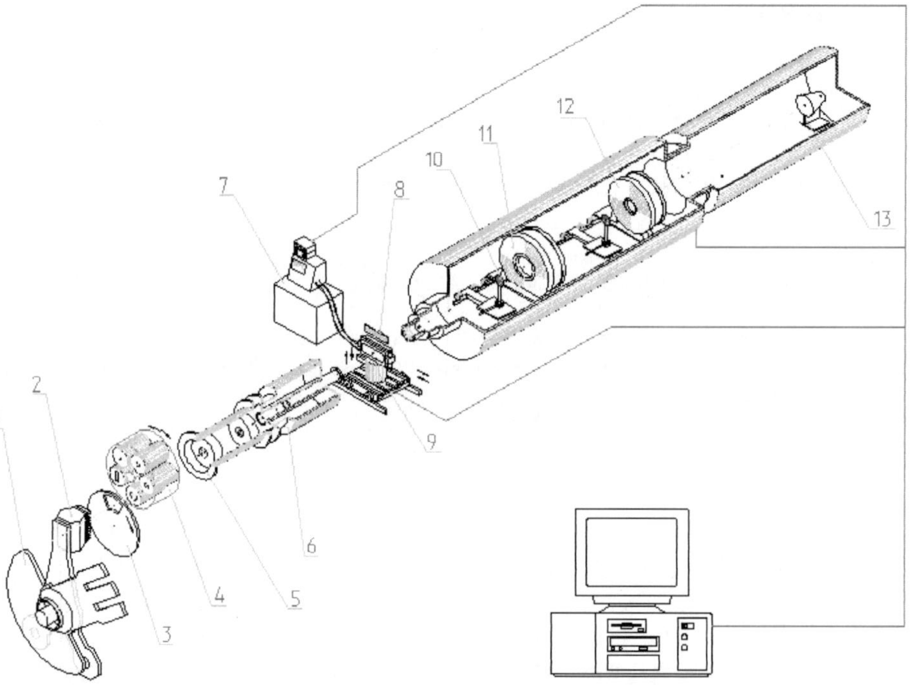

Figure 1. The setup small-angle scattering of neutrons YuMO (4-th channel IBR-2): 1 - reflectors, 2 - active zone of reactor with moderator, 3 - chopper, 4 - changeable collimator, 5 - vacuum tube with additional background collimators, 6 – justified collimator, 7 - thermostat, 8 – holder with samples on the table with thermos boxes, 9 - table of samples, 10 - the vanadium standard, 11, 12 - detectors ("OLD" and "NEW" accordingly), 13 - direct beam detector.

Figure 1 presents a scheme of the small-angle YuMO spectrometer. The neutrons which have arisen as a result of passage of reflectors 1 near to zone 2, then thermalized in moderator. Chopper cuts the background. Neutrons go through the collimation system and scatter at the sample 8 and than and comes on detectors 11, 12 and 13. More detailed description of the spectrometer operation in two-detector variant can be found in [3-4].

3. Mathematical background and realization

In general, processing with SAS includes three main stages:

1. preliminary corrections of the time spectra;
2. normalization of the time spectra over the scattering from the vanadium standard;
3. conversion time spectra into the space of neutron momentum transfer.

4. Conversion into Q-space and averaging spectra

On this final processing stage, we should convert time-channel and angle of scatter to q-space. To check conversion, we used a procedure, describing in [12]. The values of the module q-vector scattering and its dispersion are calculated by the formulas shown below.

Values of the transferred momentum q of a neutron and its dispersion can be calculated according to:

$$q = \frac{4\pi \sin(\Theta/2)}{\lambda} \approx \frac{2\pi\Theta}{\lambda}, \quad \Delta q = q\sqrt{\frac{\sigma_\Theta^2}{\Theta^2} + \frac{1}{12}\left(\frac{\Delta t}{t}\right)^2},$$

where Θ and σ_Θ – average scattering angle for the corresponding detector (ring of detector) and its dispersion, λ – neutron wavelength, t and Δt – the time channel center and its width. Average scattering angle and its dispersion can be calculate according to the:

$$\Theta = \frac{\int_{\Theta_-}^{\Theta_+} \theta F(\theta) d\theta}{\int_{\Theta_-}^{\Theta_+} F(\theta) d\theta}, \quad \sigma_\Theta = \frac{\int_{\Theta_-}^{\Theta_+} \theta^2 F(\theta) d\theta}{\int_{\Theta_-}^{\Theta_+} F(\theta) d\theta} - \Theta^2,$$

where $F(\theta)$ is the instrument resolution function,

$$\Theta_- = \Theta_{min} - d\Theta, \quad \Theta_{min} = \frac{R_{in}}{d_{sample-detector}},$$

$$\Theta_+ = \Theta_{max} + d\Theta, \quad \Theta_{max} = \frac{R_{out}}{d_{sample-detector}},$$

$$d\Theta = \underbrace{\frac{R_{collimator-1}}{d_{reactor-sample}}}_{\theta_1} + \underbrace{\frac{R_{collimator-2}}{d_{reactor-sample}} + \frac{R_{collimator-2}}{d_{sample-detector}}}_{\theta_2}.$$

An averaging procedure on interval $[Q_{min}, Q_{max}]$ with step equal to ΔQ is follows:

$$Q_n = \left(\underbrace{\sum_{i}^{7} \sum_{j=N_{begin}}^{N_{end}}}_{i,j:\left[(q_j^{(i)} - Q_{min})/\Delta Q\right]=n} \frac{q_j^{(i)}}{dI_j^{(i)}} \right) \left(\underbrace{\sum_{i}^{7} \sum_{j=N_{begin}}^{N_{end}}}_{i,j:\left[(q_j^{(i)} - Q_{min})/\Delta Q\right]=n} \frac{1}{dI_j^{(i)}} \right)^{-1},$$

$$I_n = \left(\underbrace{\sum_{i}^{7} \sum_{j=N_{begin}}^{N_{end}}}_{i,j:\left[(q_j^{(i)} - Q_{min})/\Delta Q\right]=n} \frac{I_j^{(i)}}{dI_j^{(i)}} \right) \left(\underbrace{\sum_{i}^{7} \sum_{j=N_{begin}}^{N_{end}}}_{i,j:\left[(q_j^{(i)} - Q_{min})/\Delta Q\right]=n} \frac{1}{dI_j^{(i)}} \right)^{-1},$$

$$\Delta I_n = \left(\sqrt{ \underbrace{\sum_{i}^{7} \sum_{j=N_{begin}}^{N_{end}}}_{i,j:\left[(q_j^{(i)}-Q_{min})/\Delta Q\right]=n} \frac{1}{dI_j^{(i)}} } \right)^{-1},$$

$$\Delta Q_n = \left(\underbrace{\sum_{i}^{7} \sum_{j=N_{begin}}^{N_{end}}}_{i,j:\left[(\Delta q_j^{(i)}-Q_{min})/\Delta Q\right]=n} \frac{q_j^{(i)}}{dI_j^{(i)}} \right) \left(\underbrace{\sum_{i}^{7} \sum_{j=N_{begin}}^{N_{end}}}_{i,j:\left[(q_j^{(i)}-Q_{min})/\Delta Q\right]=n} \frac{1}{dI_j^{(i)}} \right)^{-1},$$

$$n = 0\ldots(Q_{max} - Q_{min}) / \Delta Q,$$

Summation is made for nonzero values of $q_j^{(i)}$, $I_j^{(i)}$ and $dI_j^{(i)}$. Upper index (i) refers to the detector ring number (e.g. $i = 1\ldots8$ refers to the 1st detector and $i = 9\ldots16$ - to the 2nd detector). In addition, it is possible to exclude some rings from the averaging procedure for some reason.

The program provides an opportunity for non-averaging. In this case, sorting points of the spectrum over neutron momentum transfer is performed.

5. Realization and example

Since SAS [4, 13] is object-oriented, and elementary object deals with a detector ring, it does not really matter to which of two physical detectors the ring belongs. This allows one to process all the rings of both detectors simultaneously in the procedure described above. SAS does NOT operate with "OLD" and "NEW" detectors separately, and does NOT sew spectra from them. Instead, it operates with a set of their rings and produces the final spectrum in a natural way. In other words for realization in SAS [4, 13] seamless merger of curves, we shift a "vector of data treatment" from physical detector to each ring as self-sufficient unit. It means that we have spectra as a file from each of the rings and do averaging of these files.

Below, figure 2 shows the overlapping of the scattering curves from the "OLD" (green triangles) and "NEW" (blue squares) detectors. Insertion shows zoom of the overlapping region. Points with poor statistics (small angles for "OLD" and wide angels for "NEW") have no influence on the resulting curve (red circles). Also figure 2 demonstrates possibility on dynamical range of q-space by more than 100. Another scale (delta Sigma/ delta Omega - is a differential cross section) is an absolute scale. The given example is not unique. Good overlapping of the scattering curves from the "OLD" and "NEW" detectors simultaneously shows the real q-range which could be used for a part of ranges for small and wide q.

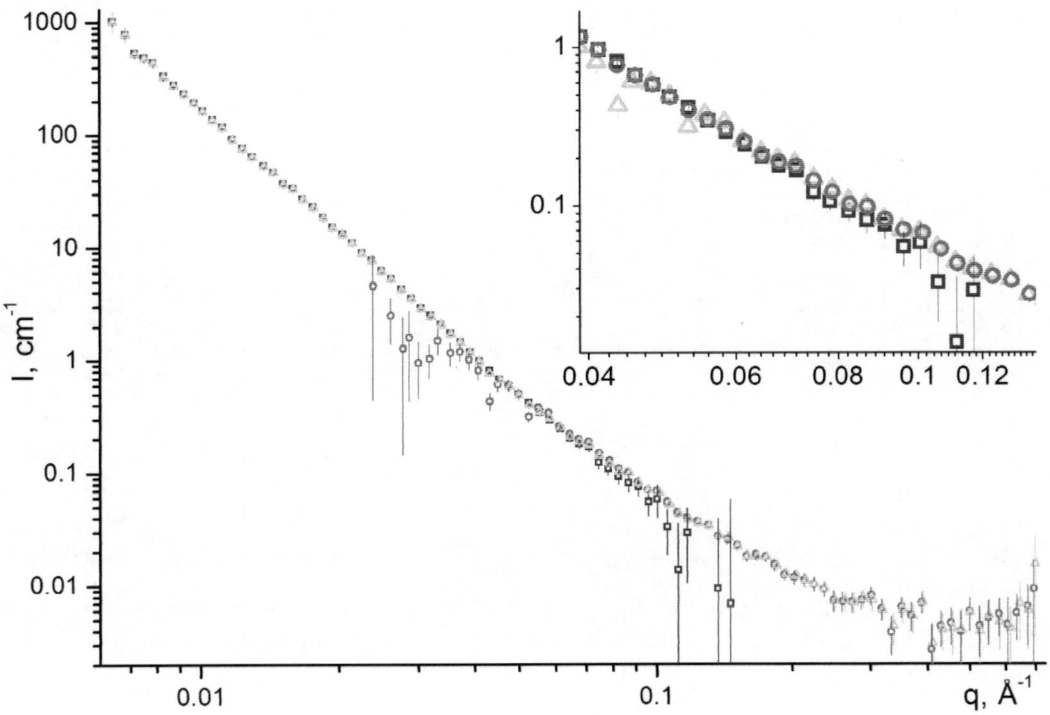

Figure 2. shows overlapping of the scattering curves from the "OLD" (green triangles) and "NEW" (blue squares) detectors. Insertion shows zoom of the overlapping region. Points with poor statistics (small angles for "OLD" and wide angels for "NEW") have no influence on the resulting curve (red circles).

To demonstrate good quality of the overlapping procedure with different detectors, in figure 3 we presented a view of monitor screen with above-described SAS program data treatment for series of silver alkanoates [14] curves measured at YuMO spectrometer with two-detector system and by using smoothing procedure [10]. As we can see, the different position of peaks does not have asymmetry in different position of q-scale. It is important to stress that only 2 rings of "New" detector "see" peaks and hence take part in averaging of curves. We specially used wide neutron wavelength range to show influence of tail of "New" detector. The stability of curves behaviour in this case has been also demonstrated.

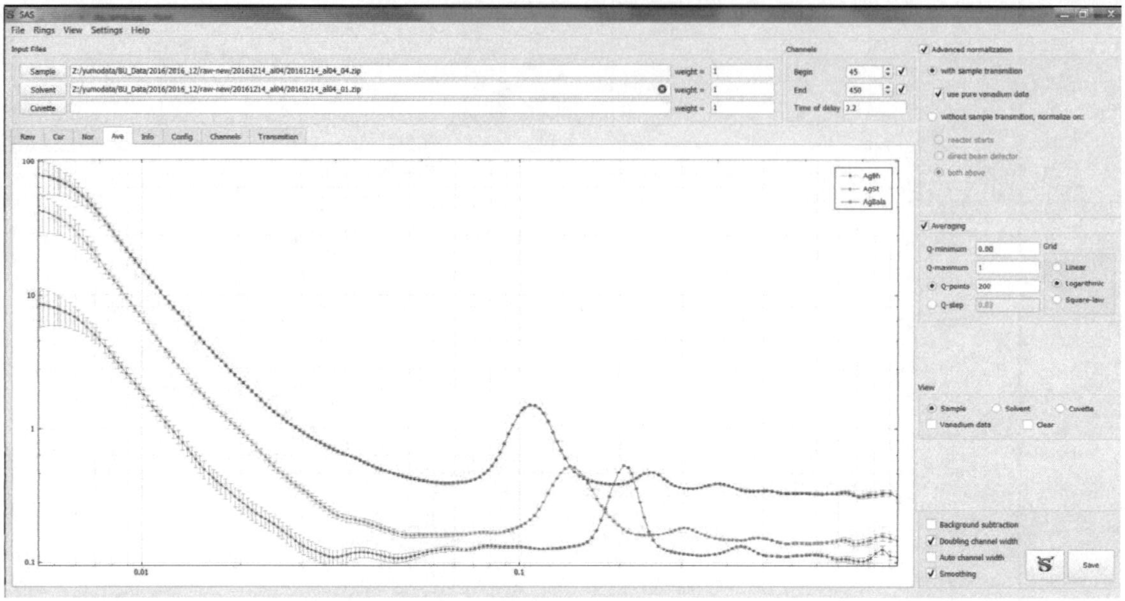

Figure 3. View of monitor screen with above-described SAS program data treatment for series of silver alkanoates [14] curves measured at YuMO spectrometer with two-detector system and by using smoothing procedure [10].

6. Conclusion

The improvement of the SAS program as an attempt to adapt it to a two-detector system for YuMO spectrometer are described. It is shown that applying a weight algorithm technique for raw data averaging procedure allows one to avoid a "sewing" problem resulting from using different detectors in rather wide q-range. In particular, choosing the time range (the wavelength range) has now a mathematical background. On the other hand, it is an additional "advise" from the curves view for maxima and minima q to use for consideration.

Acknowledgements

We thank for financial support Plenipotentiary of Chech Republic, Romania and Slovak Republic.

References

[1] Ostanevich Y M 1988 *Makromol.Chem. Macromol.Symp.* **15** pp 91-103
[2] Kuklin A I, Islamov A K and Gordeliy V I 2005 *Neutron News* **16(3)** pp 16-18
[3] Kuklin A I, Islamov A K, Kovalev Yu S, Utrobin P K and Gordelii V I 2006 *Poverkhnost'. Rentgenovskie, Sinkhrotronnye i Nejtronnye Issledovaniya* **6** pp 74-83
[4] Soloviev A G, Solovieva T M, Stadnik A V Islamov A Kh and Kuklin A I 2003 *Communication of JINR* **P10-2003-86**
[5] Kuklin A I, Soloviov D V, Rogachev A V *et al* 2011 *J. Phys.: Conf. Series* **291 No. 1** 012013
[6] Heenan R K, Rogers S E, Turner D, Terry A E, Treadgold J and King S M 2011 *Neutron News* **22(2)** pp 19-21
[7] Dewhurst C D 2008 *Measurement Science and Technology* **19(3)** 034007
[8] Klose F, Constantine P, Kennedy S J and Robinson R A 2014 *J. Phys.: Conf. Series* **528 No. 1** 012026
[9] Kuklin A I, Rogov A D, Gorshkova Yu E, Utrobin P K, Kovalev Yu S, Rogachev A V, Ivankov O I, Kutuzov S A, Soloviev D V and Gordeliy V I 2011 *Physics of Particles and Nuclei Letters* **8(2)** pp 119-128.

[10] Soloviev A G, Litvinenko E I, Ososkov G A, Islamov A K and Kuklin A I 2003 *Nuclear Inst. and Methods in Physics Research*, A. 502/2-3 pp. 498—500.
[11] Kuklin A I, Islamov A H, Balasoiu M, Murugova T N and Gordeliy V I 2012 *J. Phys.: Conf. Series* **351 No. 1** 012001
[12] Nyam-Osor M, Soloviov D V, Kovalev Y S, Zhigunov A, Rogachev A V, Ivankov O I, Erhan R V and Kuklin A I 2012 *J. Phys.: Conf. Series* **351 No. 1** 012024
[13] http://yumo.jinr.ru/software/sas
[14] Binnemans K, Van Deun R, Thijs B, Vanwelkenhuysen I and Geuens I 2004 *Chemistry of materials* **16(10)** pp 2021-2027

IOP Conf. Series: Journal of Physics: Conf. Series **848** (2017) 012021 doi:10.1088/1742-6596/848/1/012021

Nonequilibrium chemo-electronic conversion of water on the nanosized YSZ: experiment and Molecular Dynamics modelling problem formulation

A S Doroshkevich[1,2], A I Lyubchyk[3], A K Islamov[1], V A Turchenko[1,2], T Yu Zelenyak[5], A V Shylo[2], M. Balasoiu[1], A V Saprykina[4], S Ohmura[6], O S Lygina[7], S I Lyubchyk[8], T E Konstantinova[2], M V Lakusta[2], V I Bodnarchuk[1], S B Lyubchyk[7], Yu Yu Bacherikov[9], Ye Aliyeva[10], E B Asgerov[1,11], D A Suvorov[5], V R Gashimova[5], A L Gasev[5], H S Dadakhanova[5], Kh T Kholmurodov[1,5]

[1] Joint Institute for Nuclear Research, 141980, str. Joliot-Curie, 6, Dubna, Russia.
[2] Donetsk Institute for Physics and Engineering named after O O Galkin NAS of Ukraine, 03680, Nauki ave, 46, Kyiv, Ukraine.
[3] i3N/CENIMAT, Department of Materials Science, Faculty of Science and Technology, New University of Lisbon and CEMOP/UNINOVA, Campus de Caparica, 2829-516 Caparica, Portugal.
[4] Government Institution "Donetsk Institute for Physics and Engineering named after A A Galkin", 83114, R. Luxembourg str., 72, Donetsk, Ukraine.
[5] Dubna International University for Nature, Society and Man, 141980, st. Universitetskay, 19, Dubna, Moscow region, Russia.
[6] Hiroshima Institute of Technology, Hiroshima 731-5193, Japan.
[7] REQUIMTE, Chemical Department, Faculty of Science and Technology, New University of Lisbon, Campus de Caparica, 2829-516 Caparica, Portugal.
[8] CQE-Centro de Química Estrutural, Instituto Superior Tecnico, Universidade de Lisboa, Av. Rovisco Pais 1, 1049-001 Lisbon, Portugal.
[9] Institute of Semiconductor Physics NAS, Kiev, Ukraine, 03028, Nauki ave, 41, Kiev, Ukraine.
[10] Institute of Physics, Azerbaijan National Academy of Science,131 H. Javid ave., Az1143 Baku, Azerbaijan.
[11]National Nuclear Research Center, AZ-1073, Baku, Azerbaijan.

E-mail: doroh@jinr.ru

Abstract. The exothermic heterogeneous electrochemical energy conversion to the electric energy through interaction of the ZrO_2-based nanopowder system with atmospheric moisture was explored within this work. Electrical properties of the experimental samples were investigated during humidification at the conditions of molecular flux density gradient. The morphological features of the surface cross-section and aggregates of 3 mol% Y_2O_3-doped ZrO_2 nanopowder systems were investigated. Initial conditions for molecular dynamics modelling of the adsorption processes were obtained. A novel approach for developing of chemo-electronic converters based on nanoscale processes and materials with dielectric conductivity type proposed.

IOP Conf. Series: Journal of Physics: Conf. Series **848** (2017) 012021 doi:10.1088/1742-6596/848/1/012021

1. Introduction

Development of a new electricity generation techniques is the one of the most relevant tasks, especially nowadays under conditions of extreme growth in energy consumption [1 - 2]. The water is well known conventional source of electrical energy used by mankind. In liquid state, it is traditionally used for the electric energy conversion by an external thermostat of hydropower plants. However, direct conversion of water in gas state, (atmospheric moisture) to electric energy, remained unrealizable for centuries as a top scientific and practical aim. With the development of nanotechnology, the implementation of these ideas become potentially possible. In this regard with nanopowder system based on ZrO_2 is the best candidate for the direct conversion of the energy of water adsorption into electric energy.

It is known [3] that extremely nonequilibrium thermodynamically, the surface of nano-dispersed oxide systems is exist in a state of dynamic charge and adsorption equilibration. I.e., the change of quantity of adsorbates is accompanied by a changing of the total charge in the system. In the case of nanoscale powder of zirconia dioxide, on its surface are preferably water is adsorbed, and certain stages of a matter exchange between the system and external environment have an exothermic behaviour. Hence, in a cyclic mode, the nanopowder system based on ZrO2 can convert the chemical energy of water molecules adsorption to an electric form. The possibility of such processes has been shown previously for the bulk semiconductors [4 - 5].

The aim of present work is the experimental proof of concept and formulation of a problem for the following Molecular Dynamic (MD) calculations of water adsorption on the surface of nanosized yttria-stabilized zirconia (YSZ).

2. Experimental section

2.1. Specimens preparation.

As an object of study the compacts (compressed structures) of YSZ (ZrO_2 + 3 mol% Y_2O_3) nanopowder with a particle size 9 nm were used (Figure 1). As can be seen from Fig. 1, the nanoparticles are monodisperse and well distributed. Consequently, the binding of aggregates is weak and adhesion between particles of this powder is low. The technology of nanopowders production was previously published in [6] and contains several technological operations. At first, by co-precipitation method from the chloride raw materials the hydrated zirconia hydroxide was obtained. After dehydration in a specialized microwave oven (T = 120 C, t = 0.4 h) an amorphous powder was crystallized through annealing at 400°C for 2h. Then, in order to create the compacts structures, 1g of obtained powder was compressed by uniaxial pressure of P = 40 MPa until the tablets of 20 mm in a diameter and a height of 2-3 mm were formed.

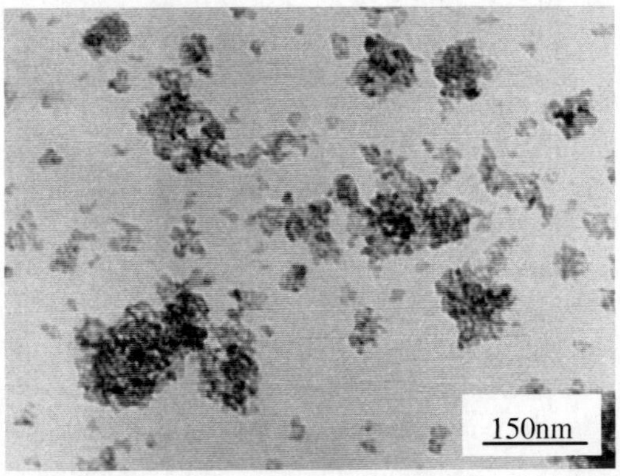

Figure 1. TEM image of ZrO_2+3mol% Y_2O_3 nanopowder prepared at, 400°C for 2h.

Before the compression in order to unify initial physical conditions the powder was pre-treated in an oven at $T = 120°$ C for 1h. After the tablets form creation, the specimens undergo compression by a high hydrostatic pressure (HHP, 500 MPa). Then, the graphite electrodes were deposited mechanically on the sides of the tablets (Fig. 2).

The pre-treatment of specimens before measurements consisted of dehydration through drying at 120 °C for 1h. After dehydration samples was placed in a climatic chamber under specific conditions of temperature $T = 21$°C and relative humidity $\eta = 70\%$. It worth noting that prior to the placement of specimens into the climatic chamber the spherical clamping silver contacts were attached to the graphite electrodes. In order to create the potential difference between the contacts during the adsorption of water molecules a gradient of physical properties within the specimen was created. Thus, the adsorption at one side of the tablet has been limited by deposited waterproof layer of paraffin.

2.2 Samples characterization

The spatial and structural characterization of the samples was performed by transmission (TEM) [7] and scanning (SEM) electron microscopies using JEM 200A and JSM640LV (JEOL) instruments respectively.

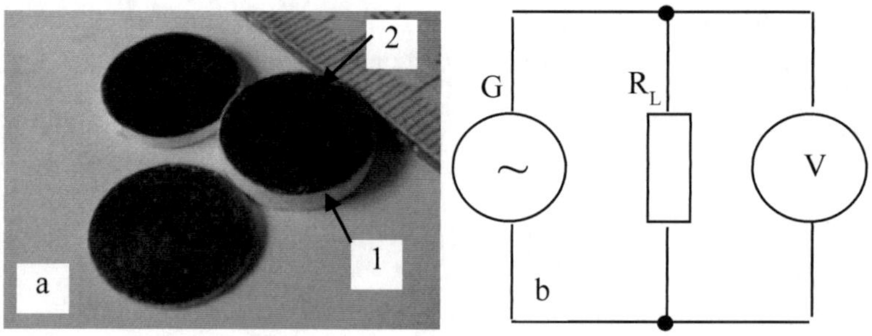

Figure 2. a) Specimen - tablets (1) with graphite electrodes (2) and b) electrical circuit of measurement, where G – specimen in generation mode, R – applied load (1MOhm), and V - digital voltmeter.

Small-angle neutron scattering (SANS) experiments were done using the time-off-light YuMO spectrometer and the IBR2 highflux pulsed reactor (Dubna, Russia). The detectors of the spectrometer were located at distances of 5.28 and 13.04 m from the sample; the range of transferred pulses q attained in the experiment was from 0.006 to 0.5 A^{-1} [8, 9]. X-ray diffraction analysis (XRD) of the samples was performed using an Empyrean diffractometer (PANalytical). Measurements of potential difference (EMF) of produced samples were carried out by digital voltmeter AM1199 (AKTACOM) under eclectic circuit shown in Fig. 2b.

3. Results

3.1. Experimental realization of the chemoelectronic conversion effect.

The variation of changes of potential differences (EMF) and mass δm of the specimen within the time is shown on Figure 3. As can be seen, the change of specimen mass increased monotonically, and reaches saturation plateau after 100 min. During this time, the sample adsorbs around 4 wt% of moisture. Herewith after 20 minutes the potential difference between the electrodes $V(t)$ increases up to 100 mV. The variation of $V(t)$ in contrast to variation of $\delta m(t)$ is not monotonic and during the next 80 minutes the potential difference between the electrodes is decreases and the polarity of voltage is inversed. The amplitude of signal of reverse polarity is approximately 25% from the total *EMF* amplitude. According to the form of the *EMF* curve (Figure 3), it can be concluded that the electrochemical process in the sample material has two stages. Initially, the electrodes are separated by an insulating layer of dewatered samples material and have fixed potentials. A slight potential difference usually takes place even in the initial state. It is induced by polarization of the sample during compaction process (electrets effect [10]). In our case, such value is around 20 mV (Figure 3, curve 2). During the humidification process of specimen, the potential of water permeable electrode increases because of localization in a bulk of adjacent space region of a free charge carriers through nonequilibrium chemoelectronic process (First step) [11], wherein the electrical resistance of layer between electrodes is still large (according impedance spectroscopy of about 10^6 - 10^7 Ohms / mm).

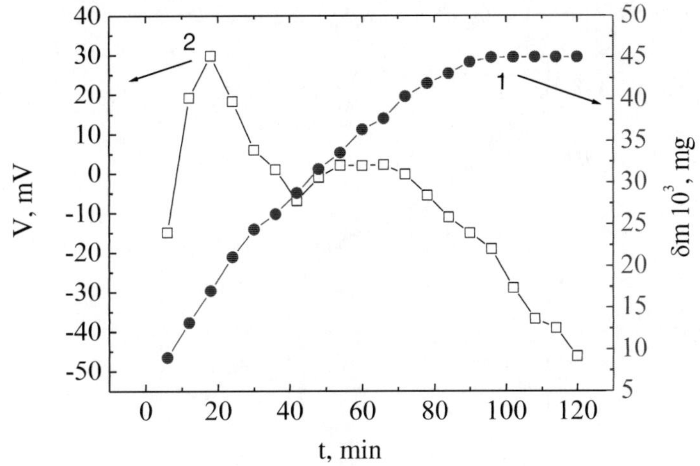

Figure 3. Variation of weight, *δm* (1) and the *EMF*, *V* (2) of the sample changes from the time.

The potential of waterproofed electrode remains constant. During the distribution of moisture within pores of the tablet conductivity between the electrodes begins increase whereas a potential difference between electrodes decreases accordingly (Second step). Increasing of the potential differences is observed at 20 min. There is an inflection on the potential curve herewith, as can be seen from curve 1, the dynamics of water absorption of tablet is monotonous. This indicates that the generation of power by tablet is limited exclusively by the geometric features of sample, rather than the nature of the mass transfer process. The nature of the minimum observed on curve 2 (Figure 3), occurred after 40 minute is not clear yet.

3.2. Investigation of physical properties of the specimens. The morphology of the specimens.
The typical SEM image of the tablet cross-section is shown on Figure 4.

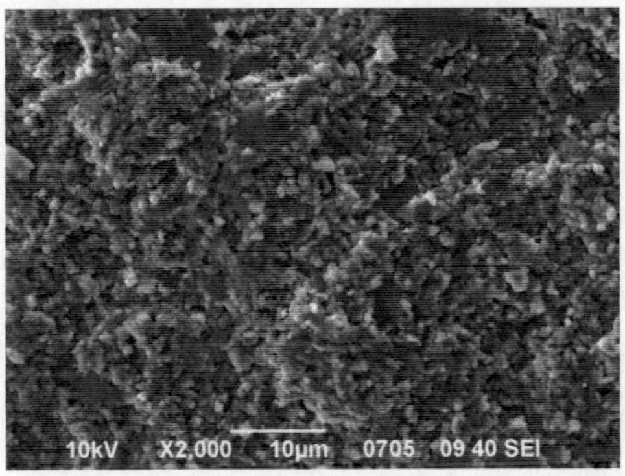

Figure 4. SEM image of the tablet cross-section.

It can be seen that the microstructure is relatively friable. The calculated density of the tables is about $\rho_c = 3.1$ g / cm^3, which constitute about 40-50% of the density of ZrO_2-based ceramics. Moreover, the branched network of relatively large (approx. 1 μm) pores is present in the bulk of specimen. Thus, water can easily seep into the bulk of tablet. Most probably, the first water molecules in the gas phase fill pores in the surface layers of tablet, and then, with a delay proportional to the resistance of the flow are penetrates to bulk.

3.3. The bulk structure of the samples. SANS - investigation.

The SANS spectra of tablets (curve 1) and nanopowder (curve 2) are presented in Figure 5.

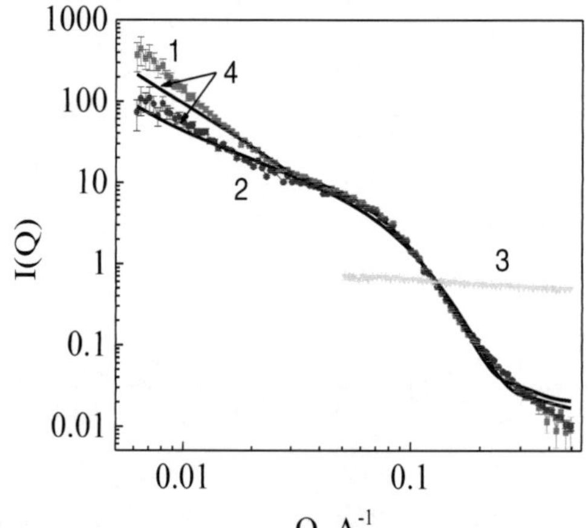

Figure 5. The results of investigation of spatial structure of sample by SANS, where 1 - the scattering of the nanopowder ZrO_2 + 3 mol% Y_2O_3 with a particle size of 9 nm; 2 - scattering of the tablet and 3 - scattering of water; 4 - fitting curves.

To compare the amount of water in the samples, scattering from pure water is shown in Figure 5 (curve 3). The scattering was divided on two parts. Low Q < 0.03 A^{-1} region was described by Guinier law and large Q > 0.03A^{-1} region was described by scattering from sphere:

$$I(Q) = A \cdot \exp(-Rg^2Q^2/3) \qquad Q < 0.03A^{-1}$$

$$I(Q) = B \cdot \left(3 \cdot \frac{\sin(QR) - QR\cos(QR)}{(QR)^3}\right)^2 + C \quad Q \geq 0.03A^{-1},$$

where: R is the radius of nanoparticles, A and B are scaling parameters and C – is the background constant, corresponding to incoherent scattering (the presence of H_2O in system).

Guinier law allows to estimate aggregation number of nanoparticles, whereas large Q region gives the information about internal structure of system, in particular - the average size of particles. In both cases (nanopowders and tablets) the scattering is well described be spheres of diameter 9nm, whereas the aggregation for powders and tablets differs significantly and amounts to 15 and 5 nanoparticles, respectively. Thus, for the compressed structures the access of moisture to the surface of samples is available at all the scales.

3.4. Assumptions on localization and delivery mechanism of charge carriers to electrodes. Relationship between effects of chemo-electronic conversion and structural features of surface layer of nanoparticle.

Earlier V.I. Alekseenko et al [12] proposed an adsorption mechanism of Tetragonal - monoclinic transformation (T-M - transformation). Considering the T-M transition from the electrophysics standpoint, he revealed that a change in the charge state of the ion of stabilizer Y causes the T-M transformation in YSZ ceramics under chemical adsorption of H_2O. Ions of stabilizer exist in material in the form of impurity - vacancy dipoles (IVD) such as $Y^{3+(-)}$ - $Vo^{-(+)}$. In the solid solution, based of Zirconium oxide impurity atoms Y^{3+} have on the one electron less than Zr^{4+} atoms and they are acceptors. The impurity oxygen vacancies formed for compensation of excess of volume and charge of inovalent impurity - are donors. Vacancy Vo donates electron to Y^{3+} ion. Thus, a chemical bond between them is formed. The ionization energy of IVD is relatively small (E ~ 0,1 ÷ 0,6 eV) [13 -14]. A shift of electron density from the bulk of the material to adsorbates occurs during chemical interaction of adsorbate with surface of ZrO_2. The binding energy in IVD near the ZrO_2 particle surface decreases, hence its chain length Zr - Vo - Y, increases, which is accompanied by the transition in structure of the solid solution system with a large volume of cell - A phase. This effect on the surface of a bulk of ZrO_2-based ceramic was observed by the method of X-ray diffraction under the creation of high-stress concentrations using indentation [15]. After indentation in a thin layer of the ceramic material at the subsurface a 2-D monoclinic crystal was appeared. The T-M transformation in ZrO_2 – ceramic is reversible and has diffusionless mechanism, i.e., mutually convertible structures transforms into one another coherently, without breaking the bonds [16].

In nano-sized objects up to now the effect of changing of structure of surface layer of nanoparticles based on ZrO_2 was not discovered due to insufficient resolution of used research methods. However, considered effect of chemoelectronic conversion as one of specific manifestations of adsorption-induced electrokinetic processes suggest that recharging of ions of crystal lattice near the surface of nanoparticles is still going on. There is reason to think that an electric charge, localized from the crystal lattice of discussed structural transformation of adsorption type can be delivered by DEL – ions to the electrode in the presence of an external electric field or gradient of electrical properties of sample in the space between electrodes through the relay mechanism [17]. Specifically, due to a relay mechanism the hydronium and hydroxyl ions have higher numbers of charge transfer [18 - 19]. Limit of mobility of hydrogen and hydroxyl ions are 349.8 and 199.2 Sm•m^2 / mole, respectively and the mobility of other ions are in the range between 40 - 80 Sm•m^2 / mole. Proton hopping from hydronium ion H_3O^+ to a certain way oriented neighboring water molecule is occurs during the relay mechanism. As a result, it becomes an H_3O^+ - ion. Then proton is passed on to the next water molecules. These jumps occur much faster than the simple movement of hydronium ion, which provides high mobility and a high transport number of hydronium ion. Similarly explained the charge transfer of hydroxyl ions. In this case proton does hopping from water molecules to adjacent OH^- ions because proton in a molecule of water is linked stronger than hydroxonium ion, it jumps in this case more difficult, and the mobility of hydroxyl ion is lower than mobility of hydronium ions. At the moment from the available experimental data it is not clear what kind of ions is charge transfer with generalized heterophase boundaries on electrodes is carried out.

A charge transfer to an external electrical circuit occurs, consequently, as a result of ions discharge at the electrodes. The excess cations are then discharged at a cathode. Similarly, the anions are discharged at an anode, and its amount is thus decreases in its anode compartment. Thus, based on the nature of electrokinetic processes on the nanoparticles surface, we have to identify by MD simulations, the two main tasks (as were stated in part 1 of the Introduction):

(1) - to calculate the adsorption energy and to show the ability of nanoparticles surface ionization, i.e., an electron exemption from the volume of nanoparticles material into a dispersion medium during the water adsorption process;

(2) - to estimate a charge state of ions of ZrO_2 nanoparticles and to demonstrate the consequent changes of the crystal structure caused through the nanoparticles surface layers (T-M transformation).

3.5. Initial conditions for mathematical modeling of electrons generation in ZrO_2 based nanopowder system.

As initial conditions for the MD calculations of the ZrO_2+3mol% Y_2O_3 - nanoparticles material the structural parameters, obtained from the X-ray analysis data [20] were used. The X-ray pattern of investigated nanopowder sample of ZrO_2 + 3%Y_2O_3 is shown on Figure 6. According to the X-ray data, the material of the studied nanoparticles has a tetragonal type of crystal lattice $P4_2$ / nmc. Calculated structural parameters of nanoparticle structures are shown in Table 1.

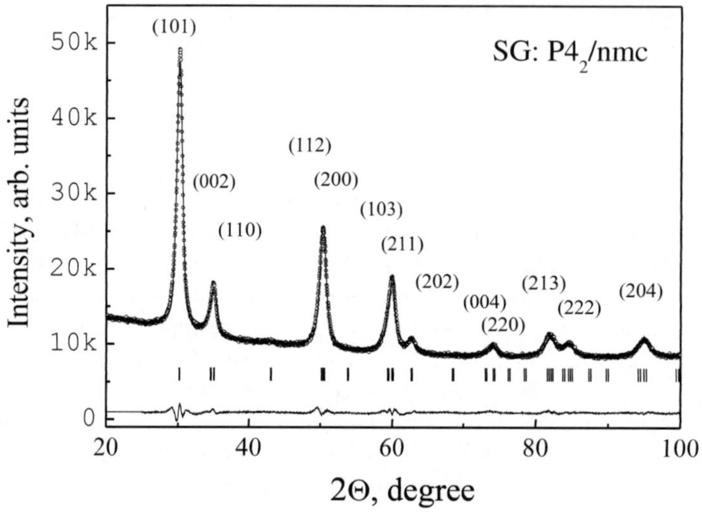

Figure 6. X-ray pattern of ZrO_2 + 3%Y_2O_3 - nanopowder sample.

Table 1. Atomic positions and isotropic thermal parameters of $Zr_{1-x}Y_xO_2$ (x = 0.05), calculated using Rietveld method from X-ray diffraction. $P4_2$ / nmc: Zr (2a) (0.75, 0.25, 0.75); O (4d) (0.25, 0.25, z)

	Atomic parameter	x = 0.05 tetragonal $P4_2$/nmc
	a, Å	3.614 (2)
	c, Å	5.167 (5)
	V, Å^3	67.48 (8)
Zr/Y	X	0.7500
	Y	0.2500
	Z	0.2500
	Biso (Å^2)	0.543
	Occupancy	0.4735/0.026
O1	X	0.25000
	Y	0.25000
	Z	0.04669
	Biso (Å^2)	0.727
	Occupancy	1.0

3.6. Quantum molecular dynamics method of the adsorption processes on a nanopowders surface based on ZrO_2.

In the MD calculations the exchange-correlation potential approximation as generalized gradient approach formulated, by Perdew, Burke and Ernzerhof (GGA-PBE), and the effective potentials of ionic cores as projector-augmented plane-wave method seem to be a preferable approach. To describe the changes in the chemical bond energy, related to a hydrogen-producing process, a bond-overlap population analysis [21-22] and extended electron wave functions of the atomic orbital basis are of common method. So far, we do have a complete filling of the $Z_i(t)$ for the ith atom and the imposition of a set of links $O_{ij}(t)$ for a pair of ith and j^{th} atoms, with dependence of time t. From $Z_i(t)$, to be easy estimate the charge of $O_{ij}(t)$ atoms, which can provide a semi-quantitative assessment of a covalent bond energy existing between the atoms. The nudged elastic band (NEB) method [23] and extend calculations to a free energy [24] of the assessment in the energy barrier in the chemical reactions and adsorption processes give us the general picture of the processes.

Method NEB provides the energy profiles of the processes at zero temperature. Calculation of free energy ensures to estimate the effect of the temperature on the outlined chemical reactions. Previously we have used a combination of these methods for the activation state calculations and estimation of the reaction parameters which are associated with the H_2O adsorption on a Al cluster surface [25 - 26].

4. Conclusions

1. It is shown that a nanopowder system based on ZrO_2 can convert chemical energy of gas phase water molecules adsorption to an electrical form. It was suggested that the mechanism of established effect is associated with chemical adsorption.
2. Experimental preparation and formulation of the problem for the molecular dynamics (MD) modelling of the moisture adsorption on ZrO_2 - nanopowder surfaces were made.

The scientific and practical importance of the work lays down in new prospects for development of nonconventional ways for generation of electricity, in particular, direct conversion of gas-phase water adsorption energy to electric energy.

Acknowledgments

A.S. Doroshkevich, O. S. Lygina and A. I. Lyubchyk acknowledges funding by the EU H2020-MSCA-RISE-2015 through the HUNTER project (grant nº 691010).
JINR-Romania Cooperation Programme Project of 2017 Order No. 219 / 55 is acknowledged.
International innovation centre of nanotechnologies of the CIS (grant №2016/101) is acknowledged.

The authors acknowledges G K Volkova, V A Glazunova, V V Burkhovetskiy from Government Institution “Donetsk Institute for Physics and Engineering named after A A Galkin” for their help in preparing and discussing the materials of the article.

References

[1] *Forecast electricity consumption in Russia in the medium term, Information-analytical materials*, 2013 (Moscow: Vnesheconombank) p 24 http://www.veb.ru/common/upload/files/veb/analytics/fld/20140122_2.pdf
[2] Brauner G, D'Haeseleer W, Gehrer W and all. 2013 *Electrical Power Vision 2040 for Europe Study by EUREL Short version* (Brussels: EUREL) p 44 http://www.eurel.org/home/TaskForces/Documents/EUREL-PV2040-Short_Version_Web.pdf
[3] Voznaya, N F Water 1979 *Chemistry and Microbiology* (Moscow: Higher School) p 340
[4] Styrov V V, Simchenko S V 2013 *Reports of the Nat. Acad. of Sci. of Ukraine* **80** № 5
[5] Styrov V V, Tyurin Yu I, 2003 *Non-equilibrium chemo-effects on the surface of solids* (Moscow: Energoatomisdat)

[6] Konstantinova T E, Danilenko I A, Glazunova V A, Volkova G K, Gorban O A 2011 *J. Nanoparticle Research* **13** № 9 4015
[7] Doroshkevich A S, Danilenko I A, Konstantinova T E 2006 *Electron microscopy and strength of materials* **13** 151
[8] http://flnph.jinr.ru/ru/facilities/ibr-2/scientific-installations.
[9] Ostanevich Yu M, Serdyuk I N 1982 *Physics-Uspekhi* **137** №1 85
[10] Gubkin AN 1987 *Electrets* (Moscow: Nauka) p 192
[11] Grankin V P, Styrov V V 1980 *JETP Letters* **31** № 7 403
[12] Alekseenko V I, Volkova G K 1999 *Technical Physics* **70** № 9 57
[13] Ptitsyn V E 2007 *Technical Physics* **77** Vol. 4 113
[14] Gurvich L V, Karachevtsev G V, Kondratyev V N 1974 *Energy of chemical bond, ionization potentials and electron affinities* (Moscow: The science) p. 351
[15] Evstratov A A 2007 Russian Chem. Journal **51** № 6 52
[16] Chevalier J and Gremillardw L 2009 *The Tet ragonal-Monoclinic Transformation in Zi rc onia:Lessons Learned a nd Future Trends* J. Am. Ceram. Soc., **92** № 9 1901 – 1920 DOI: 10.1111/j.1551-2916.2009.03278.x
[17] Golikov G A, 1988 *Guide to Physical Chemistry: A Textbook. allowance for Chem-primary process. Specialists* (Moscow: Higher school) p. 383
[18] Voronov V N, Larin B M, Senin V A 2006 *Chemical and nuclear technological modes with VVER: a manual for schools* (Moscow: MEI Publishing House) p. 390
[19] Evstratova K I, Kupina N A, Malakhov E E 1990 *Physical and Colloid Chemistry; Proc. for the farm. universities and faculties* (Moscow: Higher school) p. 487
[20] https://www.ill.eu/sites/fullprof/php/downloads.html
[21] Mulliken R S 1955 *The Journal of Chemical Physics* **23** 1833
[22] Shimojo F, Nakano A, Kalia R K, Vashishta P 2008 *Physical Review E - Statistical, Nonlinear, and Soft Matter Physics* **77** 66
[23] Henkelman G Jonsson H 2000 *Journal of Chemical Physics* **113** 9978
[24] Hass K C, Schneider W F, Curioni A, Andreoni W 1998 *Science* **282** 265
[25] Shimojo F, Ohmura S, Kalia R K, Nakano A, Vashishta P 2010 *Phys. Rev. Let.* **104** 126102.
[26] Ohmura S, Shimojo F, Kalia R K, Kunaseth M, Nakano A, Vashishta P 2011 *Journal of Chemical Physics* **134** 244702

Structural features of the ionic self-complementary amyloidogenic peptide

V V Egorov[1,2], Ya A Zabrodskaya[2], D V Lebedev[1], A N Gorshkov[2] and A I Kuklin[3,4]

[1] DMRB PNPI NRC KI, Gatchina, Russia
[2] Influenza Institute MH of Russia, Saint-Petersburg, Russia
[3] Joint Institute for Nuclear Research, Dubna, Russia
[4] Moscow Institute of Physics and Technology (State University), Dolgoprudny, Russia

E-mail: vlaegur@omrb.pnpi.spb.ru

Abstract. In this study we investigate the effect of triazavirine on ionic self-complementary (iSCM) containing peptides fibrils with transmission electron microscopy (TEM), small-angle neutron scattering (SANS) and MALDI mass-spectrometry (MS). It was shown that triazavirine is capable to dissociate iSCM amyloid fibrils. The mechanism of such an action is proposed.

1. Introduction

Conformational diseases such as amyloidoses are one of the main causes of the self-care capacity loss in the elderly. Amyloid fibril formation is the process which underlies the pathogenesis of many conformational diseases. Such fibrils have abnormal resistance to enzymes, chemical and physical influences [1].

In this study we investigate the effect of guanine nucleotide analog triazavirine [2] (Figure 4A) on the peptide GDIRIDIRIDIRG (Self-complementary Ionic, SI) fibrils dissociation. SI peptide comprises an ionic self-complementary motif (iSCM), first identified in the study of small heat shock proteins. Interaction between these proteins subunits is provided with antiparallel oriented monomer's iSCMs. It is also shown that iSCM is also involved in the assembly of crystallin oligomeric complexes – the proteins found in various organs, including the eye tissues [3]. SCM – containing peptide fibrils are subjects of nanostructured materials studies. The two-staged mechanism is proposed for iSCM amyloid fibrils formation: electrostatic attraction of protofibrils-forming monomers at the first stage and zipper-like hydrophobic interaction between protofibrils at the second mature fibrils forming stage [4], [5].

From molecular modeling of various proteins interaction with a number of model molecules – such as structural nucleotide analogs, we found that, within the model used, these substances have a tendency to interact with side chains of basic amino acid residues. This fact agrees with NMR guanosine-arginine interaction data [6]. Molecules with such a specificity of interaction may be useful for some disease treatment or prevention [7], [8].

Beta-strands are the main structures involved in the amyloid fibrils formation. Arginine-containing beta-strands are stabilized in amyloid-like fibril with hydrogen bonds and electrostatic attraction. The selective blocking of arginine residues ionic interaction from one peptide with negatively charged, i.e. aspartate, residues of another could lead to monomer repulsion, hydrogen bonds breakage and fibrils dissociation. This study is dedicated to proving the guanosine analogue triazavirine ability to dissociate amyloid-like fibrils formed with iSCM containing peptide.

2. Materials and methods

2.1. Materials
Peptide GDIRIDIRIDIRG (SI) was synthesized at NPF "Verta" Ltd., purity higher than 90%. All reagents were purchased from Sigma Aldrich, Triazavirine (2-methylthio-6-nitro-1,2,4-triazolo[5,1-c]-1,2,4-triazine-7(4I)-one) substance was a kindly gift from Ural Federal University.

2.2. Transmission electron microscopy (TEM)
TEM was performed using the standard negative-staining technique as described previously [9].

2.3. Small-angle neutron scattering (SANS)
SANS spectra were recorded with a YuMO spectrometer located in the fourth channel of the high-flux IBR-2 reactor (Laboratory of Neutron Physics, Joint Institute for Nuclear Research, Dubna). Measurements were performed in the standard geometry. A neutron beam was collimated into a 14 mm spot. The SANS data were collected on the YuMO spectrometer using two detectors [10]. The sample-detector distances were 12.96 m and 5.3 m for "New" and "OLD" detectors, respectively. Solutions of samples in D2O buffer were measured at 20ºC in 1 mm path-length quartz Hellma cells. The temperature was controlled by Lauda thermostat in special thermobox [11]. The spectrometer during the experiment was governed by SONIX+ Software program [12]. Using SAS program [13] we (i) combined the data referring to the same sample, (ii) calculated the spectrometer resolution function for given experimental conditions, (iii) carried out data correction on dead times of neutron ring wires detectors, (iv) subtracted a background signal from detector data, (v) carried out the normalization of the obtained spectrum on standard vanadium scatterer, and (vi) subtracted background sample.

2.4. Mass-spectrometry
Mass-spectrometry was performed in MALDI TOF "UltrafleXtreme" mass-spectrometer, Bruker, at positive ions registration mode using DHB matrix. Before the sample preparation samples were centrifuged at 1000 g for 10 minutes.

2.5. Fibrils preparations
SI peptide was dissolved in phosphate-buffered saline buffer pH 7.4 to the 5 mg/ml concentration and then incubated at room temperature for 1 hour with agitation. Triazavirine was added at 5x molar excess to the peptide and then the mixture was incubated for an hour at room temperature.

2.6. Molecular docking
Protein docking was performed using the Hex server software (http://hex.loria.fr/) with default parameters (electrostatic interactions included).

3. Results and discussion
The model peptide SI was studied by us previously [14]. It was found that it has all properties that characterize amyloidogenic peptides and, at the same time, is simple for computer modelling and studying of its spatial structure and oligomerization using biophysical and biochemical methods. Furthermore, the presence of only four different amino acid residues in the peptide allows us to make targeted influence on non-covalent bonds which drive protein-protein interactions in abnormal fibrils. This work aims at study the influence of the guanidine nucleotide analogue (triazavirine, TZV) on the model peptide fibrillogenesis.

(A)
(B)

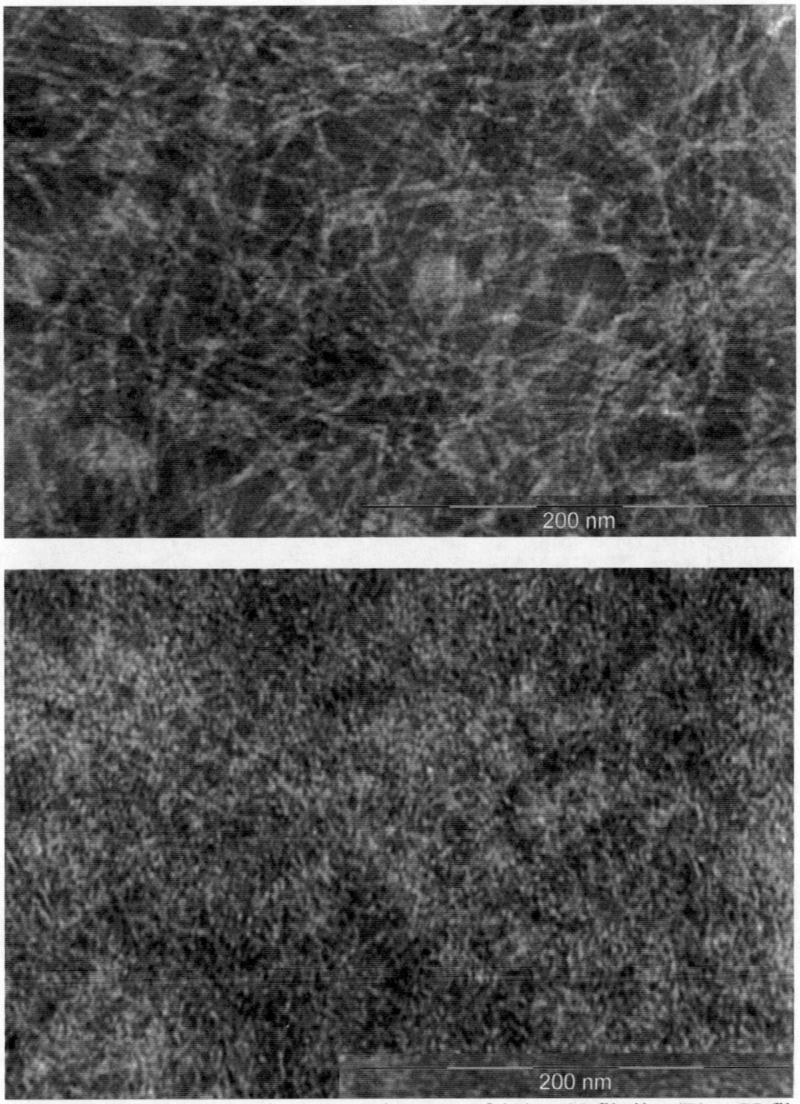

Figure 1. Transmission electron microscopy images of (A) – SI fibrils; (B) – SI fibrils with triazavirine.

As it is obvious from TEM images (Figure 1), TZV administration to the sample of SI fibrils causes shortening of fibrils and partial fragmentation of the fibrils network compared to fibrils in control conditions. Precise and reliable analysis of TZV-dependent SI fibrils structure remodeling requires additional methods in combination with TEM. Also, sorption of different species in a polydisperse sample (fibrils of different morphology, oligomers and unstructured aggregates) to the TEM grid surface may differ and could be further affected by the presence of nucleotide analog. Therefore to obtain the structural information on all of the species present in the bulk sample and for the further structure refinement we made additional experiments for SANS spectra measurements.

The SANS spectra presented in Figure 2 show substantial scattering from SI that generally followed a power law with exponent close to −2, indicating that significant SI peptide fraction forms large random-coil linear oligomeric structures or unstructured aggregates. Scattering by peptide alone was in agreement with a model of long worm-like fibrils with the diameter of approximately 3 nm and Kuhn length of ca. 12 nm (Figure 2A). The fibril diameter yielded by this approximation is close to the size of SI tetramers proposed earlier as the main prefibrillar species comprising fibrils of SI peptide [14].

In contrast with SI alone, scattering of the peptide after TZV was added to the sample could not be adequately described by worm-like model, but rather by a mixture of Gaussian coil polymer and rod-like fibrils approximately 10 nm in diameter (Figure 2B).

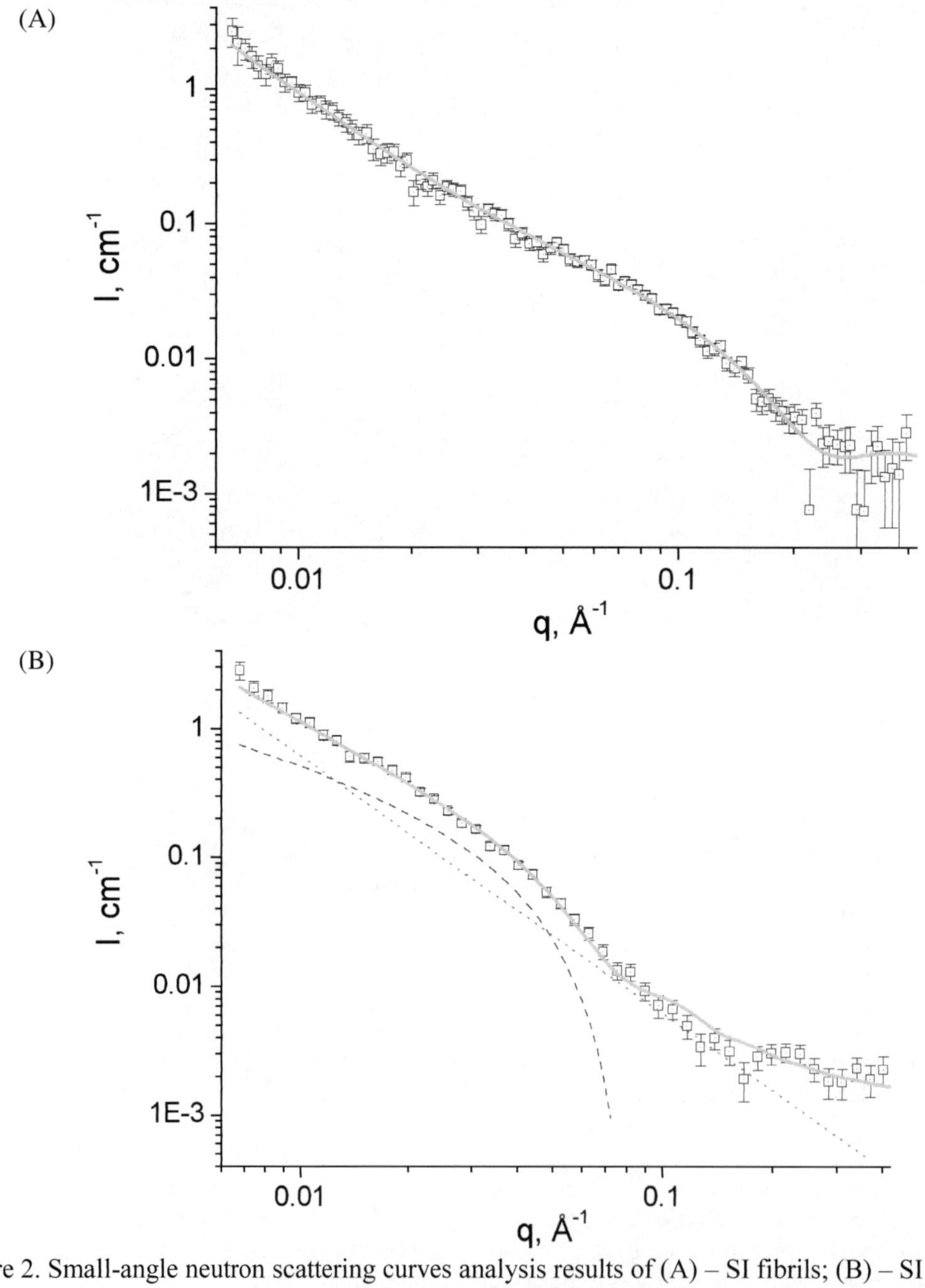

Figure 2. Small-angle neutron scattering curves analysis results of (A) – SI fibrils; (B) – SI fibrils with triazavirine. Data fitted to (A) worm-like model with fibril radius of 1.40 ± 0.04 nm and Kuhn length of 12.0 ± 0.7 nm ($\chi^2 = 0.83$, solid green line in Panel A) and (B) linear combination of random coil model (dotted red line) and long cylinders with the radii of 4.66 ± 0.14 nm (dashed blue line), $\chi^2 = 1.2$, shown in solid green line in Panel B.

The samples were further separated by centrifugation and MALDI mass-spectrometry (MS) measurement of pellets and supernatants was performed. The presence of signal in spectrum at Figure 3, A (SI peptide fibrils before centrifugation) shows that SI peptide molecules are capable for ionization ant can be detected by MALDI with the matrix chosen. The analysis showed that in SI fibrils sample there is no detectable quantity of the peptide complexes in supernatant (Figure 3B) with all peptide molecules being in the pellet (Figure 3C). After addition of triazavirine we observed no pellet in the same centrifugation conditions, but the signal which corresponding to SI peptide was detected in the supernatant (Figure 3D).

Taken together, the TEM, SANS and MS data indicate that SI fibrils were partially dissociated after triazavirine addition. In solution nearly all of the peptide appear to form long worm-like fibrils ca. 3 nm in diameter, which partially aggregate in larger structures observed in TEM images. After addition of TZV such large fibrils were no longer observed by any of the three techniques while smaller assemblies including rod-like fibrils just less than 10 nm in diameter appear to form instead. These fibrils could be identified as rod-like particles few tens of nanometres in length seen in TEM images, explaining the fact that they did not sediment during centrifugation.

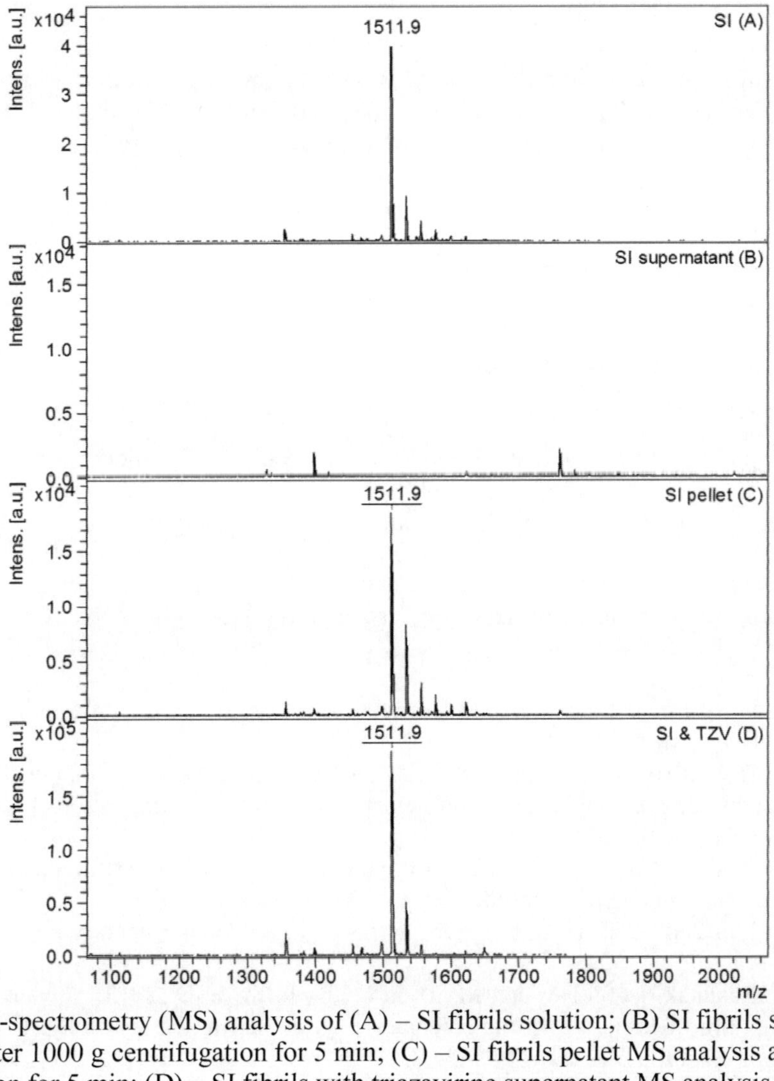

Figure 3. Mass-spectrometry (MS) analysis of (A) – SI fibrils solution; (B) SI fibrils supernatant MS analysis after 1000 g centrifugation for 5 min; (C) – SI fibrils pellet MS analysis after 1000 g centrifugation for 5 min; (D) – SI fibrils with triazavirine supernatant MS analysis after 1000 g centrifugation for 5 min.

IOP Conf. Series: Journal of Physics: Conf. Series **848** (2017) 012022 doi:10.1088/1742-6596/848/1/012022

The model of the peptide dimer was revealed by molecular dynamics simulations followed by trajectory analysis and molecular docking of the most frequent conformers [15]. There was an evident triazavirine interaction with arginine in dimer structure (Figure 4B). It is consistent with protein-nucleotide interaction data described in [6].

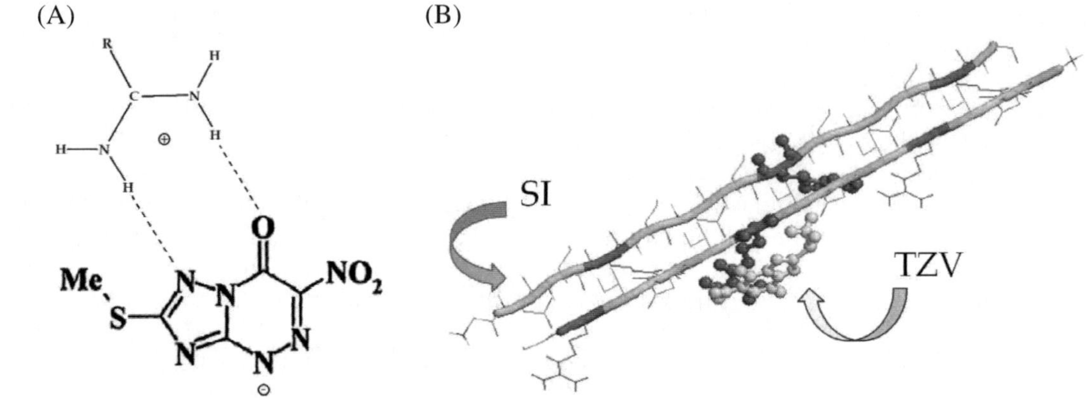

Figure 4. Hypothesis of triazavirine action on SI peptide fibrils. (A) Triazavirine can interact with arginine residue as described in [6] for nucleotide. (B) In dimer model the molecule of triazavirine (yellow) interacts with arginine residue side chain (red).

Currently substances liable to degradation of amyloid fibrils to components capable of participating in cellular catabolism are highly toxic. Therefore amyloidoses not currently be medicated. The presence of low-toxicity drugs with the potential to interact with the structural elements of amyloid fibrils holds great promise for the drug design for amyloidosis treatment. Thus, the data obtained in the study of interaction of triazavirine with iSCM-containing peptide may shed light on potential treatment of several conformational diseases.

4. Acknowledgements
The work was supported by Russian state 5-100 Russian Academic Excellence Project for enhancing the competitiveness of MIPT among the world's leading research and education centres of the Ministry of education and science and by the Russian Science Foundation (project no. 14-13-01301).

5. Conflict of Interests
The authors declare that there is no conflict of interests regarding the publication of this paper.

References
[1] Sipe J D *et al* 2014 Nomenclature 2014: Amyloid fibril proteins and clinical classification of the amyloidosis *Amyloid* **21** 221–4
[2] Karpenko I *et al* 2010 Antiviral properties, metabolism, and pharmacokinetics of a novel azolo-1,2,4-triazine-derived inhibitor of influenza A and B virus replication *Antimicrob. Agents Chemother.* **54** 2017–22
[3] Farnsworth P N and Singh K 2000 Self-complementary motifs (SCM) in alpha-crystallin small heat shock proteins *FEBS Lett.* **482** 175–9
[4] D'Auria G *et al* 2009 Self-assembling properties of ionic-complementary peptides *J. Pept. Sci.* **15** 210–9
[5] Kabiri M, Bushnak I, McDermot M T and Unsworth L D 2013 Toward a mechanistic understanding of ionic self-complementary peptide self-assembly: role of water molecules and ions *Biomacromolecules* **14** 3943–50
[6] Lancelot G, Mayer R and Hélène C 1979 Models of interaction between nucleic acids and

proteins. Hydrogen bonding of arginine with nucleic acid bases, phosphate groups and carboxylic acids *Biochim. Biophys. Acta* **564** 181–90
[7] Martellini J A *et al* 2009 Cationic polypeptides contribute to the anti-HIV-1 activity of human seminal plasma *FASEB J.* **23** 3609–18
[8] Lump E *et al* 2015 A molecular tweezer antagonizes seminal amyloids and HIV infection *Elife* **4** e05397
[9] Egorov V V *et al* 2013 Structural features of the peptide homologous to 6-25 fragment of influenza A PB1 protein *Int. J. Pept.* **2013** 370832
[10] Kuklin A I, Islamov A K and Gordeliy V I 2005 Scientific Reviews: Two-Detector System for Small-Angle Neutron Scattering Instrument *Neutron News* **16** 16–18
[11] Kuklin A I *et al* 2011 New opportunities provided by modernized small-angle neutron scattering two-detector system instrument (YuMO) *J. Phys. Conf. Ser.* **291** 12013
[12] Kirilov A S *et al* 2004 Evolution of the SONIX Software Package for the YuMO Spectrometer at the IBR-2 Reactor *Instruments Exp. Tech.* **47** 334–345
[13] Soloviev A G, Stadnik A V, Kuklin A I, Islamov A H and Solovieva T M 2003 SAS. The Package for Small-Angle Neutron Scattering Data Treatment. Version 2.4. Long Write-Up and User's Guide 22
[14] Egorov V V *et al* 2008 Self-assembling capable peptides for development of bottom-up structured materials *Innovations* **6** 84–87
[15] Altman M, Lee P, Rich A and Zhang S 2000 Conformational behavior of ionic self-complementary peptides *Protein Sci.* **9** 1095–105

SANS method study of surfactant mixture system TX-100+C16TAB in heavy water solutions

A Rajewska[1,2], **A Kh Islamov**[1], **R F Bakeeva**[3]

[1]Joint Institute for Nuclear Research, LNP, 141980 Dubna, Russia
[2]National Centre for Nuclear Research, 05-400 Świerk-Otwock, Poland
[3]Kazan State Technological University, 420015 Kazan, Russia

aldonar@jinr.ru

Abstract. The mixture system of two classic surfactants cationic and non-ionic – C16TAB (hexadecyltrimethylammonium bromide)+TX-100(p-(1,1,3,3-tetramethyl) poly(oxyethylene) in heavy water solutions was investigated at temperatures 30°, 50°, 70° for compositions 1:1, 2:1, 3:1 by the small-angle neutron scattering(SANS) method on spectrometer ('YuMO') at the IBR-2 pulsed neutron source at FLNP, JINR in Dubna (Russia). Measurements have covered Q range from $7x10^{-3}$ to 0.4 $Å^{-1}$. The SANS measurements of aqueous solutions of nonionic / cationic surfactants have shown that the mixed micelles are formed [1-2]. From the measured dependence of the scattered intensity on the scattering angle, we derived the size, shape of micelles, aggregation number at various compositions and temperatures. The size of mixed micelle is a weak function of the mixing ratio between the two components.

1. Introduction

Currently, the structures and properties of mixed solutions formed by two surfactant types, especially by nonionic and ionic surfactants are extensively studied by experimental and theoretical methods, since properties of mixed systems are in many cases better than properties of individual components. Taking into account the fact nonionic surfactants tend to ideal mixing, it can be expected that large deviations from the ideal behavior should appear for mixtures with certain interactions between two surfactant types. Although the synergistic behavior of mixed surfactant systems has long been studied, acquisition of the data on the behavior of various surfactant molecule types in mixed solutions remains an urgent problem. The aim of this work is to study mixed systems of nonionic surfactant TX-100 with cationic surfactant by small angle neutron scattering (SANS) with the determination of the micelle shape and size. The SANS and small angle X - ray scattering are efficient methods for obtaining information on the structure of mixed micelles and are often used in practice. The critical micelle formation concentration (CMC) was also determined during the study by the surface tension method.

2. Experimental section

Materials

Triton X-100 nonionic classic surfactant (p-(1,1,3,3-tetramethyl) poly(oxyethylene) with CMC= $0.2x10^{-3}$M was obtained as a gift from JINR, Dubna, (Russia), C16TAB cationic classic surfactant (hexadecyltrimethylammonium bromide (CMC= 1 mM) were purchased from Fluka(99.9 %) and was used these substances without further purification. Heavy water D_2O (99.9%) was purchased in Prikladnaya Chimia" in St.Petersburg (Russia).

2.1. *Measurement of CMC - surface tension method*

Du Nouy ring method we used for measured the CMC of investigated solutions. The CMC values for TX-100+C16TAB+D_2O micellar solutions are presented in Table 1.

Table 1. Influence of temperature on CMC for mixed system TX-100-C_{16}TAB-D_2O for different compositions.

composition of mixture $C_{TX\text{-}100}$ / C_{CTAB}	$C_{CMC}\cdot 10^{-4}$,M		
	temperature		
	30°C	50°C	70°C
1 : 1	0,915	0,941	0,790
2 : 1	0,568	0,667	0,363
3 : 1	0,465	0,470	0,474

2.2. *SANS measurements-small angle neutron scattering (SANS) method*

The SANS spectra were measured using a YuMO diffractometer by the time of flight at the IBR-2 pulsed reactor (Joint Institute for Nuclear Research, Dubna, Russia). The specific features of this diffractometer were described in details previously [3]. Neutrons in the range of 0.7–10 A^{-1} were used. Solutions were poured into tightly closed quartz cells (Hellma) 2 mm thick. The temperature within cells (30°,50°,70° ± 0.5°C) was maintained constant during measurements using a thermostat. The measured differential scattering cross section was calibrated using a vanadium reference sample. Data pre-processing was performed using the SAS program [4]. For all samples, heavy water (D_2O) was used as a solvent to achieve good contrast conditions. The experimental results were processed using the PCG Software 2.01.02 program (Austria).

3. Results and discussion

Effect of nonionic surfactant TX-100 concentration on structure of micellar solutions in mixed [5-10] system (TX-100+C16TAB+D_2O) for different compositions and temperatures

The experiment was performed at temperatures of 30°C,50°C and 70°C for three mixed solutions: (i) nonionic surfactant TX-100 (c_1 = 0.022mol/l = 1.294%) in D_2O with cationic surfactant C16TAB concentrations c =0.022mol/l (ii) nonionic surfactant TX-100(c_2 = 0.044mol/l= 2.588%) in D_2O with cationic surfactant C16TAB for concentration c =0.022mol/l = 1.294%.; (iii) nonionic surfactant TX-100 (c_3 = 0.066mol/l= 2.588%) in D_2O with cationic surfactant C16TAB for concentration c =0.022mol/l = 1.294%. The effect of nonionic surfactant TX-100 on the structure of micellar solutions of mixed surfactants system TX-100+C16TAB+D_2O was studied by small angle neutron scattering as a function of composition and temperature (Figure 1). In Table 1 shown the compositions of investigated in experiment mixture systems.

Table 2. TX-100+C16TAB+D_2O (composition of solutions)

1:1	0.022mol/l(TX-100)+0.022mol/l(C16TAB)=1.294%+0.7289%=2%
2:1	0.044mol/l(TX-100)+0.022mol/l(C16TAB)=2.588%+0.7289%=3.4%
3:1	0.066mol/l(TX-100)+0.022mol/l(C16TAB)=3.882%+0.7289%=4.6%

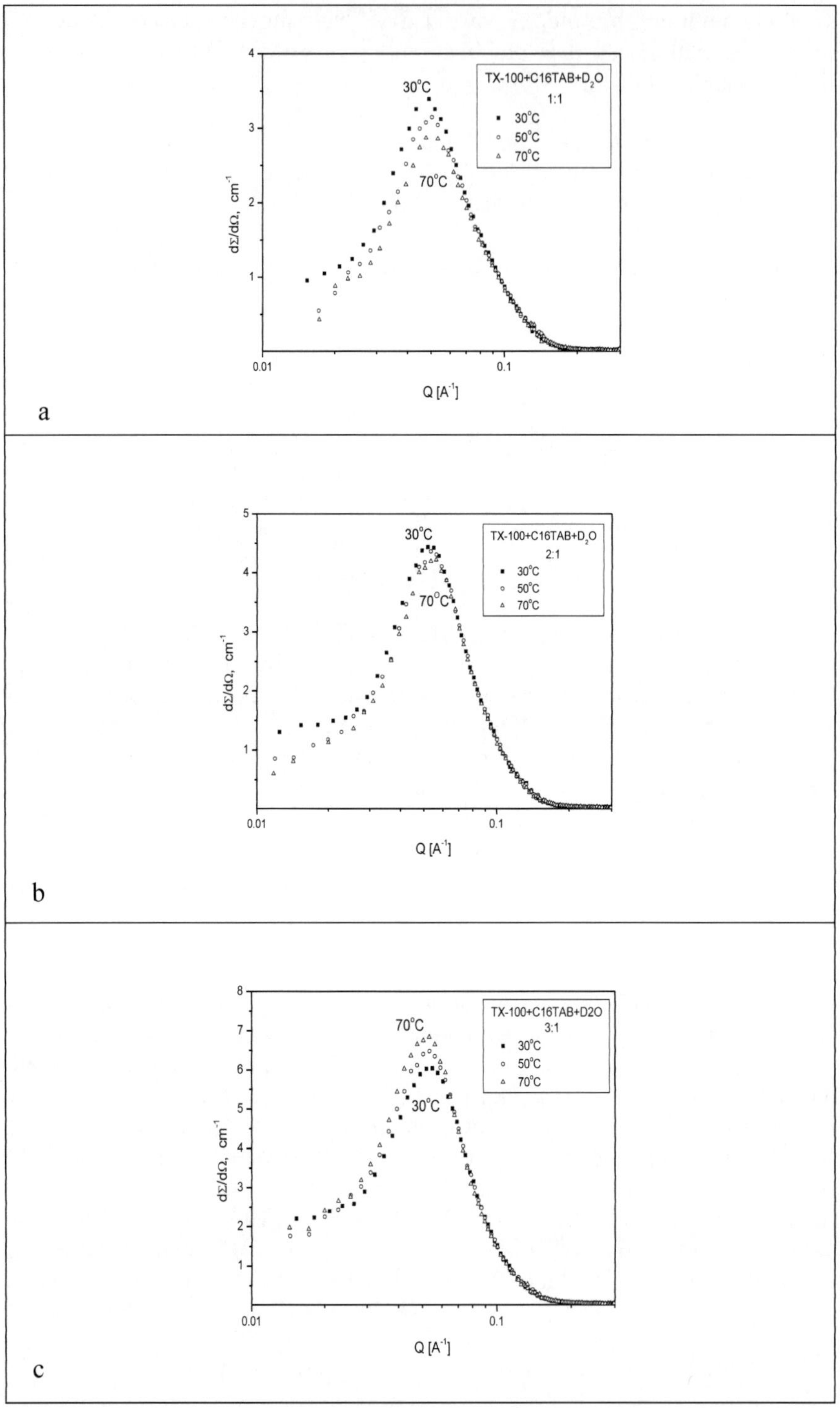

Figure 1. Intensity of neutron scattering vs scattering vector for mixture solution of TX-100 +C16TAB+D_2O for three various mix ratios 1:1(a),2:1(b),3:1(c) at temperatures 30°C, 50°C,70 °C.

Table 3. Average distance between micelles (d) and aggregation number (N_{agg}) in micellar solutions for mixture system TX-100+C16TAB+D_2O.

composition	average distance between micelles (1) aggregation number (2)					
	30°C		50°C		70°C	
	1	2	1	2	1	2
1:1	128.1 Å	72	124.5 Å	73	124.5 Å	77
2:1	121.2 Å	79	117.7 Å	80	117.8 Å	81
3:1	117.7 Å	74	117.8 Å	90	117.8 Å	107

The range of measured temperature was 30°C - 70 °C. The changes of the scattering curves were very small but tendencies for 1:1, 2:1 and 3:1 compositions of the mixed systems were opposite. In case of 1:1 composition the position of maximum curves for increases temperatures shifts to larger Q values for higher temperatures 50°C and 70°C and the maximum height decreases. The experimental curves for composition 2:1 shown that the position of the maximum is approximately the same at all three temperatures (slightly higher at 30 °C) but intensity of the scattering neutrons increases which points to the increasing micellar size. From these results the conclusion is that with increasing ratio between concentration of nonionic surfactant TX-100 and cationic C16TAB from 1:1 to 3:1 (increase the amount of non-ionic surfactant) the system changes properties from ionic like to nonionic like. In micellar solutions ionic like the micelles become smaller with increasing temperature but in nonionic like mixture system micelles increases in solutions.

The intensity of scattering neutrons for the mix ratio 3:1 is higher than for the mix ratios 1:1 and 2:1 and highest experimental curve is at temperature 70°C. The maximum of the scattering neutrons intensity increases with increase of temperature for mix ratio 3:1 and shift to the smaller Q value. The experimental results for CMC values for all investigated solutions shown that CMC changes with the composition of solutions. In mixed system solutions are mixed micelles. The intensity of scattering neutrons for the mix ratio 3:1 is higher than for the mix ratios 1:1 and 2:1 and highest experimental curve is at temperature 70°C. The maximum of the scattering neutrons intensity increases with increase of temperature for ratio 3:1 between concentrations. In Table 3 are the values of the average distance between micelles d and aggregation number N_{agg}.

Conclusions

The preliminary results obtained in this study can be summarized as follows. For all the studied aqueous solutions with constant amount of cationic surfactant C16TAB+D_2O (0.7289%) with additions of non-ionic surfactant TX-100, the existence of micelles was detected. It was found that spherical micelles were formed in mixed system TX-100+C16TAB+D_2O in all studied solutions. The aggregation number for the solutions with composition 1:1 is smaller at temperature 70°C than at 30°C. For mix ratio 2:1 the aggregation number was constant for all temperatures (30 °C,50 °C,70°C).

The aggregation number for mixed system with the highest concentration (3:1) increases with increase of temperature. The average distance between micelles in investigated micellar mixed systems at three different concentrations (at constant cationic C16TAB surfactant concentration) decreases with increasing concentration of the non-ionic TX-100 surfactant. There are only preliminary experimentally results and the next paper is ready for these mixed micellar solutions.

References

[1] Clint J 1975 *J. Chem.Soc. Faraday Trans.***1** 1327; Cantu L, Corti M, Degiorgio V, 1990 *J.Phys. Chem.* **94** 793
[2] Rathman J, Scamehorn J F, 1984 *J. Phys. Chem.* **88** 580
[3] Ostanievich Yu M 1988 *Makromol. Chem. Macromol. Symp.* **15** 91
[4] Solov'ev A G, Solov'eva T N, Stadnik A V, Islamov A Kh, Kuklin A I 2003 *Soobshch.*

Ob'edin. Inst. Yad. Issled. Dubna **86** 10
[5] Rajewska A, Mędrzycka K, Hallmann E 2014 *Physics of the Solid State* **56** 125
[6] Rajewska A, Mędrzycka K, Hallmann E 2007 *Kristallografiya* **52** 811
[7] Almgren M, Garamus V M 2005 *J. Chem. Phys.* **109** 11348
[8] Scamehorn J R, Schechter R S, Wade W H 1982 *J.Dispersion Sci. Technol.* **3** 261
[9] Puvvada S, Blankschtein D 1992 *J. Phys. Chem.* **96** 5567
[10] Zemb T, P. Charpin P 1985 *J. Phys. (Paris)* **46** 249

IOP Conf. Series: Journal of Physics: Conf. Series **848** (2017) 012024 doi:10.1088/1742-6596/848/1/012024

CEM V based special cementitious materials investigated by means of SANS method. Preliminary results

A C Dragolici[1], M Balasoiu[1,2*], O L Orelovich[2], L Ionascu[1], M Nicu[1], D V Soloviov[2,3,4], A I Kuklin[1,4], E I Lizunov[2], F Dragolici[1]
[1]Joint Institute for Nuclear Research, Dubna, Russia
[2] "Horia Hulubei" National Institute for Physics and Nuclear Engineering, Bucharest, Romania
[3]Institute for Safety Problems of Nuclear Power Plants of the Ukrainian NAS, Kyiv, Ukraine
[4]Moscow Technical Physics Institute, Dolgoprudnyi, Russia

[*]E-mail: balas@jinr.ru

Abstract. The management of the radioactive waste assume the conditioning in a cement matrix as an embedding, stable, disposal material. Cement matrix is the first and most important engineering barrier against the migration in the environment of the radionuclides contained in the waste packages. Knowing how the microstructure develops is therefore desirable in order to assess the compatibility of radioactive streams with cement and predict waste form performance during storage and disposal. For conditioning wastes containing radioactive aluminum new formulas of low basicity cements, using coatings as a barrier between the metal and the conditioning environment or introducing a corrosion inhibitor in the matrix system are required. Preliminary microstructure investigation of such improved CEM V based cement matrix is reported.

1.Introduction

Cementitious materials are much more complex than it might seem. Cement paste contains many crystalline and non-crystalline phases in various ranges of sizes. New analysis tools for concrete study, which allows the understanding of the mechanisms regulating the processes needed in different applications, have been used lately [1-11]. One of these analysis tools is the small angle neutron scattering method (SANS). Small angle neutron scattering represents a powerful tool for the characterization of the inner structure of technically relevant porous materials, e.g. hydrating cement pastes. Various parameters can be obtained by this non-destructive method, allowing repeated measurements as well as long term investigations, such as particle size distribution, specific surface of grains, etc. [3, 11]. The QENS technique was also applied and showed, among other things, that the hydrated cement paste contains C-S-H in a gel form and C-H in the form of colloidal particles imbedded in the C-S-H matrix. The water molecules incorporated in the colloidal particles appear immobile while the water molecules dispersed in the C-S-H gel matrix behave as interfacial water showing some slow dynamics, quite different from that of bulk water. Over time, the interfacial water in C-S-H gel penetrates into the colloidal particles [9].

The management of the radioactive waste assume the conditioning in a cement matrix as an embedding, stable, disposal material. Cement matrix is the first and most important engineering barrier against the migration in the environment of the radionuclides contained in the waste packages. Chemically, after mixing with water and additives, cement produces a reacting matrix with a porous microstructure. Knowing how the microstructure develops is therefore desirable in order to assess the compatibility of radioactive streams with cement and predict waste form performance during storage and disposal. Adjusting the phase fractions of the individual compounds of dry cement is one of the few ways to change the structural and mechanical properties of the matrix.

In the present paper, results on SANS measurements accomplished on CEM V samples for identification of microstructural changes induced in the special prepared cement samples are reported.

2. Experimental

2.1. Sample description

The wastes produced as result of nuclear activities are divers and sometimes incompatible with ordinary Portland cement based encapsulation matrices. A type of radioactive waste that cannot be embedded in Portland cement matrix is the radioactive aluminum [11, 12]. New formulas of low basicity cements, using coatings as a barrier between the metal and the conditioning environment or introducing a corrosion inhibitor in the matrix system [12, 13] are required.

Using CEM V with chemical composition given in Table 1, following special cementitious samples were prepared (see Table 2).

Table 1. Chemical composition CEM V

Formula	Z	Concentration
CaO	20	54,65%
SiO2	14	27,12%
Al2O3	13	9,45%
MgO	12	2,81%
Fe2O3	26	4,14%
K2O	19	0,96%
TiO2	22	0,56%
MnO	25	0,22%
SrO	38	0,05%
ZnO	30	0,02%
ZrO2	40	0,01%

Table 2 Description of the cement samples

Sample	Sample description
Sample 1	CEM V + H_2O
Sample 2	CEM V + H_2O + Al(powder)
Sample 3	CEM V + H_2O + Al(powder) + $Al_2(SO_4)_3$ + citric acid + pantarhol
Sample 4	CEM V + H_2O + Al(powder) + $Al_2(SO_4)_3$ + citric acid + pantarhol + $LiNO_3$

For SANS measurements the cement samples were prepared in the form of round plates (Fig.1).

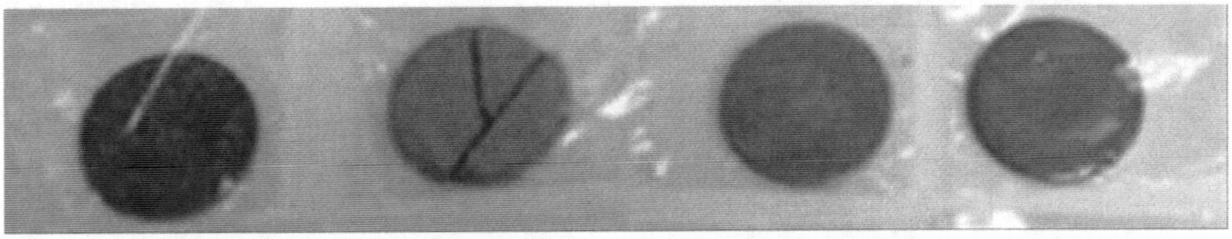

Figure 1 Cement samples: 1, 2, 3, 4 (see Table 2)

2.2 Scanning electron microscopy (SEM)
For the investigation of the morphology of the samples, a JEOL type JSM-840 scanning electron microscope was used.

2.3 *Small angle neutron scattering (SANS)*

SANS measurements on cement sample gives an intensity of scattered neutrons, I(Q) as a function of the momentum transfer $Q = (4\pi / \lambda)\sin(\theta / 2)$, with the scattering angle being θ; and λ the wavelength. In the case of cement paste studies, it was shown that for populations of approximate similar small pores and/or particles the Guinier approximation could give the pore size and shape [15]. Analysing the SANS intensity using the power-law approach very interesting features of cements can be distinguished [1-3].

Small angle neutron scattering (SANS) experiments were performed at the time-of-flight YuMO spectrometer [16] in function at the high flux pulse IBR-2 reactor, JINR Dubna. The Sonix+ software control system provides spectrometer operation [17]. The experiments were carried out at a sample-to-detector distances of 5.28 m and 13.04 m, resulting in a Q range of 0.007÷0.2 A^{-1}. The sample diameter and thickness in the beam were 14 mm and respectively 1 mm. The measured neutron scattering spectra are corrected relative to the transmission and thickness of the sample, background scattering due to the experimental scotch tape support and the vanadium reference sample using the SAS software [18], providing a neutron scattering intensity in absolute units of cm^{-1}.

Further, in the present paper, first steps in the SANS investigation of CEM V cementitious compounds are presented.

3. Results and discussions

In Figure 2, scanning electron microscopy (SEM) images of CEM V series samples are presented. The recorded images have the same magnification (x500) and evidence quite different morphology.
The presented images give the possibility to observe changes in the microstructure of matured cement paste as a consequence of adding the following components one at a time: Al(powder), $Al_2(SO_4)_3$ + citric acid + pantarhol and $LiNO_3$. In the first image (Fig.2a), clumps of quite large grains of CEM V can be noticed. In the next image (Fig.2b), agglomeration of smaller particles is visualized. Next (Fig.2c) we remark combined agglomeration of big grains and fine needle-shaped particles. In the final image (Fig.2d), some morphological similarity with first image (Fig.2a) is observable.

IOP Conf. Series: Journal of Physics: Conf. Series **848** (2017) 012024 doi:10.1088/1742-6596/848/1/012024

In Figure 3 SANS curves of samples 1, 2, 3, 4 measured at 20 ^{0}C temperature are depicted. From the beginning, a likeness between the curves aliasing is evident for samples 1 and 4, and respectively for samples 2 and 3. In the Q domain up to the value of 0.1 $A^{-1}$the power law behavior of the curves for all the samples (1-4) suggests the fractal approach to be used for the data treatment. More information will be obtained after detailed analysis of de experimental data.

In Figure 4 SANS curves of sample 2 measured at several values of the temperature (20 ^{0}C, 30 ^{0}C, 40 ^{0}C, 50 ^{0}C, 20 ^{0}C) are given. The curves present different features in the Q range higher the 0.1 A^{-1}. This fact may be due to the different content of water molecules that with the increase of the temperature could evaporate from the sample. Detailed analysis is in progress.

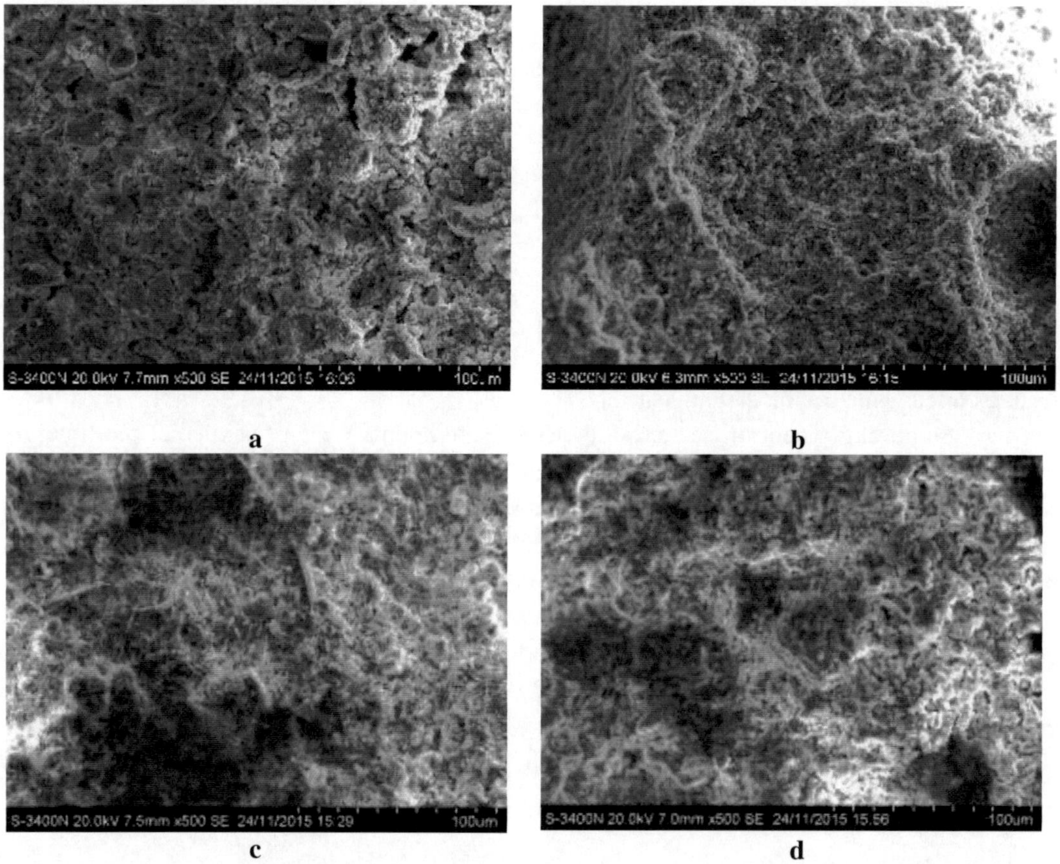

Figure 2 Scanning electron microscopy images of CEM V cement series samples: (a) Sample 1; (b) Sample 2; (c) Sample3; (d) Sample 4.

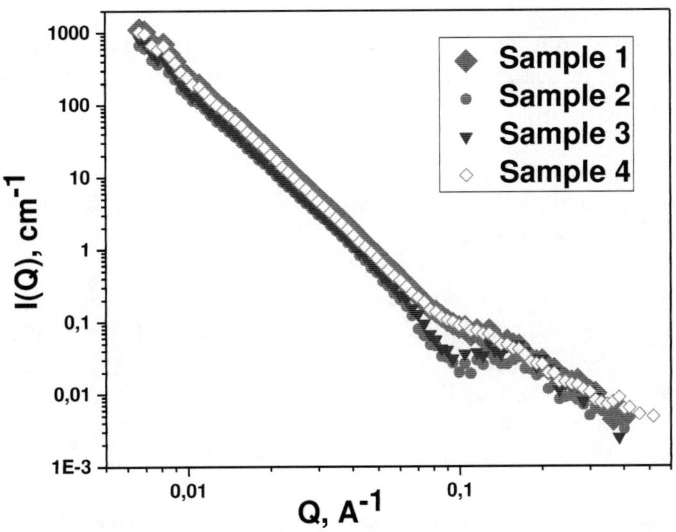

Figure 3. SANS curves of: (a) samples 1, 2, 3; (b) inset of figure 1: SANS curves for samples 1,2, 3, 4 (see the table 2).

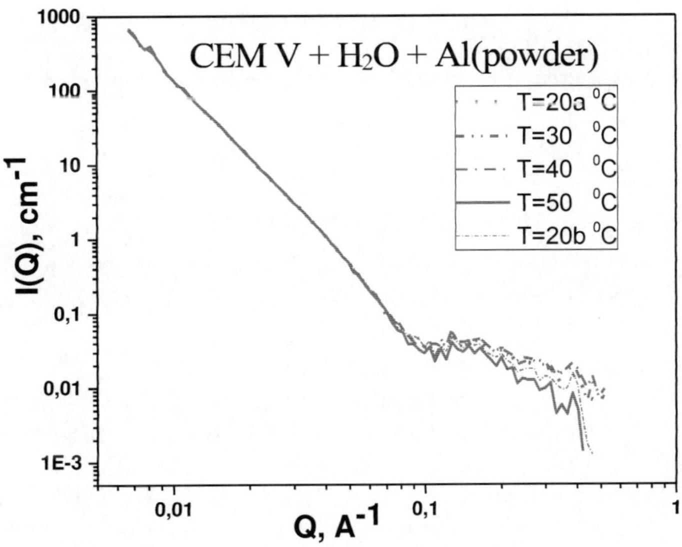

Figure 4. SANS curves of sample 2 (CEM V + H2O + Al(powder)), measured at several temperature: 20a ^{0}C; 30 ^{0}C; 40 0C; 50 ^{0}C; 20b ^{0}C. a – means that temperature is increasing; b- means that temperature is decreasing.

4. Conclusions

By admixture of additives during the cementitious material preparation the microstructure is changing at different length scales. Both scanning electron microscopy and small angle neutron scattering gives useful information about the forming processes.

Acknowledgements

The authors acknowledge the RO-JINR Grant No.94/15.02.2016 item 10 and RO-JINR Projects Nos. 219/10.04.2017 and 220/10.04.2017 item 89.

References

[1] Eichhorn E, Häussler F, Baumbach H, *Journal de Physique IV* C8 (Supplement of Journal de Physique I) **3** 369-372 (1993)
[2] Heinemann A, Hermann H, Wetzig K, Häussler F, Baumbach H, Kröning M, *Journal of Materials Science Letters* **18**(17) 1413–1416 (1999)
[3] Häussler F, Hempel M, Baumbach H, and Tritthart J, *Advances in Cement Research*, **13**(2) 65-73 (2001)
[4] Fitzgerald S A, Thomas J J, Neuman D A, Livingston R A, *Cement and Concrete Research* **32** 409–413 (2002).
[5] Fratini E, Chen S H, Baglioni P, Bellisent Funel M C, *Phys. Rev.* E **64** 020201 (R) (2001).
[6] S. H. Chen, C. Liao, F. Sciortino, P. Gallo, P. Tartaglia, *Phys. Rev.* E **59** 6708 (1999).
[7] Fratini E, Chen S H, Baglioni P and Bellisent, Funel M C, *J. Phys. Chem.* B **106** 158 (2002).
[8] Faraone A, Chen S H, Fratini E, Baglioni P, Liu L and Brown C, *Phys. Rev.* E **65** 040501(R) (2002)
[9] Padureanu I, Aranghel D, Rotarescu Gh, Dragolici F, Turcanu C, Brzozowski R, Stepinski M, Szalanski P J, Kozlov Zh A, Semenov V A, *Romanian Journal of Physics* **50**(5–6) 551–559 (2005)
[10] Arandigoyen M, Alvarez J I, Cement and Concrete Research **37**(5) 767–775 (2007)
[11] Rogante M, Domanskaya I K, Gerasimova E S, Vladimkrova E, *Universal Journal of Applied Science* **1**(1) 11-17 (2013)
[12] Ionascu L, Nicu M, Turcanu C, Dragolici F, Rotarescu Gh, *Romanian Reports in Physics* **65**(4) 1512–1518 (2013)
[13] Nicu M, Ionascu L, Turcanu C, Dragolici F, Rotarescu Gh, *Romanian Journal of Physics* **59** 360-368 (2014).
[14] Dragolici C A, Balasoiu M, Ionascu L and Nicu M, *Journal of Physics: Conference Series* **746** 012045(5) (2016)
[15] Guinier A & Fournet G 1955 *Small-Angle Scattering of X-Rays* (John Wiley and Sons Inc., New York)
[16] Kuklin A I, Islamov A K, Gordeliy V I 2005 *Neutron News* 16 (3) 16
[17] Kirilov A S, Litvinenko E I, Astakhova N V, Murashkevich S M, Petukhova T B, Yudin V E, Gordelii V I, Islamov A Kh, Kuklin A I 2004 *Instruments and Experimental Techniques* 47(3) 334
[18] Soloviev A G, Solovieva T M, Stadnik A V, Islamov A Kh, Kuklin A I 2003 SAS The Package for Small Angle Neutron Scattering Data treatment. Version 2.4. Long write-Up and User's Guide (Russian) JINR Communication P10-2003 86; http://wwwinfo.jinr.ru/programs/jinrlib/sas/indexe.html

Application masking liquid technique for SAXS study of natural bio-nanocomposites

Yu V Larichev[1]
Boreskov Institute of Catalysis, Siberian Branch of Russian Academy of Sciences, Prospekt Akademika Lavrentieva 5, Novosibirsk, 630090, Russia
Novosibirsk State University, Pirogova Street 2, Novosibirsk, 630090, Russia
E-mail: ylarichev@gmail.com

Abstract. Natural bio-nanocomposites (rice husk, sapropel, peat and brown coal) have been studied by TEM, XRD and SAXS. Particle size distributions for the mineral matter in these systems were determined by SAXS with using full contrast technique. In the rice husk silica nanoparticles are forming compact particles with bimodal distribution and mean size is equal 4.7 nm. In the case of peat and brown coal mineral matter in organic matrix are forming more large particles with compact shapes and mean sizes 14.1 nm and 14.3 nm respectively. In the sapropel inorganic matter are forming elongated structures with lengths are equal 20-70 nm and widths are equal 2-10 nm.

1. Introduction

Natural bio-resources (such as rice husk, peat, sapropel and other) have a big perspective as a source for preparation different types of low-cost carbon and carbon-mineral adsorbents and functional materials [1-5]. The most significant hindrance for wide application of these materials in industry it is unsatisfactory reproducibility of main properties due to preparation these materials. The base reasons of this problem are complex composition of natural materials, amorphous structure and respectively insufficient information about particle sizes and shapes of mineral parts of these materials. These data are needed for understanding and controlling process of formation porous structure due to calcination and activation such bio-resources. Removing of mineral matter (totally or partially) could noticeably increase specific surface and pores volume of these materials [3, 5]. Depends on the mineral matter content and their particle shape and sizes in bio-composites it is possibly to consider carbon porous structure with inorganic inclusion or vice versa porous mineral matrix covered by carbon-contain films. Nevertheless information about inorganic particle sizes and their shapes in initial bio-materials practically is absent and preparation condition of functional materials are needs to select in each case at that its depend on the batch of raw materials these conditions could be different [4, 5]. Information about sizes and shapes of mineral particles in bio-composites should help in solution such problem about reproducibility. One of the methods is suitable for obtaining data about particle sizes in bio-materials is small angle X-ray scattering (SAXS) with using special masking liquid for masking parasitic scattering signal from organic/carbon matrix of samples [6, 7]. In this work we report about results of analysis mineral components (particle size distribution and particle shapes) in rice husk, peat, sapropel and brown coal.

[1] To whom any correspondence should be addressed. E-mail: ylarichev@gmail.com

2. Experimental section

Sample preparation

For study have been used four different bio-composites. Rice husk was grown in the Krasnodar region of Russia. According to [8], this rice husk contained 19–25 % lignin, 34 – 42 % cellulose, and 17–22 % hemicellulose. The ash content of the husk was 19.5%, and the SiO_2 content in the ash was above 96%. Sapropel has been mined from Gorkoe Lake (Omsk region of Russia). The ash content of the sapropel was 64.3%. Ash has a complex composition: 63.4% Si, 7.1% Al, 8.8% Ca, 5.6% Mg, 3.4% Fe and other minor elements.

Peat has been mined from Taganskoe peat field (Tomsk region of Russia). The ash content of the peat was 29.0%. Ash consists from next elements: 31.3% Si, 15.2% Al, 18.2% Fe, 25.9% Ca, 3.5% Mg and other minor elements.

Brown coal (lignite) has been mined from Kansk-Achinsk coal basin (Krasnoyarsk region of Russia). The ash content of the brown coal was 11.1%. Ash consists from next elements: 57.1% Si, 22.1% Al, 7.5% Fe, 4.2% Ca, 2.1% Mg and other minor elements.

Sample characterization

X-ray diffraction

XRD investigations were carried out at an ARLX'TRA diffractometer (Thermo, Switzerland) with a Si(Li) solid-state detector and CuKα radiation (wavelength λ is equal to 1.5418 Å). The measurements were carried out in the 2θ range of 10–90° with a step of 0.05°. Storage time in point is a 5 seconds.

True density measurement

True densities of bio-composites were measured pycnometrically. Volume of the liquid displaced by a porous sample was measured. The measurement was made at least three times for each sample, and the obtained values were averaged. The liquid used in these experiments was heptane. The measurement error was less than 4 %.

Transmission electron microscopy

Electron microscopy examination of the samples was performed using a JEM-2010 transmission electron microscope (JEOL, Japan) with accelerating voltage 200 kV and resolution 0.14 nm. The samples for the TEM study were prepared by ultrasonic dispersing in ethanol and consequent deposition of the suspension upon a "holey" carbon film supported on a copper grid. The linear sizes of not less 300 particles from mineral part of bio-composites were measured for each sample to obtain particle size distributions ($D_N(R)$). For correct comparison with SAXS data TEM histograms were recalculated as $D_v(R) = D_N(R)*R^3$.

Small-angle X-ray scattering

SAXS data were acquired using an S3 MICRO (HECUS) diffractometer with point collimation and Cu Kα radiation (wavelength is equal to 1.541 Å). The electrical power of generator for X-ray tube is 50W (1.0 mA at 50 kV). The diffraction patterns were measured in the vector q ranges from 0.01 to 0.6 $Å^{-1}$, where $q = 4\pi\sin\theta/\lambda$. The samples were placed in a 1.5 mm quartz capillary. For selective extraction of the SAXS pattern of dispersed mineral phase from the total SAXS pattern from whole sample we developed an original full-contrast technique with using fluorocarbon compounds as masking liquids [6, 7]. A sample in the capillary was filled with a special liquid having a density, which roughly corresponds to the density of organic or a carbon matrix in the samples. Whereas electron density and pycnometric density of materials are close to each other the last value has been used for convenience. For amorphous or disordered samples are very difficult to measure electron density value instead pycnometric density. For this reason in practical applications is better to use pycnometric density and make additional density corrections if it need in experiment. Processing of experimental data and calculation of the particle size distributions were performed by GNOM software from the ATSAS package [9].

3. Results and discussion

IOP Conf. Series: Journal of Physics: Conf. Series **848** (2017) 012025 doi:10.1088/1742-6596/848/1/012025

Figure 1 are shows XRD patterns from biomaterials. Reflections observed for RH can be attributed to the cellulose structure [10, 11]. No other reflections corresponding to silica are observed. At the same time there are lot reflections for a sapropel. More intensive reflections are corresponding to α-quartz (SiO_2) structure. Also other reflections are corresponding to albite ($NaAlSi_3O_8$), muscovite ($KAl_2(AlSi_3O_{10})(F,OH)_2$), leucite ($K[AlSi_2O_6]$) and hematite ($Fe_2O_3$). In generally it is typical phase composition for the sapropels [3]. XRD patterns for peat and brown coal have also typical views for such minerals [12, 13]. In the case of peat and brown coal also are found reflections from α-quartz structure. Reflections from α-quartz structure have narrow lines and particle sizes estimated from Scherrer equation are equal approximately 100 nm. The other possible compounds are not give noticeable reflexes. Moreover XRD patterns also contain broad peaks (halo) with the maximums at 21° and 41° approximately. These broad peaks can be assigned to disordering carbon with graphite-like structure and amorphous silica too [8, 14]. According XRD data a lot part of matter in bio-composites are presented in amorphous and/or disordered states.

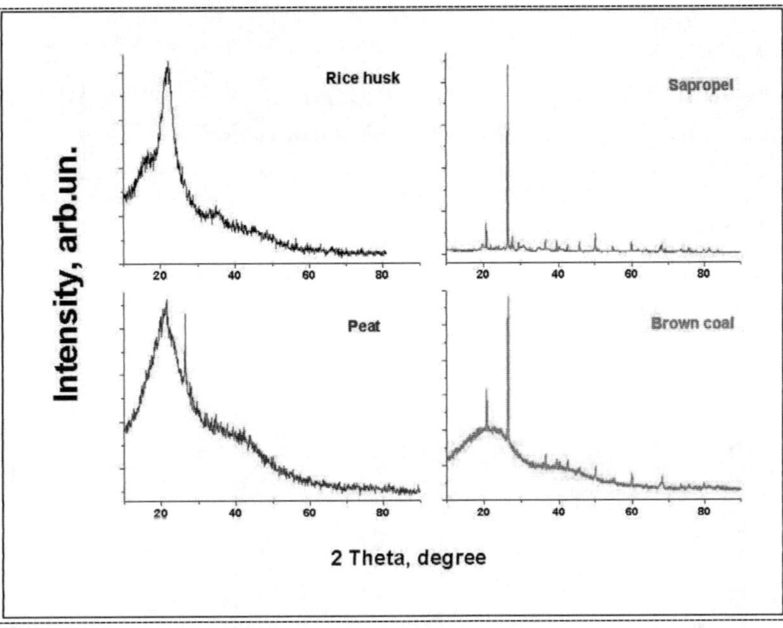

Figure 1. X-ray diffraction patterns for rice husk, sapropel, peat and brown coal.

Figure 2 displays TEM images from such biomaterials. According TEM data in the case of rice husk their mineral matter are forming particle with compact shape and typical sizes 2-10 nm. In the sapropel mineral phase are forming elongated structures with typical lengths are equal 20-70 nm and widths are equal 2-10 nm. The average ratio length/width for these aggregates is about 7.5. For the peat and brown coal has been found that mineral matter in both cases are forming compact particles with typical sizes about 10-30 nm. For the all TEM data (except sapropel) have been calculated volumes particle size distributions for comparing with corresponding SAXS data.

The SAXS curves of bio-composites are contained complex mixture information about pores, mineral particles and organic/carbon matrix. The current q range is restricting SAXS spatial resolution up to 65 nm. Therefore huge particles of quartz with sizes about 100 nm will be unavailable for SAXS analysis. To the other hand the all amorphous or disordered inorganic matter with less sizes in bio-composites must give scattering signals in the SAXS patterns.

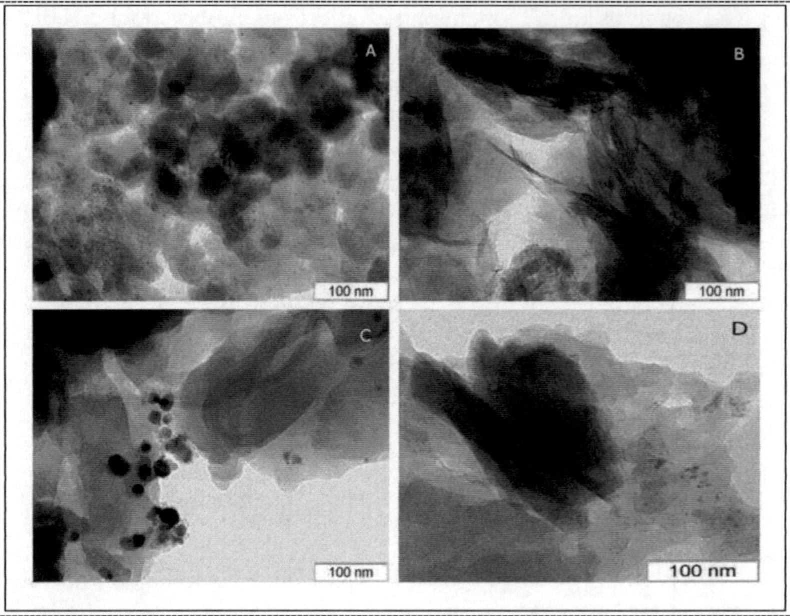

Figure 2. Typical TEM images for rice husk (A), sapropel (B), peat (C) and brown coal (D).

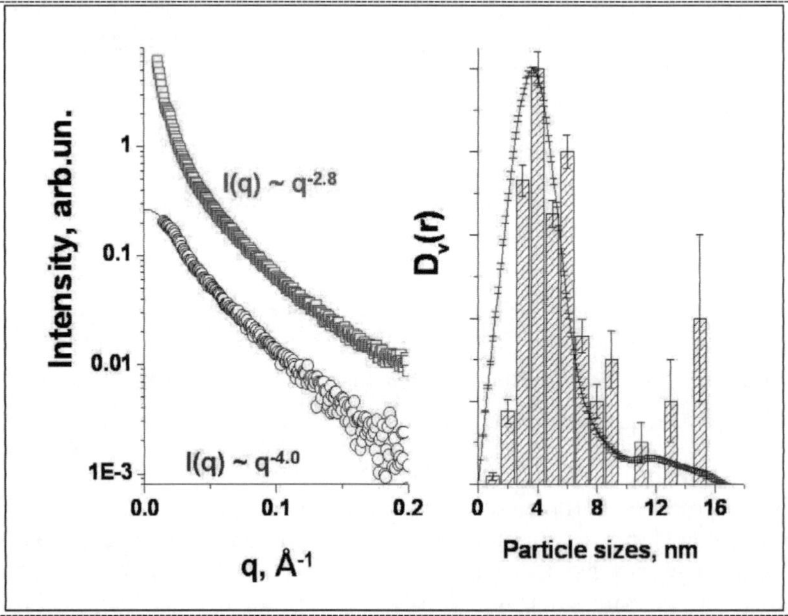

Figure 3. (Left) SAXS curves for initial rice husk (red) and rice husk after impregnation with masking liquid (blue). Black line is scattering curve from model of polydisperse spheres. (Right) Volume particle size distributions for silica particles in rice husk calculated from corresponding SAXS curve (blue line) and TEM distribution (bar chart).

Figure 3 presents SAXS data for the rice husk. The slope of the initial scattering curve I(q) for the rice husk is proportional to $\sim q^{-2.8}$. Hence, the particle form-factor cannot be determined strictly in this case, and it is difficult to find the particle size distribution. Such slope of the scattering curve can be typical,

IOP Conf. Series: Journal of Physics: Conf. Series **848** (2017) 012025 doi:10.1088/1742-6596/848/1/012025

for example, of fractal aggregates, which may consist of cellulose fibers. The application of the masking liquid technique makes it possible to suppress virtually completely the scattering signal of the organic phase. When rice husk was impregnated with the masking liquid having the density of 1.4 g/cm^3, the SAXS signal corresponding mainly to the silica phase was obtained for the sample. As it is shown in Figure 3, the scattering curve for the contrast matched sample has much lower intensity and a radically different slope ($I(q) \sim q^{-4.0}$). Such slope of the scattering curve is typical for particles with compact (sphere-like) shape [15]. Using spherical form-factor for modeling curve has been get volume particle size distribution for silica particles (Figure 3). This distribution has a small particle fraction with sizes from 1 to 7 nm and large particle fraction with sizes from 8 to 16 nm. Mean particle size in this distribution is equal 4.7 nm. This result has good qualitative agreement with TEM distribution (Figure 3). For quantitative comparing TEM and SAXS distributions there are need to have much more statistical data for TEM distribution. Typical quantity 300-350 particles for TEM histogram in our case is allow making only qualitative comparing with SAXS data. Moreover that wider particle size distribution than more number of particles for good statistics are needed. So SAXS with masking liquid technique is allow to high quality information about particle sizes of inorganic matter in such biomaterials.

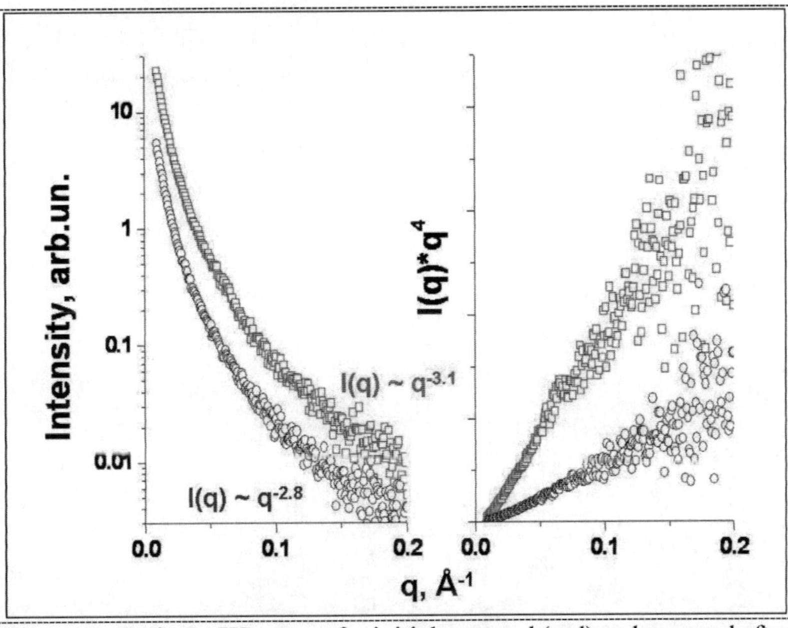

Figure 4. (Left) SAXS curves for initial sapropel (red) and sapropel after impregnation with masking liquid (blue). (Right) Such SAXS curves in coordinates $I(q)*q^4$; q.

In the case of sapropel scattering curve I(q) is proportional to $\sim q^{-3.1}$ (Figure 4). The application of the masking liquids with density of 1.35 g/cm^3 practically does not influence on this slope value. The slope value of SAXS curve after using masking liquids is increased up to -2.8. Such slope value (-2.8) for SAXS curve with using contrast is indicate that mineral matrix in sapropel are forming rod-like or fractal-like particles [15]. This assumption is agreed with TEM image of sapropel (Figure 2). The value of radii gyration (R_G) for mineral particles in sapropel is equal 16.2±0.5 nm. This value is approximately corresponding to R_G values of elongated particles on the TEM image (Figure 2). For example model solid ellipsoid with such semi-axes as (a = 5 nm, b = 5 nm, c = 37 nm) has a value of radii gyration is equal 16.8 nm [16]. Indefinite form-factor of particles in this case does not allow getting particle size distribution from SAXS data. Likewise such indefinite shape of particles in sapropel is impeding to calculate of TEM distribution too. Application of masking liquid with density of 1.21 g/cm^3 for the peat is changed slope value of the SAXS curve from -2.9 to -4.0 (Figure 5).

IOP Conf. Series: Journal of Physics: Conf. Series **848** (2017) 012025 doi:10.1088/1742-6596/848/1/012025

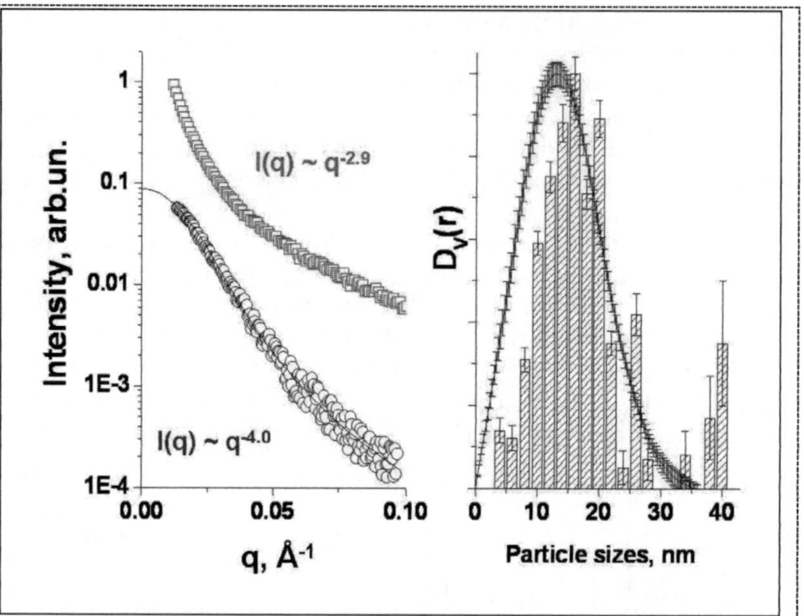

Figure 5. (Left) SAXS curves for initial peat (red) and peat after impregnation with masking liquid (blue). Black line is scattering curve from model of polydisperse spheres. (Right) Volume particle size distributions for mineral part in the peat calculated from corresponding SAXS curve (blue line) and TEM distribution (bar chart).

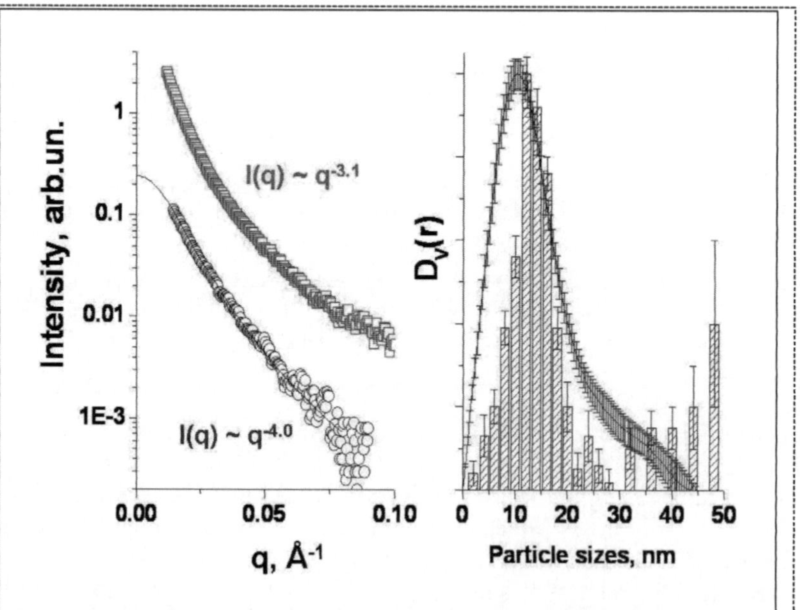

Figure 6. (Left) SAXS curves for initial brown coal (red) and brown coal after impregnation with masking liquid (blue). (Right) Black line is scattering curve from model of polydisperse spheres.

This liquid mixture is decreasing parasitic scattering from organic matrix of peat and we observed scattering intensity from mineral particles in the peat only. Such slope (-4.0) for SAXS curve is typical for particles with compact (sphere-like) shape. Volume particle size distributions for mineral part in the brown coal calculated from corresponding SAXS curve (blue line) and TEM distribution (bar chart). The volume particle size distribution for mineral matter of peat has been get with using spherical form-factor (Figure 5). This distribution has monomodal character with mean size is equal 14.1 nm. This particle size distribution from SAXS data are agreed with similar TEM particle size distribution. Some discrepancy between SAXS and TEM data are determined insufficient TEM statistics.
In the case of brown coal there is similar picture comparing to peat. After application masking liquid with density of 1.30 g/cm^3, the slope value of SAXS curve is changed from -3.1 to -4.0 (Figure 6). After using spherical form-factor for fitting SAXS curve from mineral particles in brown coal matrix has been got particle sizes distribution with mean size is equal 14.3 nm and bimodal character. TEM particle size distribution has practically the same particle sizes and also there is quality agreement between SAXS and TEM data (Figure 6). The near sizes for mineral matter in the peat and brown coal could be as simple coincidence or indicated on the similar formation conditions and transformation of such bio-composites on our planet. In any case it is need more data for precise conclusion.
Finally in Table 1 are shows some properties of bio-composites and particle sizes of mineral components. The rice husk has mineral particles with minimal sizes. Peat and brown coal are containing more large particles with similar mean sizes. Although in brown coal are presenting as less small and more large particles while in peat mineral particles has more homogeneous distribution.

Table 1. Physical properties of bio-composites and particle sizes of mineral parts therein.

Samples	Density, g/cm^3	Ash content, wt%	R_G, nm	D_{av}, nm*
Rice husk	1.54	19.5	5.8±0.1	4.7
Sapropel	2.29	64.3	16.2±0.5	-
Peat	1.62	29.0	8.4±0.2	14.1
Brown coal	1.47	11.1	11.0±0.3	14.3

* - $\langle D_{av} \rangle = \sum n_i * d_i / \sum n_i$ where d_i the diameter of i-particles on the SAXS histogram, N the number points on the SAXS histogram.

In sapropel sample mineral particles have a maximal sizes and elongated shapes. Due to high percentage of mineral phase in sapropel these particles are aggregated and formed elongated structures. Probably that for sapropels with less quantity of mineral phase their particle sizes will be close to such particles in the peat and brown coal.

4. Conclusions

The SAXS method with liquid masking technique has been used for study mineral matter in biomaterials. In the case of rice husk silica nanoparticles are formed bimodal distribution with mean sizes particles are equal 4.7 nm. In the case of sapropel mineral matter of this material are formed elongated and/or fractal-like particles with mean R_G value is equal 16.2 nm. In the case of peat mineral matter of this material are formed nanoparticles with monomodal distribution and mean particle sizes are equal 14.1 nm. In the case of brown coal mineral matter of this material are formed nanoparticles with bimodal distribution and mean particle sizes are equal 14.3 nm. There is a good agreement between TEM and SAXS results.

Acknowledgement

IOP Conf. Series: Journal of Physics: Conf. Series **848** (2017) 012025 doi:10.1088/1742-6596/848/1/012025

The author is grateful to P.M. Yeletsky, O.I. Krivonos and E.V. Maltseva for samples provision and D.A. Zyuzin, A.V. Ishchenko for assistance in the investigations of the samples by XRD and TEM. This work was supported by the Russian Science Foundation (grant no. 14-23-00037).

References

[1] Pietrzak R, Wachowska H, Nowicki P, Babeł K 2007 *Fuel Processing Technology* **88** 409
[2] Yeletsky P M et al 2009 *Microporous and Mesoporous Materials* **121** 34
[3] Kovalenko G A, Adeeva L N 2010 *Chemistry for Sustainable Development* **18** 181
[4] Virla L D et al 2016 *Microporous and Mesoporous Materials* **234** 239
[5] De Gisi S et al 2016 *Sustainable Materials and Technologies* **9** 10
[6] Larichev Yu V, Tuzikov F V 2013 *J. Appl. Cryst.* **46(3)** 752.
[7] Larichev Yu V, Tuzikov F V 2013 *Kinetics and Catalysis* **54(5)** 632
[8] Larichev Yu V et al 2015 *Journal of Physics and Chemistry of Solids* **87** 58
[9] Konarev P V, Petoukhov M V, Volkov V V, Svergun D I 2006 *J. Appl. Cryst.* **39** 277
[10] Park S, Baker J, Himmel M, Parilla P, Johnson D 2010 *Biotechn. Biofuels* **3**:10.
[11] Lu P, Hsieh Y L 2012 *Carbohydrate Polymers* **87** 564
[12] Asapo E S et al 2012 *J. Miner. Mater. Characteriz. Engineer.* **11** 709
[13] Takagi H, Maruyama K, Yoshizawa N, Yamada Y, Sato Y 2004 *Fuel* **83** 2427
[14] Larichev Yu V et al 2008 *Catalysis Letters* **120(3-4)** 204
[15] Feigin L A, Svergun D I 1987 *Structure Analysis by Small-Angle X-Ray and Neutron Scattering* (New York: Plenum Press) *335*
[16] online SAXS calculator: *http://scattering.tripod.com/xitami/java/iq.html*

IOP Conf. Series: Journal of Physics: Conf. Series **848** (2017) 012026 doi:10.1088/1742-6596/848/1/012026

Structural analysis of aqueous ferrofluids with cobalt ferrite particles stabilized with lauric acid and sodium n-dodecyl sulphate

A -M Balasoiu-Gaina[1,2,3], M Balasoiu[1,4*], O Ivankov[1,5,6], D Soloviov[1,5,6], S Lysenko[7], C Stan[8], N Lupu[9], D Creanga[10], A Kuklin[1,5]

[1]Joint Institute for Nuclear Research, Dubna, Russia
[2]CMCF, Moscow State University, Moscow, Russian Federation
[3]West University of Timisoara, Timisoara, Romania
[4]"Horia Hulubei" National Institute of Physics and Engineering, Bucharest, Romania
[5]Moscow Technical Physics Institute, Dolgoprudnyi, Russia
[6]Institute for Safety Problems of Nuclear Power Plants of the Ukrainian NAS, Kyiv, Ukraine
[7]Institute of Technical Chemistry, Ural Branch of RAS, Perm, Russian Federation
[8]"Politehnica" University of Bucharest, Faculty of Applied Sciences, Romania
[9]National Institute of Research and Development for Technical Physics, Iasi, Romania
[10]Alexandru Ioan Cuza University, Iasi, Romania

*E-mail: balas@jinr.ru

Abstract. Small angle neutron scattering (SANS) experiment on $CoFe_2O_4$/lauric acid/sodium dodecyl sulphate/H_2O ferrofluid was performed at the time-of-flight YuMO spectrometer. Concentration effects on the structure variation of the investigated ferrofluid are presented. Using the zero concentration approximation, the structure factors of samples with different concentration were determined. It was shown that the structure factor of values higher than 1 is present in samples with particle volume concentration equal and greater than 0.5%.

1. Introduction

Ferrofluids reveal a wide range of applications in many technical and industrial fields, as well as in medicine and biotechnology. Water-based ferrofluids present the most complex microstructural behaviour and their properties improvement represents an important challenge in this field [1-7]. Use of nano-ferrites in wide technical and biomedical areas, as for example, magnetic recording, permanent magnets, ferrofluids, drug delivery, hyperthermia, etc., make these materials very interesting to researches for improving their properties, magnetization and coercitivity on one hand, and particle dimensions and shapes, on the other [8, 9].

Small angle neutron scattering is a well-known useful method for investigating such systems [10].

Preliminary results on the CoFe2O4/lauric acid/ sodium dodecyl sulphate/H2O ferrofluid sample of 3% particle volume concentration were reported earlier [11]. TEM images have revealed that the sizes and the geometrical forms are not uniform and even that some particles show rhomboidal shape. The size distribution is found to be lognormal with mean value of particle diameter 8.48 nm.

The model fit of SANS data gave triaxial ellipsoid and ellipsoidal shell, as objects characterizing the investigated system [11]. μSR technique [12, 13] was applied to the same sample and the results evidenced that the system is formed from monodomain nanoparticles of an average dimension about 85 A and a high anisotropy coefficient [14]. Further, a multifractal approach [15] was applied to the TEM experimental data and the existence of multifractal properties in the sample was demonstrate [16].

In the present paper concentration effects on the structure variation of a non-ionic $CoFe_2O_4$/lauric acid/ sodium dodecyl sulphate/H_2O ferrofluid investigated by means of small angle neutron scattering are presented.

2. Experimental

2.1. Samples preparation

The ferrofluid was prepared by the coprecipitation method of $Fe(OH)_3$ and $Co(OH)_2$. Further step in the preparation was the adsorption of lauric acid on ferrite particles and peptisation of hydrophobic precipitate in aqueous solution with sodium n-dodecyl sulphate [11].

Before SANS measurements the sample was centrifuged in Biofuge 15R (Heraeus instruments) 2 times 6000 rot/min for 60 min. After that, the obtained sample of 3% particle volume concentration was diluted with double distillated water and obtained the following concentrations: 1%; 0.5%; 0.25%; 0.14%; 0.07%; 0.04%.

2.2. Neutron scattering experiment

Small angle neutron scattering (SANS) experiments were performed at the time-of-flight YuMO spectrometer [17] in function at the high flux pulse IBR-2 reactor, JINR Dubna. The Sonix+ software control system provides spectrometer operation [18]. The experiments were carried out at a sample-to-detector distances of 5.28 m and 13.04 m, resulting in a Q range of 0.007÷0.2 A^{-1}. The sample diameter and thickness in the beam were 14 mm and respectively 1 mm. The measured neutron scattering spectra are corrected relative to the transmission and thickness of the sample, background scattering due to the experimental cuvette and the vanadium reference sample using the SAS software [19], providing a neutron scattering intensity in absolute units of cm^{-1}.

2.3. Small angle neutron scattering method

Small-angle neutron scattering (SANS) has been shown to be a valuable technique for studying the properties of magnetic fluids (e.g. [20-23]), which characteristic structural features lies mostly in the interval of 1-100 nm.

The scattered intensity on an absolute scale for any interacting particulate systems of scatters can be expressed as [24, 25]:

$$I(Q) = \phi P(Q) S(Q) \qquad (1)$$

where:

Q is the modulus of the scattering vector defined as $Q = (4\pi / \lambda)\sin(\theta / 2)$, with the scattering angle being θ ; $\phi = N / V_0$ is the density of particles in the volume V_0 of the sample, $P(Q)$ characterize a particle and is related to its form factor, $F(\vec{Q})$, by the relation $P(Q) = \left\langle \left| F(\vec{Q}) \right|^2 \right\rangle$

with

$$F(\vec{Q}) = \int_{\text{Volume of particle}} (\rho - \rho_0) \exp(i\vec{Q}.\vec{r}) dV ;$$

ρ and ρ_0 are the coherent lengths densities of the particle and respectively of the polymer matrix.

$S(Q)$ is the interparticle term-the structure factor-related to the spatial distribution of the centres of mass:

$$S(\vec{Q}) = \left\langle \sum_{\alpha,\beta} \exp i\vec{Q}(\vec{R}_\alpha - \vec{R}_\beta) \right\rangle$$

where $\langle .... \rangle$ denotes a statistical average which taken over the available positions and orientations of the particles.

In case of materials with magnetic properties, the neutron scattering intensity contains a part representing the contribution of the magnetic scattering [23]:

$$I(Q) = I_n(Q) + I_m(Q)$$

It is known, that if the nanoparticles in ferrofluid are dispersed in H_2O, the nuclear contribution mostly dominates the magnetic one:

$$I_m(Q) \ll I_n(Q)$$

and in the case of non-polarized neutron scattering it is correct the following approximation:

$$I(Q) = I_n(Q) + I_m(Q) \approx I_n(Q)$$

Further analysis of the experimental results is based on this approximation.

3. Results and discussion

3.1. Concentration series analysis

SANS data sets for 1%, 0.5%, 0.25%, 0.14%, 0.07% and 0.04% particle vol. concentration samples were obtained (Fig. 1).

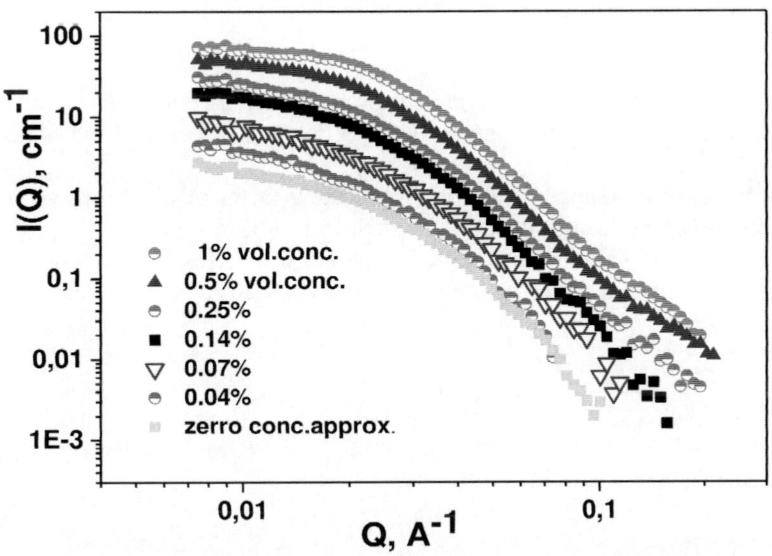

Fig.1 SANS experimental curves for 1%, 0.5%, 0.25%, 0.14%, 0.07% and 0.04% particle vol. concentration samples and the zero concentration approximation curve.

For a qualitative analysis of a polydisperse system accounting particle interaction, a ratio between the scattering curves for concentrated and zero concentration diluted solutions can describe the effective structure-factor [26]:

$$S_\phi(Q) \equiv \frac{I_\phi(Q)}{I_{\phi\to 0}(Q)}$$

In figure 2 the representations of these ratios for every volume particle concentration are given.

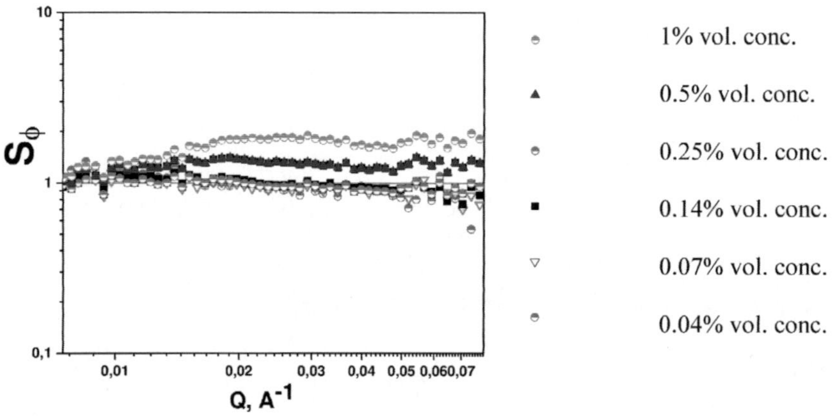

Fig.2 Effective structure-factor in dependence on 1%, 0.5%. 0.25%, 0.14%, 0.07% particle volume concentration samples.

Values of structure factor higher than 1 are determined for volume particle concentrations equal and greater than 0.5%.
For the data processing, the PRIMUS software was used [27].

4. Conclusion

Small angle neutron scattering (SANS) experiments on CoFe2O4/lauric acid/DDS-Na/H2O ferrofluid samples of 1%, 0.5%, 0.25%, 0.14%, 0.07% and 0.04% particle vol. concentrations were performed at the time-of-flight YuMO spectrometer.

Using the zero concentration approximation the structure factor of the samples was determined. It was shown that the structure factor of values higher than 1 is present in samples with particle volume concentration equal and greater than 0.5%. In ferrofluid samples based on nonpolar and polar liquid base with magnetite particles it was determined that the structure factor value in not higher than 1 for volume particle concentration up to 1% [28].

Acknowledgements

The authors acknowledge the RO-JINR Projects Nos. 219/10.04.2017 and 220/10.04.2017 item 45, 62, 64 and the RO-JINR Grant No.218/10.04.2017 item 16. The PRIMUS software is acknowledged (https://www.embl-hamburg.de/biosaxs/primus.html).

References

[1] Bica D, Vekas L, Avdeev M V, Marinica O, Socoliuc V, Balasoiu M, Garamus V M 2007 *J Magn Magn Mater* **311** 17

[2] Balasoiu M, Avdeev M V and Aksenov V L 2007 *Crystallography Reports* **52**(3) 505

IOP Conf. Series: Journal of Physics: Conf. Series **848** (2017) 012026 doi:10.1088/1742-6596/848/1/012026

[3] Puscasu E, Sacarescu, L., Lupu, N. N. Lupu, M. Grigoras, Oanca G, Balasoiu M, Creanga D 2016 *J Sol-Gel Sci Technol* **79** 457(9) doi:10.1007/s10971-016-3996-1

[4] Puscasu E, Sacarescu L, Domocos A, Leostean C, Turcu R, Creanga D, Balasoiu M 2016 *Romanian Journal of Physics* **61**(5–6) 946

[5] Oprica L, Nadejde C, Andries M, Puscasu E, Creanga D, Balasoiu M 2015 *Environmental Engineering and Management Journal* **14** (3) 581

[6] Nadejde C, Puscasu E, Brinza F, Ursu L, Creanga D, Stan C 2015 *U Politeh Buch Ser A* **77** 277

[7] Almasy L, Creanga D, Nadejde C, Rosta L, Pomjakushina E, Ursache-Oprisan M 2015 *Journal of the Serbian Chemical Society* **80**(3) 367

[8] Bala T, Sankar C R, Baidakova M, Osipov V, Enoki T, Joy P A, Prasad B L V, Sastry M 2005 *Langmuir* **21**(23) 10638

[9] Houshiara M, Zebhib F, Razia Z J, Alidousta A, Askaria Z 2014 *J. Mag. Mag. Mater.* **371** 43

[10] Avdeev M V, Aksenov V L 2010 *Physics Uspekhi* **53**(10) 971

[11] Balasoiu M, Ivankov O I, Soloviov D V, Lysenko S N, Yakushev R M, Balasoiu-Gaina A -M, Lupu N 2015 *J Optoelectron. Adv. Mater.* **17**(7-8) 1114

[12] Balasoiu M, Barsov S G, Bica D, Vekas L, Vorob'ev S I, Gritsaj K I, Duginov V N, Zhukov V A, Komarov E N, Koptev V P, Kotov S A, Mamedov T N, Petrescu C, Shcherbakov G V 2008 *JETP Letters* **88**(3) 210

[13] Balasoiu M, Aksenov V L, D. Bica D, Duginov V N, Gritsaj K I, Mamedov T N, Tripadus V, Vekas L, Zhukov V A 2008 *Magnetohydrodynamics* **44**(1) 61

[14] Mamedov T N, D.S. Andrievskii D S, Balasoiu M, Gritsaj K I, Duginov V N, Getalov A L, Komarov E N, Kotov S A, Moroslip A E, Pavlova I I, Scherbakov V G, Vorob'ev S I, Stan C 2015 *J Optoelectron. Adv. Mater.* **17**(7-8) 1086

[15] Stan C, Balasoiu M, Ivankov A I, Cristescu C P 2016 *Romanian Reports in Physics* **68**(1) 270

[16] Stan C, Balasoiu M, Buzatu D, Cristescu C P 2017 *Journal of Computational and Theoretical Physics* **14** 2017 (in print)

[17] Kuklin A I, Islamov A K, Gordeliy V I 2005 *Neutron News* **16** (3) 16

[18] Kirilov A S, Litvinenko E I, Astakhova N V, Murashkevich S M, Petukhova T B, Yudin V E, Gordelii V I, Islamov A Kh, Kuklin A I 2004 *Instruments and Experimental Techniques* **47**(3) 334

[19] Soloviev A G, Solovieva T M, Stadnik A V, Islamov A Kh, Kuklin A I 2003 SAS The Package for Small Angle Neutron Scattering Data treatment. Version 2.4. Long write-Up and User's Guide (Russian) *JINR Communication* P10-2003 86; http://wwwinfo.jinr.ru/programs/jinrlib/sas/indexe.html

[20] Grabcev B, Balasoiu M, Tarziu A, Kuklin A I and Bica D 1999 *J. Mag. Mag. Mater.* **201** 140

[21] Avdeev M V, Balasoiu M, Torok Gy, Bica D, Rosta L, Aksenov V L, Vekas L 2002 *J. Mag. Mag. Mater.* **252** 86

[22] Balasoiu M, Avdeev M V, Aksenov V L, Hasegan D, Garamus V M, Schreyer A, Bica D, Vekas L 2006 *J. Magn. Magn. Mater.* **300** e225

[23] Balasoiu M, A.I. Kuklin A I 2016 *Romanian Journal of Physics* **61**(3-4) 473

[24] Guinier A and Fournet G 1955 *Small Angle Scattering of X-rays* (New York: John Wiley)

[25] Feigin L A and Svergun D I 1987 *Structure Analysis by Small-Angle X-Ray and Neutron Scattering* (Plenum Press: Springer)

[26] Petoukhov M V, Franke D, Shkumatov A V, Tria G, Kikhney A G, Gajda M, Gorba C, Mertens H D T, Konarev P V, Svergun D I 2012 *J. Appl. Cryst.* **45** 342

[27] Konarev P V, Volkov V V, Sokolova A V, Koch M H J and Svergun D I 2003 *J. Appl. Cryst.* **36** 1277

[28] Balasoiu M, Avdeev M V, Kuklin A I, Aksenov V L, Bica D, Vekas L, Hasegan D, Torok Gy, Rosta L, Garamus V, Kohlbrecher J 2004 *Magnetohydrodynamics* **40**(4) 359

IOP Conf. Series: Journal of Physics: Conf. Series **848** (2017) 012027 doi:10.1088/1742-6596/848/1/012027

Study of carbon-supported bimetallic PtCu nanoparticles by ASAXS

Bulat N.V., Avakyan L.A, Pryadchenko V.V., Srabionyan V.V., Belenov S.V., Bugaev L.A.

Southern Federal University, Rostov-on-Don, Russia

E-mail: laavakyan@sfedu.ru

Abstract. Bimetallic platinum-copper nanoparticles on carbon support are studied as a perspective electrochemical catalyst by anomalous small-angle X-ray scattering near the Pt absorption L_3-edge. The simultaneous fitting of several diffraction patterns measured at different photon energies lead to a satisfactory agreement between experimental and model curves in the assumption of core-shell structure of the particles with Pt-rich shell and Cu-rich core. It is shown that the average size of as prepared nanoparticles is about 6 nm with distribution spread of about ±2 nm and with thickness of Pt-rich shell approximately 1.6 nm. After annealing at 350°C the average size of the particles increased by two times with additional enlargement of the Pt-rich shell thickness.

1. Introduction

Platinum-containing materials, especially nanocomposites, are widely used as a catalysts of many practically important reactions, in particular, hydrogen electro-oxidation in low-temperature fuel cells [1]. High cost of platinum catalysts severely restrains their commercial applications, so that the search of cheaper and more durable materials is very important. One of the promising ways is the synthesis of catalytic layer containing bimetallic nanoparticles of core-shell architecture, with a core of transition metal atoms and a shell of platinum atoms, in combination with new types of carbon supports [2, 3]. There were synthesized a series of bimetallic nanoparticles with platinum shell and different core metals [4], such as silver [5, 6], ruthenium, gold or copper [7, 8]. Additionally, such catalysts are expected [4] to be more stable and durable than monometal ones.

Such an increasing interest to complex multi-metallic nanosized systems requires the development and application of techniques for structural diagnostics which should be sensitive to the details of atomic structure at a scale from one to hundreds of nanometers. Particular interest is in the size distribution of nanoparticles, the composition of inner and shell regions, thickness and continuity of the near-surface area. The small-angle scattering diffraction (SASD) is proved to be a powerful tool for studying of the structure of materials at an intermediate scale range between the atomic scale, where spectroscopic methods (XAFS, XES, EELS, etc.) are applicable, and mesoscopic scales where direct observations (TEM, SEM, etc) could be performed. The small-angle scattering of X-rays (SAXS) near the absorption edge of a particular element is called anomalous SAXS or ASAXS. The dependence of scattering contrast from the energy in ASAXS experiments allows to consider the distribution of selected element in more details. This technique was successfully applied for the studies of platinum-copper alloyed particles in carbon [9], gold-silver core-shell particles in glass [10] and even more complex metal-oxide systems [11, 12, 13].

In the present paper the structure of platinum-copper bimetallic particles on the carbon support (PtCu/C) is studied by means of ASAXS diffraction with photon energies at Pt L_3-absorption edge. Section 2 presents the details of sample preparations and data acquiring, Section 3 describes the fitting strategy and the results of ASAXS analysis are summarized in Section 4.

2. Experimental

The considered samples were synthesized by sequential (PtCu-seq/C) chemical reduction of copper and platinum ions of corresponding salts ($CuSO_4$ and H_2PtCl_6) and deposition to a carbon suspension (Vulcan XC-72, Cabot) [14]. At the first step the nanostructured material Cu/C was obtained by the reduction of copper ions using $NaBH_4$, and at the second step aqueous solution of H_2PtCl_6 containing an excess of NH_3 and $NaBH_4$ was added to the solution where Cu/C were already formed. In order to boost stability of the catalyst it was annealed at 350°C for 1 hour.

The ASAXS experiments were performed near Cu K-edge (8697 eV) and near Pt L_3-edge (11558 eV) using the μSPOT beamline at the synchrotron BESSY II at Helmholtz-Zentrum Berlin (HZB) [15]. Three energy points were used for ASAXS data collection, within -10, -100 and -300 eV before each edge. The q-scale is calibrated using a silver behenate standard sample, measured at each photon energy. The integration and calibration of the detector data was done by using the software GSAS-II [16].

3. ASAXS data analysis

The scattering intensity of dilute system of dispersed particles with the number density distribution $N(R)$ can be described as: $I(q) \propto \int_0^\infty N(R)|F(q,R)|^2\, dR$ where $q = 4\pi\sin\theta/\lambda$ is the scattering vector, R – particle radius, $F(q,R) = \int \Delta\eta(\vec{r})e^{i\vec{q}\cdot\vec{r}}d^3r$ is the scattering length (scattering amplitude), depended on how scattering density $\Delta\eta(\vec{r}) = \eta(\vec{r}) - \eta_{\text{matrix}}$ is distributed in the particle. Conventionally the particle volume $V(R)$ and averaged scattering contrast $\Delta\eta$ are extracted from scattering length, so that remained factor $\Phi(q,R)$ which contains only information about the shape (form-factor): $F(q,R) = \Delta\eta \cdot V(R) \cdot \Phi(q,R)$. For spherical particle:

$$F_{\text{Sp}}(q,R) = \Delta\eta \cdot \left(\frac{4}{3}\pi R^3\right) \cdot \left(3\frac{\sin(qR) - qR\cos(qR)}{(qR)^3}\right) = \Delta\eta \cdot V_{\text{Sp}}(R) \cdot \Phi_{Sp}(q,R), \quad (1)$$

and for core-shell object with total particle radius R, core radius $R_c = \nu R$, core contrast $\Delta\eta_c = \xi \cdot \Delta\eta$ and shell contrast $\Delta\eta$:

$$F_{\text{CS}}(q,R) = \Delta\eta\,[V_{\text{Sp}}(R)\cdot\Phi_{\text{Sp}}(q,R) - (1-\xi)V_{\text{Sp}}(\nu R)\cdot\Phi_{\text{Sp}}(q,\nu R)]\,. \quad (2)$$

The only energy-dependent term is the scattering contrast ($\Delta\eta$ and ξ for core-shell), which is calculated with account for anomalous contribution to atomic factor as it implemented in GSAS-II:

$$\eta \propto \rho \sum c_a(Z_a + f'_a(E)) \Big/ \sum c_a M_a\,, \quad (3)$$

where ρ is the density, c_a is the number of atoms with charge Z_a and mass M_a in the unit cell, $f'_a(E)$ is a real part of anomalous correction (tabulated). Figure 1 compares the energy dependence of experimental SASD curves at arbitrary chosen value of $q = 0.1$ Å^{-1} of two samples and calculated behavior of squared scattering contrast $|\Delta\eta|^2$ of pure platinum or copper with respect to carbon matrix. It can be seen, that experimental and theoretical curves are in good agreement near Pt L_3-edge, but near Cu K-edge anomalous copper corrections are not visible. The exact reasons of non-anomalous Cu K-edge behavior is not clear, so that current consideration was limited by Pt L_3-edge.

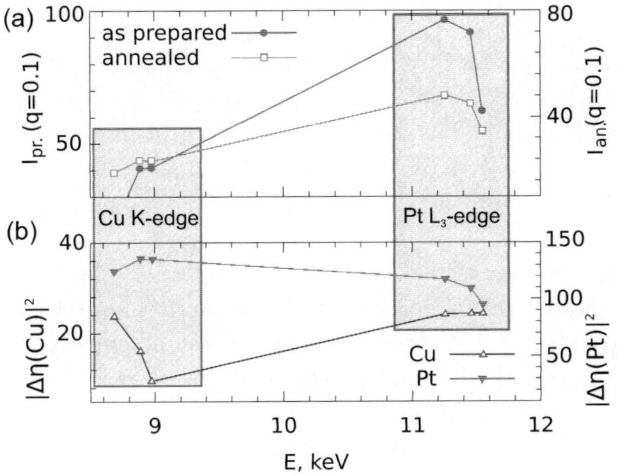

Figure 1. The energy dependence of: a) experimental scattering intensity of as prepared (●) and annealed (□) samples at $q = 0.1$ Å^{-1} and b) theoretical contrast of pure copper (△) and platinum (▼) with respect to carbon matrix.

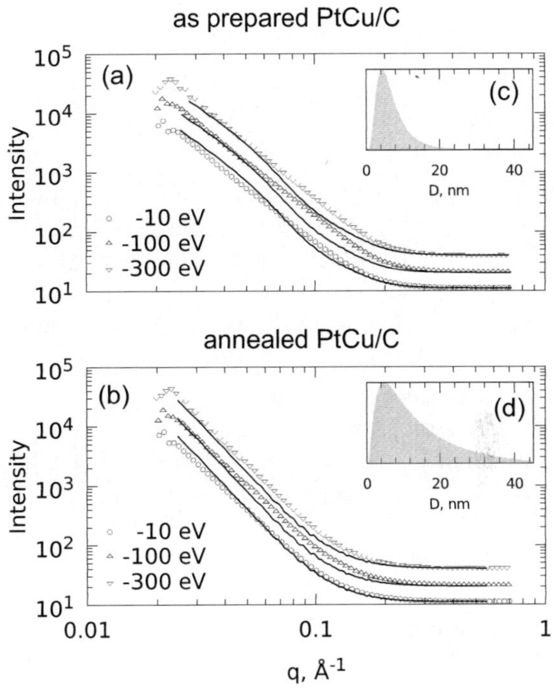

Figure 2. The comparison of experimental data measured at energies within 10 (○), 100 (△) and 300 (▽) eV before Pt L_3-edge and calculated curves (lines) of: a) as prepared and b) annealed samples. The insets c) and d) shows obtained size distributions.

According to EXAFS study of bimetallic catalysts [8, 6], prepared by the same technique of sequential deposition, the structure of averaged particle is expected to be close to core-shell architecture with platinum shell and copper core. This allows to use form-factor of core-shell particle (2) with pure platinum shell and copper core for the calculation of scattering intensity. The radii R of the particles were distributed according to log-normal law [10], with two free parameters controlling average value (μ) and dispersion (σ). The common scale parameter and the fraction of core radius $\nu = R/R_{\text{core}}$ were varied. The background contributions are clearly distinguished and were fixed to observed values. The contrast of the core and shell with respect to the matrix was not varied and was set to calculated by eq.(3) values for platinum and copper at different photon energies.

Table 1 presents the results of simultaneous fitting of the samples together with goodness of fit parameter (R-factor) and Figure 2 illustrates the fitting quality. For clarity the curves at Figure 2ab are shifted along vertical axis. The extension of the fitting model by inclusion of free dispersed platinum particles to the core-shell particles produced very small volume fraction of pure particles (< 10 %), with approximately the same parameters of core-shell particles. The average size ($D_{\text{av.}} \approx 2\mu$) of core-shell particles is reasonably [17] increased from 6 nm to 12 nm as a result of annealing procedure. The observed distribution spread is increasing too,

IOP Conf. Series: Journal of Physics: Conf. Series **848** (2017) 012027 doi:10.1088/1742-6596/848/1/012027

Table 1. Results of simultaneous fitting of the ASAXS curves near Pt L_3-edge.

Sample	R_w-factor, %	μ, nm	σ	$\nu = R/R_{\text{core}}$
as prepared	19	3.0	0.6	0.5
annealed	22	6.4	0.9	0.3

as it illustrated on the insets of Fig. 2. The average shell thickness of Pt-rich shell region is estimated as $\mu(1-\nu) = 1.6$ nm before and 4.7 nm after annealing. However, the estimations of the number of platinum and copper atoms corresponding to the core and shell volumes does not agree with total composition of the materials estimated by EDX, so that more detailed analysis of the chemical compositions of core and shell regions is required.

4. Conclusions

Pt L_3-edge ASAXS data of as prepared and annealed PtCu/C can be satisfactory explained by a model of log-normally distributed core-shell particles with Cu-rich core and Pt-rich shell. The performed study showed that annealing at 350° resulted in i) the increase of the particle size from ~ 6 nm to ~ 12 nm; ii) the increase of the distribution spread from $\sigma \approx 0.6$ to $\sigma \approx 0.9$ and iii) enlarging of the average Pt-rich shell thickness from ~ 1.6 nm to ~ 4.7 nm.

Acknowledgments

The study was performed with the support of Grant #213.01-07.2014/10ПЧВГ of Southern Federal University. The authors are also grateful to the BESSY-II for beamtime at μSPOT beamline, Dr. Ivo Zizak and Dr. Armin Hoel for their help during measurements.

References

[1] Antolini E 2003 *J. Mater. Sci.* **38** 2995–3005
[2] Sankar M, Dimitratos N, Miedziak P J, Wells P P, Kiely C J and Hutchings G J 2012 *Chem. Soc. Rev.* **41**(24) 8099–8139
[3] Xu Y, Chen L, Wang X, Yao W and Zhang Q 2015 *Nanoscale* **7**(24) 10559–10583
[4] Guterman V E, Lastovina T A, Belenov S V, Tabachkova N Y, Vlasenko V G, Khodos I I and Balakshina E N 2014 *J. Solid State Electrochem.* **18** 1307–1317
[5] Li C and Yamauchi Y 2013 *Phys. Chem. Chem. Phys.* **15**(10) 3490–3496
[6] Pryadchenko V V, Srabionyan V V, Mikheykina E B, Avakyan L A, Murzin V Y, Zubavichus Y V, Zizak I, Guterman V E and Bugaev L A 2015 *J. Phys. Chem. C* **119** 3217–3227
[7] Wei Z, Feng Y, Li L, Liao M, Fu Y, Sun C, Shao Z and Shen P 2008 *J. Power Sources* **180** 84 – 91
[8] Bugaev L A, Srabionyan V V, Pryadchenko V V, Bugaev A L, Avakyan L A, Belenov S V and Guterman V E 2016 *J. Phys: Conf. Ser.* **712** 012048
[9] Yu C, Koh S, Leisch J E, Toney M F and Strasser P 2009 *Faraday Discuss.* **140**(0) 283–296
[10] Haug J, Kruth H, Dubiel M, Hofmeister H, Haas S, Tatchev D and Hoell A 2009 *Nanotechnology* **20** 505705
[11] Krycka K L, Borchers J A, Salazar-Alvarez G, López-Ortega A, Estrader M, Estradé S, Winkler E, Zysler R D, Sort J, Peiró F, Baró M D, Kao C C and Nogués J 2013 *ACS Nano* **7** 921–931
[12] Bogomolova A, Hruby M, Panek J, Rabyk M, Turner S, Bals S, Steinhart M, Zhigunov A, Sedlacek O, Stepanek P and Filippov S K 2013 *J. Appl. Crystallogr.* **46** 1690–1698
[13] Raghuwanshi V S, Harizanova R, Tatchev D, Hoell A and Rüssel C 2015 *J. Solid State Chem.* **222** 103 – 110
[14] Guterman V, Belenov S, Pakharev A, Min M, Tabachkova N, Mikheykina E, Vysochina L and Lastovina T 2016 *Int. J. Hydrogen Energy* **41** 1609 – 1626
[15] Hoell A, Zizak I, Bieder H and Mokrani L 2006 patent DE 10 2006 029 449
[16] Toby B H and Von Dreele R B 2013 *J. Appl. Crystallogr.* **46** 544–549
[17] Srabionyan V V, Bugaev A L, Pryadchenko V V, Makhiboroda A V, Rusakova E B, Avakyan L A, Schneider R, Dubiel M and Bugaev L A 2013 *J. Non-Cryst. Solids* **382** 24–31

Institute of Physics
Dirac House, Temple Back
Bristol BS1 6BE UK

ISSN: 1742-6588
ISBN 978-1-5108-4242-7